The Human Imperative

Power, Freedom and Democracy in the age of Artificial Intelligence

By
Paul Nemitz and Matthias Pfeffer

The Human Imperative: Power, Freedom and Democracy in the age of Artificial Intelligence
By Paul Nemitz and Matthias Pfeffer

This book first published 2023
Ethics International Press Ltd, UK
British Library Cataloguing in Publication Data
A catalogue record for this book is available from the British Library

Print Book ISBN: 978-1-80441-195-7
eBook ISBN: 978-1-80441-196-4

Enhanced English edition, updated and adapted from the German language original. The copyright of the German language original is with Dietz Verlag, Bonn, ISBN 978-3-8012-0565-2. The German original was published in July 2020 under the title *"Prinzip Mensch – Macht, Freiheit und Demokratie im Zeitalter der Künstlichen Intelligenz"*.
Translation into English by the Authors and Barrister Nyasha Weinberg, London.
The translation into English was supported by the Friedrich Ebert Foundation, Berlin, the Brost Foundation, Essen, and the Mercator Foundation, Essen.

Reactions to the German language version of this book with the title "Prinzip Mensch"

"The book Silicon Valley won't like."
Janosch Delcker, Politico Europe, AI Decoded.

"Tech without democracy and rule of law cannot work. This book shows us a clear path forward that will allow us to integrate tech into our social fabric. The US Congress must act, for if America does not reign it in, others will."
Professor Alexander "Sandy" Pentland, MIT Media Lab, Boston, Author of "Social Physics".

"Ein großer Wurf zu einem wichtigen Thema unserer Zeit." (A big hit on an important topic of our time.)
Alexander Kluge, Filmmaker and Author, Munich.

More reviews and reactions at www.PrinzipMensch.eu.

Contents

New Foreword for the English Language Edition

The worldwide hype about Chat GPT, generative AI and its problems shows that the analyses and demands of this book have become even more pressing.

Some readers of the German version - in particular, journalists - told us that we should have put the last chapter of the book first. It sets out the key suggestions for political action to create a legal and policy framework for AI and the digital economy. Those working on legislation for AI and the digital economy in Washington[1] and elsewhere may want to read this chapter first.

More generally, we hope that this book injects European thinking into the global debate on how we want to live with AI in the future. The European Union has adopted or proposed several legislative measures in order to give a democratic framework to AI and the digital economy.

In our book, we discuss in particular the General Data Protection Regulation (GDPR), which is the basis for the interim interdiction in Italy of ChatGPT by the Italian Data Protection Authority.[2]

The AI Act proposed by the European Commission in 2021 is still under negotiation between the European Council and the European

---

[1] Washington vows to tackle AI, as tech titans and critics descend, Cat Zakrzewski, Washington Post 8 April 2023, https://www.washingtonpost.com/technology/2023/04/08/washington-artificial-intelligence-regulation/

[2] Decision of the Italian Data Protection Authority ("Garante") of 31 March 2023, Nr. 112 (Doc. Web 9870832), https://www.garanteprivacy.it/web/guest/home/docweb/-/docweb-display/docweb/9870847#english

Parliament.[3] This book contains much of the underlying thinking for the AI Act which has been inspired by our work.[4] And the need to act in the face of fast advancing disruptive technology is greater than ever.

Immediately after generative speech AI began its breath-taking triumphant march in early 2023 a worldwide six-month moratorium on AI was put into play. OpenAI founder Elon Musk and more than a thousand other experts warned of an impending catastrophe if the unchecked development was not temporarily halted to develop safety measures against unintended consequences of the technology.[5] A moratorium in the sense of a pause for reflection would be a sensible step in principle. But it is completely unrealistic. Hardly anyone would adhere to it in the global race for the most powerful AI. And the appeal focuses too much attention only on one point, the dangers of an approaching uncontrolled superintelligence. But the danger lies in the fact that even non conscient AI can cause immense damage, precisely because it cannot know what it is doing.

The more important point however is the constantly evolving, "autonomous" i.e., self-learning AI and how it accumulates power and can be used to exercise power over both individuals and societies. It is this question of technology induced power that this book focuses on. And in the context of this power question, we need to examine the control problem. The control problem exists even if we do not assume that an Artificial General Intelligence (AGI) will

---

[3] A European approach to artificial intelligence | Shaping Europe's digital future (europa.eu), https://digital-strategy.ec.europa.eu/en/policies/european-approach-artificial-intelligence

[4] See already Paul Nemitz, Constitutional democracy and technology in the age of artificial intelligence, 15 October 2018, Philosophical Transactions of the Royal Society A, https://royalsocietypublishing.org/doi/full/10.1098/rsta.2018.0089

[5] https://futureoflife.org/open-letter/pause-giant-ai-experiments/

emerge soon. Because it has two sides: One the one hand, the danger of the loss of human control over technology. And on the other hand, the opposite danger of concentration of technological power in the hands of a few and the resulting problem of lack of democratic control and lack of accountability.

ChatGPT represents another open-air experiment by GAFAM companies: the profits will once again be reaped by these big companies. The technology will be trained and improved by people, the users, for purposes these companies determine alone, and which guarantee them maximum profits. The negative consequences, however, are once again borne by society, which is completely unprepared for this technology.

The Large Language Models (LLMs) used in the case of Chat GPT embody a new special form of power technology because people organize their thinking, knowledge, and actions through language. This is why forging AI into logos is particularly problematic, even if it still seems rather clumsy, like mathematical intelligence without good judgment and wisdom.

AI has a deceptive nature. These programmes try to imitate human intelligence and pretend that they have really achieved it. LLMs, in combination with ML and generative AIs, are particularly "good" at this because, in addition to the appearance of intelligence, they also produce the appearance of creativity.

The nature of this type of "creativity" needs to be better understood: Since AIs still have no reference to the world, they infer meaning only from the syntactic use of words, which they find in gigantic quantities in training files, composed of all published texts in this world. Many of these texts are protected by copyright which LLMs simply ignore. They appropriate the langue of these texts and its value without making explicit reference to them like any scientist would have to do nor pay their copyright dues. The language models train in this way on texts that are not published with the intent to be used subsequently as fragments to be completed with the help of probability calculations and to be the basis of "self-creation" by machines in the process.

The AIs then gather feedback and become "better" at calculating the effect of their new texts. But not with a view to their truth and relevance. They are trained purely on the effect of the texts on people who, as the AI very quickly finds out, are best controlled by emotions. By providing users with what they are most likely to want to hear, rather than what they should know, or need to know, AI focuses on emotions and avoiding cognitive dissonance, thus leading into a dangerous narrowing of thinking.

AI extrapolates, averages the data available to it, data from the past, but not creative solutions for the new, the surprising, for the challenge posed by an open future.

LLMs perform a stringing together of words according to statistical calculations. Probability takes the place of truth, which thus becomes an appearance only. Above all, AIs produce the appearance of a truthful intelligence. The "ought" is reduced to the stochastically calculated "having-been". From data of the past probabilities of occurrence are calculated, which can block alternative really creative and therewith better solutions.

Let us not forget: Only humans can account for the truthfulness of texts and take responsibility for their ideas. It is for this reason that copyright does and should only protect human creators. And it is humans alone who have the ability to imagine the non-existent or not yet existent, thus, through criticism of the existing and dreams and imagination of the new and better, create the really new which goes beyond just extrapolation of the past or a new remixing of the already existing.

But the texts produced by GPT look convincing. This is what makes them so dangerous, in addition to the opacity of the programme and of the operating company.

The problem: OpenAI has stated that neither the algorithms nor the training data of the software will be published. Officially for security reasons. The company is thus 're-arcanizing' knowledge; making it more, not less, arcane - an enormously significant decision in the prevailing information economy. If, in today's information economy, not only significant parts of data processing, but also of language and knowledge generation are carried out according to criteria and algorithms that are not accessible to the public, a new arcane regime is de facto created that unites decisive information and decision-making power. The "Open" in OpenAI has thus become a farce.

While the initial aim was to make the blessing of the revolutionary new technology of self-learning AI available to as many people as possible, any demands for transparency of the software are now rebuffed by pointing out how dangerous this technology is in the wrong hands. What is worse, the texts do not even contain the footnotes which every writer needs to insert if using texts of others. Thus, OpenAI steals texts and ideas without referencing. A major sin in modern writing, an act of anti-enlightenment. The law thus needs to oblige such programmes not only to make their code inspectable to the public, but also to set out in every text their sources.

With Chat GPT, the apparent is taking over the reins of our information ecosystem, while the public does not know the rules according to which this new "knowledge" is generated. On what basis, then, will informed decisions be made in the future? And by whom?

We cannot absolve the companies that develop and use this technology of the responsibility to be completely open and transparent and to respect the rules of reason and rules of law which society imposes on them. And we cannot absolve the democracies of this world and their legislators from the duty to adopt the necessary, strict rules for AIs which are necessary to ensure that these programmes do not harm to people and societies, do not undermine democracy and fundamental rights and the rule of law, now and in the future.

We must impose tough and binding rules on AIs, their producers and users, because of the tremendous consequences and the existential risks to humanity.

The thesis that technology will automatically improve people's lives and solve their problems is the great illusion of our time. If we are to see through this solutionism as a dangerous fallacy and ensure that the benefits of AI outweigh the harms, we must regulate extensively. Regulation has never been more needed than it is today.

Regulation means developing clear rules for transparency and liability. This applies to AI just as it does to nuclear technology, bioengineering and genetics.

If we do not regulate, we allow the development of the biggest chaos machine imaginable for our information ecosystem, inventing new facts in places where they are not expected, producing lies and disinformation that feel very likely right but are actually totally wrong. These machines will persuade readers instead of convincing them with arguments.

ChatGPT enjoys too great popularity. The next stages of the chatbot will undoubtedly lead more and more to the readers of the texts loving the inventions of the machine, rather than texts written by critical and protesting humans, telling them the truth they need to know to face new challenges as autonomous beings. Humans unfortunately are lazy, and prefer entertainment and jokes to thinking and political engagement.

That would not be the *Human Imperative* which we describe in this book, but its mockery.

The basic theses and questions of the book gain considerable weight through the introduction of large language models: Will generative language AI undermine the language in which humans must formulate not only their knowledge of the world and their moral convictions, but which is altogether fundamental for any culture and civilization? The Philosopher of Language Wilhelm von Humboldt told us that understanding of the world only functions via language. Will language AI rule over our understanding of the world? The human as *zoon logon echon*, as the animal endowed with language and thus with thinking, reason and spirit, is in the process of outsourcing language production and thus its most powerful tool, to an inscrutable black box. Humans thereby promise themselves higher insights but could just as well forfeit the ability of reasonable self-determination, public discourse, mutual understanding and compromise, so necessary for peaceful co-existence and democracy. This technology thus becomes a potential direct threat to peace and democracy; just looking at the immense disinformation it makes possible is enough to confirm this thesis. And we have not even mentioned yet the use of AI for hot war, which is now back in Europe with the Russian aggression against the Ukraine.

The impact of AI on individual autonomy and democracy is the focus of this book. Since the publication of the German edition in

2020, numerous international groups and institutions have dealt with the demands derived from the threat potential of AI.

We want to highlight the EU AI Act,[6] which will be the world's first comprehensive law to regulate Artificial Intelligence. It combines a risk-based classification of the powerful technology with a new ex ante approach that enables forward-looking design of AI based on clear obligations. And it creates a new governance system for AI. A separate legislative proposal will establish an appropriate civil liability regime.[7]

The near-exponential growth of technology must be mastered by the intentionally slow deliberation and decision-making processes of democratic institutions. They are based on the need to take the people along and on that power is not concentrated, as it is in dictatorships like China or in the platform economy, but shared, critiqued, and decisions legitimized in a democratic deliberative way. The fact that the GAFAM companies Google, Amazon, Facebook and Microsoft immediately lobbied massively against key obligations of the AI Act will not surprise the reader of our book.[8]

The fact that the Italian data protection authority referred to violations of the GDPR in its ban of ChatGPT shows once again how important binding law is to prevent negative consequences of technology.

But the new technology also raises other questions. Never before has political action and legislation been so challenged by technology as it is today.

---

[6] Text of draft legislation proposed: https://eur-lex.europa.eu/legal-content/EN/ALL/?uri=celex:52021PC0206

[7] https://eur-lex.europa.eu/legal-content/EN/TXT/?uri=CELEX:52022PC0496

[8] https://corporateeurope.org/sites/default/files/2023-02/The%20Lobbying%20Ghost%20in%20the%20Machine.pdf

The results of the Transatlantic working group on Democracy and the Rule of Law, which published a manifesto on June 14, 2021, calling for democratic regulation of AI and the digital economy on both sides of the Atlantic, can provide further inspiration.[9] As can the EU proposal of a regulation for Internet-based advertisement in election campaigns.[10] A complete list of all official texts on policy and ethics globally can be found at CAIDP.[11]

Let's get back to the control problem: The control problem of AI is about who decides and who will be in control in the future. It is about power. An increasingly powerful AI raises the control problem, and on two levels: Should a super intelligent AI be created, which experts increasingly believe is possible, humanity faces the problem of whether this AI could still be controlled by humans at all, and if so, how. If the answer to this is no, the question of the possibility of human freedom is also decided negatively, because an uncontrollable AI wins every game against humans. The risk of such a superintelligence being created may still be low today. But since its consequences would be so massive, it must be investigated, and the developments of today's AI must be steered accordingly so that the risks remain as controllable as possible.

But the control problem exists even without creating a superintelligence: A technology that changes itself at lightning speed and makes decisions without our understanding at the time poses existential risks without requiring consciousness or our own will: it is enough that unintended side effects occur that can have

---

[9] Paul Nemitz, Democracy through law: The Transatlantic Reflection Group and its manifesto in defence of democracy and the rule of law in the age of "artificial intelligence", 01 December 2021, European Law Journal, https://onlinelibrary.wiley.com/doi/full/10.1111/eulj.12407 and https://doi.org/10.1111/eulj.12407

[10] https://eur-lex.europa.eu/legal-content/EN/TXT/?uri=celex:52021PC0731

[11] https://www.caidp.org/resources/

existential consequences for people in a highly interconnected world. The negative side effects of an originally effective tool of the advertising industry, micro targeting, on social media users and thus on societies worldwide have given us a taste of this.

Artificial Intelligence is a high-risk technology; its use must not result in future generations no longer being able to live in freedom or not being able to decide the important matters of their lives themselves. The restriction of freedom through digital technologies can be as irreversible as climate change. This is because the possibilities of surveillance and manipulation could rule out a return to freedom, just as the heating of the earth beyond a certain tipping point could hardly be slowed down.

Such a process is by no means without alternatives - not yet. A comprehensive technology impact assessment must therefore explore the consequences of this powerful technology for future generations today and draw clear boundaries: The consequences of AI must not destroy institutional conditions for living in freedom and democracy today and in the future. To this end, red lines must be drawn on the development and use of AI, just as they are drawn on the growing power of digital corporations.

For a humane life in the digital future, it will be indispensable to solve the control problem of AI. It must be solved technologically by finding ways and means that make AI controllable and manageable, even as a potential superintelligence. And it must be solved through processes and institutions that ensure that AI is aligned with the law of the land and its constitutional values.[12] And the political problem of control over an immense technological-economic power, which is

---

[12] See also https://www.sarahspiekermann.com/blog-en/6-months-ethical-diet-for-chatgpt-with-iso-iec-ieee-24748-7000

owed to digital technologies and AI, combined with the brute force of capitalism, must also already be solved today. The question is how the immense power of surveillance and manipulation made possible by AI systems in the hands of a few mega corporations in Silicon Valley and governments can be democratically controlled. The answer can only be law and strong institutions to enforce the law.

It is questionable how effective such control can be when power accumulation is allowed to the extent it has been during the past three decades, in which the Internet was not sufficiently regulated, and power concentration remained largely unchecked. The right lesson to learn from this concentration of power due to underregulating and underenforcement is to introduce the principle of separation of powers, well known in Democracies, in technology and networks as well. The question must not be what we can do with technology, but what we should do to use its effects for the general good. To this end, democratically legitimized legislation must demand responsibility for this powerful technology through clear rules and tough sanctions. Self-regulation of the powerful is an illusion, as the failure of AI Ethics and Ethics Boards set up by companies demonstrate over and over again.

The dangers and necessary limits of AI must be addressed in a broad public discussion, across all disciplines and borders. This is what this book is calling for.

If the positives and benefits of this technology are to be seen at all, the negative effects must be clearly recognized and mitigated through regulation. We must take the necessary political decisions today in order to be able to still decide as citizens and in democracy tomorrow. We can still do that now. And this book shows how.

Paul Nemitz and Matthias Pfeffer, Brussels/Munich, April 2023

Part A

Introduction: Why this Book?

This book is about power in the age of Artificial Intelligence. It looks at what the new technical powers that have accrued over the last decades mean for the freedom of people and for our democracies. Our starting point is that AI must not be considered in isolation, but rather in a very specific context: the concentration of economic and digital-technological power that we see today. Analysis of the effects of AI requires that we take a holistic view of the business models of digital technologies and of the power they exercise.

Technology, economic and political power are entering into ever closer symbiosis. Digital technologies and their corporate masters now know more than people know about themselves or governments know about the world. These technologies accumulate more and more decision-making powers. Taken together this leads to a massive asymmetry of knowledge and power in the relationship between man and machine.

The classical models of action and decision-making in democratic societies are being gradually undermined by such developments. In a new way, the question of the control of technical power arises.

Who will decide in future? And who decides, who decides?

The power to shape reality through technology is currently undergoing a fundamental change. This is a real stress test for the fundamental intellectual and cultural models on which modern societies are based. And it is unavoidable. Since we are already experiencing the second stage of the digital revolution with the current upheavals caused by AI and quantum computers on the horizon, it is worth taking a look back at the beginning of the digital age - to understand and learn why the great hopes that were associated with it have largely not been fulfilled.

In this second phase of the revolution we are currently in, we can no longer afford to make mistakes like those made in the early days of digital technology and the global Internet. Technology and knowledge seem to be exploding. Some speak of an exponential increase. In the near future this could turn into a whole new - uncontrollable - dimension.

This is contrasted by the intentionally slow processes of change in deliberative democracies. The reason for this is that experience has shown how important it is to include reflection and discussion in questions of the human exercise of power in democracies before opinion-forming and decision-making. Another consequence of this insight is the separation of powers with the checks and balances between the arms of the state.

If technology creates facts and if it develops faster than democracies decide, then that means in this game of hare and hedgehog, technology is sure to win. Does technology even have its own developmental logic, a logic that could at some point be immune to democratic control? Today, technology creates facts at a pace that could decide the question of power in its favour just because of this tempo.

We believe that the question of who will rule in the future and who will make the decisions must be asked today. Whoever wants to answer it in the spirit of democracy must bring together the representatives of technology and democracy into a new discussion.

We know that we are standing on the shoulders of giants: Immanuel Kant, Jürgen Habermas and, yes, also Edward Snowden, without whose courage the world might never have known how intensively the most powerful digital surveillance machines ever conceived by humans are already observing and manipulating us all today. The title of this book also reveals that we are drawing from Hans Jonas and Ernst Bloch, two exiles in life and thought. With their works *The Imperative of Responsibility*[13] and *The Principle of Hope*[14], they shaped the authors of this book in their youth. This book wants to analyse, orient and appeal for considered political thinking and action relating to technological change in the age of AI. It therefore concludes with a series of concrete recommendations for political action necessary to safeguard and further develop liberty, democracy and the rule of law in times of AI.

---

[13] Jonas, Hans. The Imperative of Responsibility: In Search of an Ethics for the Technological Age University of Chicago Press, 1985.

[14] Bloch, Ernst. The Principle of Hope: V. 1. Reprint edition. Cambridge, Mass: MIT Press, 1995.

Part B

Critique of Technological Power

> *Creon: Orders are manifold. Yet: Who orders?*
> *Antigone, Sophocles*[15]

It is important to understand how technological power is acquired and used today. Whether and how we - both individuals and the decision-making systems of market and democracy - are dominated by this power determines the degree of our freedom and, in the long term, the prosperity of our circumstances.

In a simpler way, you could say: The control of technological power is a central function of democracy (B1). But with the domination by economic power through *its* technology, which we are already experiencing today, we are still before the phase in which technology itself takes over: namely the domination by a strong AI that can set itself goals. We urgently need to deal with both forms of domination today. It is still unclear when exactly systems of AI, i.e., technology alone, will be able to exercise power for their own purposes directly, due to a loss of control by humans. But we must address the problem of the centralization of technical power, which seeks to escape democratic control, which is already dramatic today (B2).

We need a whole system view of the technically dominated future (B3). This begins with an individual analysis of the ten central technologies of digital power that connect to the Internet (B4). The adoption of such a holistic view of these technologies demonstrates their potential power (B5).

---

[15] Brecht's Antigone, 116, see Brecht, Bertold. Collected Plays: The Antigone of Sophocles. The Days of the Commune. Turandot or the Whitewasher's Congress (Methuen 2003).

We then turn to the problem of the concentration of power in the hands of the digital-technological-economic complex (B6).

The technology companies *Google, Amazon, Facebook, Apple* and *Microsoft,* the so-called Terrible Five[16] collectively named GAFAM, are now the richest, most powerful companies in the world. Their business models influence in one way or another all of our lives (B7). How they gain their power and how they use it is demonstrated in an analysis of the eight sources of their power including the ideology of total technical feasibility, the *Californian Ideology* (B8).

[16] Manjoo F, 'Tech's Frightful Five: They've Got Us' The New York Times. 10 May 2017 https://www.nytimes.com/2017/05/10/technology/techs-frightful-five-theyve-got-us.html

B1 The control of technical power as a central function of democracy

Our interest in the critique of technological power stems from concerns about the functioning of democracy in the technological age.

We must first understand what technology actually has to do with democracy and make clear the base of our concerns that technological developments could become a problem for the functioning of democracy.

As we have reflected on the power of technology companies and the use of technology by governments, especially the Chinese government, we thought it necessary to re-read what has been written about power and domination.

1.1. Power and rule - what is that actually?

Max Weber, Hannah Arendt[17], Parsons Talcott[18], Heinrich Popitz[19] and Jürgen Habermas[20] have all written about power and domination.[21]

Max Weber said:

---

[17] Arendt, Hannah. 1970. On Violence. Houghton Mifflin Harcourt.

[18] Parsons, Talcott. 'On the Concept of Political Power'. Proceedings of the American Philosophical Society 107, no. 3 (1963): 232–62.

[19] Popitz, Heinrich. Phenomena of Power: Authority, Domination, and Violence. Columbia University Press, 2017

[20] Habermas, Jürgen. 1986. Theory of Communicative Action, Volume 1: Reason and the Rationalization of Society: Reason and the Rationalization of Society Vol 1. New Ed edition. Boston: Polity.

[21] For a single overview (in German) of theories of power see: Anter, Andreas. Theorien der Macht zur Einführung. Hamburg: Junius Verlag GmbH, 2012.

> *"Power is any opportunity to assert one's will within a social relationship, even against resistance."*[22]

For Max Weber, rule is an increase in power in the sense that the observance of one's own will by others is secure because institutions and law ensure it.[23]

Hannah Arendt's definition of power, on the other hand, is communicative:

> *"Power corresponds to the human ability not just to act but to act in concert. Power is never the property of an individual; it belongs to a group and remains in existence only so long as the group keeps together".*[24]

This model of communicative understanding is not aimed at a power that is an end in itself and can be instrumentalized. According to Habermas' understanding of Arendt the goal of power is not just any success, but possession of reasonable validity of universalizable claims. The recognition of these claims to validity is motivated by rational reasons, not by violence and coercion:

> *"Convictions can be manipulated, but not the rationality claim from which they subjectively draw their force."*[25]

We keep these distinct understandings of power as an opportunity and power as domination, in order to ensure that others obey one's own will, in mind when we turn to consider the relationship between technology, power, domination and democracy.

---

[22] Weber, Max. Economy and Society, Edited by Johannes Winckelman, 5th ed. 1985, p. 28.

[23] Weber, Max. Economy and Society, Edited by Johannes Winckelman, 5th ed. 1985, p. 28.

[24] Arendt, Hannah. 1970. On Violence. Houghton Mifflin Harcourt.p. 45.

[25] Arendt, Hannah. Hannah Arendt: Critical Essays. SUNY Press, 1994. p.213.

Hans Jonas, in his work "The Imperative of Responsibility, in Search of an Ethics for the Technological Age", spoke of the combined power of technical innovation and economic interests which, united, can push aside all other interests in a society.

He reminded us of the special nature of human power and the responsibility associated with it:

> *"Great is the power of tigers and elephants (...), greater still that of bacteria and viruses. But it is blind and unfree (...). Only in man is power emancipated from the whole through knowledge and arbitrary will and only in man can it become fatal to him and to itself. His capacity is his fate, and increasingly becomes the general fate. In him, therefore, and in him alone, there arises out of the willing itself the "ought" as the self-control of his consciously exercised power (...)".*[26]

Man, according to Jonas becomes "*executor[s] of a trust which only [w]e can see, but did not create. [...] Since in [man] the principle of purposiveness has reached its highest and most dangerous peak through the freedom to set himself ends and the power to carry them out, he himself becomes the first object of his duty not to ruin, as he well can do, what nature has achieved in him by the way of his using it".*[27]

---

[26] Jonas, Hans. Ontological Grounding of a Political Ethics: On the Metaphysics of Commitment to the Future of Man in Schurmann, Reiner. The Public Realm: Essays on Discursive Types in Political Philosophy (SUNY Press) which takes text from Jonas, Hans. The Imperative of Responsibility: In Search of an Ethics for the Technological Age (University of Chicago Press, 1985).

[27] Jonas, Hans. The Imperative of Responsibility: In Search of an Ethics for the Technological Age (University of Chicago Press, 1985). p.130

1.2. Hans Jonas and the precautionary principle as an element of policy-making

Jonas' *The Imperative of Responsibility* is the basis of the precautionary principle, which has now become a constitutional principle of European law.[28] It states that if a technology is likely to have long-term, serious consequences for human beings or the environment, we must take tough decisions to avoid with certainty these serious negative consequences, even if we are not certain they would ever arise. To do this, we must invest in the science of technology assessment. And we must put ourselves in the emotional and political position to be able to make the necessary decisions about the future of technology today.

The reception of Jonas' book shows how democracy and law are capable of learning, and that is precisely why Jonas' thesis of the dominance of technical innovation and economic interests does not always hold true. He has probably actually succeeded in changing not only thinking but also the political reality in Europe.

In democracy, there is a constant debate about what should be done in the interest of the common good. Today it is conceivable, but not inevitable, that in this debate the interests of economic and technological innovation will prevail. This is particularly likely if interested companies are able to demonstrate that the innovation they finance and produce serves the common good and does not harm other public interests, such as environmental protection, data protection, fundamental rights and democracy.

---

[28] Article 191, Consolidated version of the Treaty on the Functioning of the European Union, PART THREE - UNION POLICIES AND INTERNAL ACTIONS, TITLE XX – ENVIRONMENT https://eur-lex.europa.eu/legal-content/EN/TXT/?uri=celex:12016E191

As long as the debate about the common good is conducted with factual arguments and based on the truth, such a result can be legitimate if the sovereign spheres of law, democracy and human dignity are preserved. Scientific technology assessment is the traditional means by which the Jonas' precautionary principle is applied.[29]

It is however not legitimate for economic power and technology to be used as a means of manipulation and pressure.

With complex technology that is becoming increasingly difficult to understand, with a dubious attitude towards the law, indeed a culture of lies in some GAFAM companies - according to the Harvard Business School professor Shoshana Zuboff[30] - and of course with the enormous money in the pockets of these companies, the temptation to do just that is growing.

Democracy and freedom need innovation and economic success, not only to compete with dictatorships like China, but also to avert political radicalization and populism. The history of Germany before the Second World War teaches us that unemployment, poverty and precarious employment conditions are the basis for a slide into populism and fascism.

Recent history also teaches us, especially the election victory of Donald Trump in the United States and the Brexit in Great Britain,

---

[29] European Parliament, 'The Precautionary Principle: Definitions, Applications and Governance - Think Tank'. n.d. https://www.europarl.europa.eu/thinktank/en/document.html?reference=EPRS_IDA(2015)573876 and currently the report of the *Data Ethics Commission* of the Federal Government of 23 October 2019, https://www.bmi.bund.de/EN/topics/it-internet-policy/data-ethics-commission/data-ethics-commission-node.html and the Enquete Commission on Artificial Intelligence of the German Bundestag, https://dserver.bundestag.de/btd/19/237/1923700.pdf

[30] See below chapter 8.2 and Zuboff, p. 103f, 112f.

that the sense of loss of control and of being dominated by a system of technological and economic power, and a political establishment linked to this system, also fuels populism and anti-democratic tendencies.

A vibrant democracy that wants to remain so is therefore well advised to enable innovation and economic growth, securing prosperity and avoiding poverty and unemployment, without, however, placing the political shaping of society in the hands of powerful economic interests or even those who run the technological systems. The people feel that democracy has been replaced by a system in which they and their interests no longer play an appropriate role and in which manipulation, half-truths and lies dominate the debate.

There is no question that technical innovations such as the Internet and AI have great potential for the common good. We all use these technologies, and depend on them, some of us even love them. Imagine the search for information today without Internet search services. The Internet and AI are technologies that have the potential to increase productivity in many areas and thus serve the common good. However, the increase in productivity through the Internet and digitalization has not yet been proven in macroeconomic terms. This is explained by the fact that productivity gains are highly concentrated in some large companies and in others, especially small and medium-sized enterprises, they do not occur at all, or only come much later.[31]

That power is exercised through technology is beyond question and is nothing new in itself. The development of weapons technology

[31] Brynjolfsson, Erik, Daniel Rock, and Chad Syverson. 'Artificial Intelligence and the Modern Productivity Paradox: A Clash of Expectations and Statistics.' Working Paper. Working Paper Series. National Bureau of Economic Research, November 2017. https://doi.org/10.3386/w24001

is the best historical example, but communications systems and networks are also instruments of power.

Eben Moglen, Professor of Legal History at Columbia University and the legal mind of the Free Software movement, begins his *Snowden lectures*[32] with a reference to the first technical network of power: the Roman road network. Not only did the Roman legions march along these roads, but information was also transmitted faster and more reliably than ever before due to the innovation of signal posts. Even in the Roman Empire, networks and information transmitted on them was power.

The question that we have to ask ourselves in terms of democracy today is whether the GAFAM are already ruling in a way that is no longer compatible with democracy, or whether we will soon be facing such rule, because the development of the technologies and business models that connect to the Internet will enable rule in the hands of powerful corporations.

1.3. The sociologist Heinrich Popitz and the phenomena of power

That the development of technology can lead to a basically unlimited increase in power, and that therefore a central task of democracy is to control the development of technology was explained by the sociologist and political scientist Heinrich Popitz in his book "Phenomena of Power" from 1992:

---

[32] Moglen, Eben, 'Eben Moglen: Snowden and the Future.' http://www.snowdenandthefuture.info/

> *"In a high-tech society the objective, objectified conditions of human existence change radically with the speed at which we turn the pages of a calendar. The person who today decides the technical shaping of our environment of existence, who has the power of data constitution, can within a very short lapse of time exercise an immense amount of power over an immense number of humans, and perhaps (as with the building of a nuclear power station) over immensely long stretches of time. We can look back and detect the technical progress. But if we look forward, we cannot decide how long and how extensively the efficiency of technical action will further increase. Our experience so far does not suggest any principle from which to derive such a prognosis. Technical action seems to constitute an intrinsically open-ended human faculty. […] We cannot know in which nameless regions the potential of social power can further build up. If technical action is in principle open-ended, so is the potential dangerousness of humans for humans as well.*
>
> *One ought to bear in mind this lack of knowledge – a truly fundamental lack – if one seeks to predict future societal developments. But, with this, the problems of power control become more and more difficult to solve. At the same time it becomes more and more certain that in modern society, the control over technical action constitutes the key aspect of all control over power."*[33]

Within technical action, the control of which he declares to be the pivotal point of power control, Popitz distinguishes three forms of technology: weapons technology, the techniques of power exercise and "data-setting technology". In his words this third category involves the "creation of technically perfect facts", which do not belong to either of the two other categories, but which nevertheless determine our lives.

[33] Popitz, Heinrich. Phenomena of Power: Authority, Domination, and Violence. Columbia University Press, 2017. p.180

The development of weapons technology, such as self-controlling drones, combat robots and digital programs of Internet warfare through AI and quantum technology have made huge leaps in development.[34] We do not go into detail here of the implications of these technologies. In the second category, namely the techniques of the exercise of power, Popitz saw not only techniques of "bondage of the subjugated" such as electric fences and minefields, but also "electronic data processing" and the "centralization of supplies, for example of electricity, which "ties even the simplest life processes to central distributors."

Whether the Internet, against the background of these definitions, falls into the group of techniques for exercising power or the group of data-setting techniques can be left open here.

It is clear however that Popitz' statements about the necessity of power control make sense in the context of the Internet and the new digital technologies. The high speed of the development from ever faster computers to quantum computers, the rapid geographical, worldwide expansion of the Internet, but also the rapid rise of applied technologies, such as image processing and self-learning algorithms, confirm Popitz' findings about the open innovation capacity of man.

Many examples demonstrate that power over people and systems is connected with these technologies. This starts with the use of the Internet by secret services for the purpose of mass surveillance, as

[34] The Economist, "Artificial Intelligence Is Changing Every Aspect of War." https://www.economist.com/science-and-technology/2019/09/07/artificial-intelligence-is-changing-every-aspect-of-war. Hitchens, Theresa, and Egozi, Arie. 'How AI Could Change The Art Of War.' Breaking Defense (blog). https://breakingdefense.com/2019/04/how-ai-could-change-the-art-of-war/.

Edward Snowden has revealed,[35] and extends to the use of facial recognition in states all over the world, such as China, to control the whole population.

Shoshana Zuboff described the "instrumentarian power" of behaviour manipulation, which GAFAM obtain through the mass collection and analysis of personal data. According to research by the UK House of Commons, the vote on Brexit in the UK has been significantly influenced by the use of this "instrumentarian power", in this case by the company Cambridge Analytica, which in turn used *Facebook* data to make a targeted and individual appeal to people and their preferences in its covert campaign for Brexit.[36] The same techniques have been used in US elections and helped Trump come to power.

---

[35] MacAskill, Ewen, Dance, Gabriel, Feilding, Cage, Chen, Gren and Popovich, Nadja. 'NSA Files Decoded: Edward Snowden's Surveillance Revelations Explained'. The Guardian. 1 November 2013. https://www.theguardian.com/world/interactive/2013/nov/01/snowden-nsa-files-surveillance-revelations-decoded

[36] See the documentary "The Great Hack", by Karim Amer; Jehane Noujaim, 2019, https://www.netflix.com/lu-en/title/80117542; and the House of Commons, Digital, Culture, Media and Sport Committee, "Disinformation and 'fake news': Interim Report', and the final report of 18 February 2019, with statement of the Select Committee Chair, https://publications.parliament.uk/pa/cm201719/cmselect/cmcumeds/1791/1791.pdf

B2 Rule by technology comes before the rule of technology

> *"We believe that modern technology platforms, such as Google, Facebook, Amazon and Apple, are even more powerful than most people realize (...), and what gives them power is their ability to grow – specifically, their speed to scale. Almost nothing, short of a biological virus, can scale as quickly, efficiently or aggressively as these technology platforms and this makes the people who build, control, and use them powerful too." Eric Schmidt, ex-CEO Google*[37]

Technologically enabled concentration of power and private rule threaten the functioning of the essential control systems of our society: democracy and the market.

Both are based on the assumption that free individuals can reach the best decisions through communication and rational motives if they organize themselves through the institutions of the market and of democracy. However, this assumption only works if the will that is formed by free individuals as well as in institutions is not distorted or dominated, either by manipulating information, by manipulating institutions or by directly manipulating individuals in such a way that we can no longer speak of free individuals.

It is not enough to simply take a look at the potential of future technologies that is already practically foreseeable today. It is also not enough to speculate about the theoretically conceivable potential of the technologies that lie further in the future. Both are important,

---

[37] Schmidt, Eric and Cohen, Jared. The New Digital Age: Transforming Nations, Business and Our Lives, New York, Vintage, 2014, p. 9

and of course it is right, as a first step, to assess the theoretical potential and the risks of a technology separately, that is, separately from other technologies and from the use of the technology within the framework of certain business models or by governments.

2.1 The coercive character of technology today

Today's technology has already taken on a coercive character even without AI, especially in its connection with economic and state power. Since we are already completely dependent on technology in all areas of life, even without AI, the further development of technology inevitably follows in order to solve the problems it has itself created. Since the economy and therefore jobs, prosperity and ultimately democracy depend on this constant development, even without AI, the impression is created that there is no alternative to this development and that democracy is weak in the face of it. One does not have to assume an "autonomous induction"[38] in technological "evolution", as presented by Singularity mastermind and Google development boss Ray Kurzweil[39]. But we must ask the question of how the power of technology is already growing inherently and systemically due to the colonization of the world by technology that has already taken place.

[38] This is a term that derives from the philosophy of technology, which describes the "inherent law" of technology: "This doctrine sees technical development as a blind automatism to which people are helplessly at the mercy." Günter Ropohl. Allgemeine Technologie : eine Systemtheorie der Technik, Karlsruhe 2009, p. 287; see also Gehlen, A. Man in the Age of Technology. Reprinted Ed edition. New York: Columbia University Press, 1989.

"The connection between science, technical application and industrial evaluation has long since also formed the superstructure, which is itself automated and ethically completely indifferent."

[39] Kurzweil, Raymond. 2006. The Singularity Is Near. London: Duckworth. op. cit. and below chapter C 1 d.

The new question in the age of AI is indeed whether technology, when it is self-learning and self-controlling, can also exercise power itself, i.e., whether it is no longer people who exercise power through technology, but technology which gains power over people, perhaps even developing a will to power itself.

Is it technically conceivable that technology could rule over humans without a human will to power using this technology? And if so, do we want to allow technology to rule over humans in this way? It is important to separate the potential of technology to rule from the question of whether technology is used to exercise power and dominion.

The temporal relevance of these two questions is different: the question of power and domination through technology is currently being raised against the background of the Internet-based concentration of power and technological innovation in the hands of GAFAM, and in view of political power that already uses the new technologies for domination purposes, most dramatically in China.

The question of the domination of technology over human beings becomes relevant to the extent that technologies are developed and disseminated which can learn and evolve by themselves in such a way that they themselves may at some point determine the purposes of their activities and thus no longer serve the purposes of their human producers and masters, but self-defined purposes.

In place of the domination *through* technology we then may see domination *by* technology and with it the control problem described by the computer scientist Stuart Russell of Berkeley University. His theory is as follows: even if the AI could perhaps set its own purposes, wherever value decisions have to be made or the application of principles weighed against each other, AI must always return to the human being and ask: What now? Because AI alone cannot and must not weigh up and apply principles in its decision making.[40] The weighing of principles shall always remain in the hands of humans. Stuart Russell rightly points out that data and faster processing alone, such as by quantum computers, will only produce false results faster unless we dramatically improve the quality of data and algorithms, and ensure that where value judgements need to be made that the machines have to ask people to make those and feed them back to the machines.[41]

Today, it can no longer be ignored that, due to the power structures in the Internet and with regard to the development and operation of modern technologies, there is a considerable risk of undermining the functioning of the market and democracy. Impact assessment of technology today must therefore always take into account the real power structures that determine the purposes of developing, deployment and use of the technologies.

We therefore separate the analysis of *power through technology*, which asks about the power and the potential for domination of the powerful tech companies, their technologies and business models, and the analysis of the *power of technology*, which asks whether these technologies will one day rule over people without serving other interests, thus whether they will have become autonomous.

2.2. When does the general AI or a super intelligence take over power?

More and more prominent writers, and thinkers now claim that we humans are on the way to losing control over our own lives to a

---

[40] Russell, Stuart: Human Compatible, Artificial Intelligence and the Problem of Control, 2019.

[41] Russell, Stuart: Human Compatible, Artificial Intelligence and the Problem of Control, 2019, p. 37 and p.149

technical system of AI.[42] While the specific AI is always theoretically limited to a special field, even when learning independently, a general AI should have the ability to transfer what has been learned in one field to another.

Whether and when the technology of AI will become so powerful, however, remains unclear despite many efforts at predicting its advance. There are some experts who claim that within five to ten years AI will win at "every game". The games that AI could win against humans include speculation on the stock market and even elections, according to a prominent AI researcher who does not want to be quoted by name.

Others claim that the possibilities of current forms of AI, i.e., neural networks and deep and reinforcement learning, have already been exhausted. From a scientific point of view, no breakthroughs in the form of significant further developments are to be expected with regard to these methods; their practical applications are largely known. However, in the field of cognitive computing and probabilistic programming, there are serious possibilities for enormous further developments.[43] With these methods human level intelligence could be achieved and surpassed in many areas.

Ray Kurzweil, the chief developer of Google, says there is a 50 percent probability that we will reach the general AI by 2029.[44] And he predicted that by 2045, our brains will be connected to the cloud,[45]

---

[42] See, among others, Bostrom, Nick. Superintelligence: Paths, Dangers, Strategies. Oxford: OUP Oxford, 2014.

[43] MIT, "MIT Probabilistic Computing Project." http://probcomp.csail.mit.edu/

[44] Ford, Martin. Architects of Intelligence: The Truth about AI from the People Building It. 1st edition. Packt Publishing, 2018.

[45] Sabin, Dyani. "Ray Kurzweil Has a Prediction About Linking Our Brains to the Cloud." Inverse. https://www.inverse.com/article/33373-ray-kurzweil-singularity-thoughts

- if it still exists, given the technology trend towards decentralizing the processing of data at the edge of the Internet.

Australian IT researcher Toby Walsh has identified the year 2062 as the year in which AI will be equal in processing power to us humans, according to his survey of scientists.[46] His book contains a list of everything that's wrong with the Internet and the digital economy. He ends by calling for action, through legislation and the transformation of society, to prevent the dystopian vision of rule by technology from materialising. We broadly agree with his proposals. According to Stuart Russell, a British researcher teaching computer science at Berkeley University, the general AI similar to intelligence of human beings will take at least another 80 years. But he also warns that, as with nuclear fission, which was long thought impossible, it could arrive tomorrow.[47]

One thing is certain, however: Even before such software, which may set itself goals, is put into operation, tests and appropriate precautions must be taken to ensure that AI can never take command of itself and control over humans. Otherwise, this would be the last invention of mankind following the beliefs of the British mathematician Irving John Good[48] and the Swedish philosopher Nick Bostrom.[49] As early as 2015, scientists led by Steven Hawking and Stuart Russell issued a warning and drafted an urgent research agenda to ensure that AI remains human-centric.[50]

---

[46] Walsh, Toby. 2062: The World That AI Made. La Trobe University Press, 2018.

[47] Russell, Stuart: Human Compatible, Artificial Intelligence and the Problem of Control, 2019. p. 77.

[48] Irving, John Good, "Speculations Concerning the First Ultraintelligent Machine." The Edward A. Feigenbaum Papers - Spotlight at Stanford. 1966. https://exhibits.stanford.edu/feigenbaum/catalog/gz727rg3869

[49] Bostrom, Nick. Superintelligence: Paths, Dangers, Strategies. Oxford: OUP Oxford, 2014.

[50] Future of Life Institute, 'Scientists' warning cry 2015 - Stephen Hawking, Elon Musk, Max Tegmark, Stuart Russell' https://futureoflife.org/ai-open-letter with annex on research priorities https://futureoflife.org/data/documents/research_priorities.pdf?x69354

On the other hand, there are countless promises of salvation in the tech literature. It is claimed that the global technical system will solve all the problems of this world better than we humans can, because it processes more information faster and is therefore *more intelligent*. It will, according to the vision, exercise control over the earth and mankind. We will have to get used to being dominated by this technical system. Decisions that humans have made so far will be made by self-programming general AI.[51]

Microsoft has formed a billion-dollar alliance with *Open AI*[52] the research and development laboratory founded by Elon Musk, to develop such strong, general Artificial Intelligence for the common good and, above all, to advance the meaningful creation and understanding of language.[53] Bill Gates once said that we tend to overestimate the development of the next two years and underestimate the developments in the next 10 years. If *Microsoft* is investing billions in the development of general AI, then in the light of Bill Gates' philosophy and given the state of development of AI today, this could be a sign that a lot can be expected in this area within the next 10 years. In particular, *Open AI's* research into artificial, argumentative, persuasive language creation that cannot be recognized as machine speech should be closely monitored. The ability of AI to communicate in language without humans realizing they are communicating with a machine touches the core of human autonomy and the democratic system.[54]

---

[51] Lovelock, James. Novacene, The Coming Age of Hyperintelligence, London, 2019;

[52] See the website of the research project: 'https://openai.com/'. Tegmark, Max. Life 3.0: Being Human in the Age of Artificial Intelligence. 01 edition. Allen Lane, 2017.

[53] Piper, Kelsey. 'Microsoft Wants to Build Artificial General Intelligence: An AI Better than Humans at Everything'. Vox. 22 July 2019. https://www.vox.com/2019/7/22/20704184/microsoft-open-ai-billion-investment-artificial-intelligence. And https://openai.com/blog/microsoft/'.

[54] GPT-2: 6-Month Follow-Up, in: https://openai.com/blog/gpt-2-6-month-follow-up/

B3 System view of the technically dominated future

Applying Max Weber and Hannah Arendt, we now speak of technological power where technologies provide the ability to control the will and communication and thus the actions of people, technical systems or the development and behaviour of nature.

3.1. The power of networks combined with other technologies

The networking power gives those who know how to use it access to all regions of the world. In the future, it will also provide access to human's inner worlds through bio-physical systems and nanotechnologies. Using sensors and observations of the communication and behaviour of people, it penetrates into the most private spaces and most private relationships, into the thinking and feeling, the hopes and dreams, in short, the soul of the human being. It ranges from high-resolution satellites for earth observation, cable laying and telecommunication services to social networks and search engines, cookies and other technologies that observe the behaviour of people, machines and nature and bring them together via the Internet.

The processing power is the power of analysis of collected information and data. It is the power of the self-learning algorithm, which in the future will be executed by quantum computers. It is the power to draw conclusions, to recognize affinities and probabilities and then, via the Internet, to play these back to humans and connected machines as commands - or invitations to act - in more or less manipulative ways. Shoshana Zuboff calls this power "instrumental power", because it makes it possible to turn people, by remote control, into de facto instruments of corporate profit or into compliant party soldiers of communism, as the Communist Party in China is striving to do.[55] But it is also the power to understand and influence the functioning and changes of the global system, from global technical and social systems to individual societies.

The more elements of these forms of technical power are concentrated a few hands, the less control is exercised over these new forms of power, the greater the dangers of this concentration of power for the freedom of individuals and the functioning of the market and democracy.

3.2. The Internet as the central networking medium

The most powerful medium of power of our time is the Internet. It creates the technical foundation for the exercise of a centralized form of rule, through the control and manipulation of people and decentralized technical and economic processes.

Behind the harmless concept of networking, as used by Mark Zuckerberg, and the promise of freedom of the Internet, which gives us seemingly unlimited access to information and the supposed possibility to speak to and be heard by all, there is huge potential for centralizing control. The Internet is a multi-way medium which can also be used by governments and corporations to monitor, store and evaluate all communication and activities of each individual citizen. The Internet allows states and corporations to control and manipulate the masses, just as it allows data from the Internet of

[55] Parton, Charles. 'China's big brother', Chatham House, The World Today, August & September 2018, https://www.chathamhouse.org/publications/the-world-today/2018-08/chinas-bigger-brother

Things (IoT) to be brought together and technical and economic systems to be controlled and manipulated.

The control and management potential of the Internet connected with AI is increasing: with its geographical spread, its growing transmission capacity and the progressive networking of devices in the IoT, as well as the willingness of people, or their compulsion, to be in constant contact - *always on* - up to the momentary or permanent bio-physical connection.

The creation of a quantum Internet, already an aim in technology policy today, will combine the enormous processing potential of quantum technology with a correspondingly enormous transmission capacity.[56] This project is an example of how necessary it is to consider not only the consequences of individual technologies, but of all technologies based on the Internet together within the framework of an overall assessment method. In the *Californian Ideology*, the Internet is becoming a global neuronal system that extends from outer space to the biology of the Earth, and even to the psyche of each individual human being, and enables central control and management by networking everyone with everyone and everything.

[56] European Commission, Technical agreement signed for a European plan on quantum communication infrastructure, https://ec.europa.eu/digital-single-market/en/news/technical-agreement-signed-european-plan-quantum-communication-infrastructure

B4 The ten technologies of power connected to the Internet

Connected to the Internet as we know it and the World Wide Web as Tim Berners Lee developed it, there are ten technologies that need to be considered together to obtain a precise understanding of consequences of the concentration of power among a few private and state actors.

These are technologies that serve partly to improve and extend networking and partly to process information more quickly and better. Developments such as genetic manipulation and nanotechnology are also associated with the system of technological power described here.

Analysis and descriptions of each of these technologies fills shelves of literature by themselves. What we want to do here is to give a short introduction to the technologies of power and to develop an understanding for the necessary synoptic view of these technologies and their potential uses and consequences.

4.1. Big data

Big data, first of all, describes a situation in which there is a great deal of data that possibly cannot be processed meaningfully by classical computers and classical programs. Big Data therefore initially describes a challenge for technological and computer science.

But then Big Data is also an assertion in theory: If you only collect all the data you can get hold of, then you will always be able to do something with it later, for example recognize patterns that are of commercial interest or even produce scientific findings that are in the public interest. The more data one collects, the greater the probability of learning something from it, is the claim.

The theory of Big Data underlies GAFAM's business model and has become an ideology that attempts to remove any limitation on data collection and hides the profit motive behind claims of public interest motivation. The European Union (EU) General Data Protection Regulation, the GDPR, limits the collection of personal data to what is absolutely necessary for the respective, legal purpose of the collection. According to the principles of purpose limitation and data minimization, at the core of GDPR,[57] no more than that, thus only the strict minimum of personal data necessary may be collected.[58] Conversely, however, this also means that data that is not personal data can be collected without restriction under GDPR.

Big Data is often the basis for training self-learning programs, the so-called machine learning, also called AI. The GDPR must also be observed in this context when personal data from Europe are processed, whether inside Europe or outside, e.g., in the US.

Where Big Data does not identify or allow to identify a person, such as in the case of many machine operating data, large system data and environmental data without personal information, the economic returns from processing can be very high, and the GDPR does not apply.

---

[57] Regulation (EU) 2016/679 of the European Parliament and of the Council of 27 April 2016 on the protection of individuals with regard to the processing of personal data, on the free movement of such data and repealing Directive 95/46/EC (General Data Protection Regulation) (Text with EEA relevance) OJ L 119, 4.5.2016, p. 1-88,

[58] Article. 5 (1) b and c GDPR

4.2. The Cloud - the place of processing and power

Technically, the consolidation of data in the cloud enables the decentralization of power, as the data may be distributed to many different data centers and no longer stored in one place. This increases data security and ensures access without delay, without the so-called latency. At the same time, however, the market for cloud services is extremely concentrated at a number of American companies, especially *Amazon, Google* and *Microsoft*[59]. Furthermore, the accumulation of data in the cloud by a few companies gives them potentially enormous power, especially if they take the right to view the data and to evaluate and exploit the content. This is what Google and others explicitly do under their own terms and conditions, at least when using the free cloud options:

"When you upload, submit, store, send or receive content to or through our Services, you give Google (and those we work with) a worldwide license to use, host, store, reproduce, modify, create derivative works (such as those resulting from translations, adaptations or other changes we make so that your content works better with our Services), communicate, publish, publicly perform, publicly display and distribute such content. The rights you grant in this license are for the limited purpose of operating, promoting, and improving our Services, and to develop new ones. This license continues even if you stop using our Services (for example, for a business listing you have added to Google Maps). Some Services may

[59] Dignan, Larry. 'Top Cloud Providers in 2020: AWS, Microsoft Azure, and Google Cloud, Hybrid, SaaS Players | ZDNet.' 11 May 2020. https://www.zdnet.com/article/the-top-cloud-providers-of-2020-aws-microsoft-azure-google-cloud-hybrid-saas/ Original German source: Herrmann, Wolfgang. Die Konzentration im Cloud-Markt nimmt weiter zu, 25. April 2018, Computerwoche, https://www.computerwoche.de/a/die-konzentration-im-cloud-markt-nimmt-weiter-zu.3544812

Walker, Peter.'Concentration Risk Rises as FS Moves to the Cloud'. FStech. https://www.fstech.co.uk/fst/Cloud_Provider_Competition_Concentration_Risk.php

offer you ways to access and remove content that has been provided to that Service. Also, in some of our Services, there are terms or settings that narrow the scope of our use of the content submitted in those Services. Make sure you have the necessary rights to grant us this license for any content that you submit to our Services.

Our automated systems analyse your content (including emails) to provide you personally relevant product features, such as customized search results, tailored advertising, and spam and malware detection. This analysis occurs as the content is sent, received, and when it is stored."[60]

Since Snowden, we have known that data transfers over the Internet and into the cloud are also subject to industrial espionage in the US.[61] After Snowden, the US declared that the NSA does not engage in industrial espionage in the sense that it passes on to the US economy information that it previously collected as part of its national security intelligence.[62] But not many people believe that.[63]

Basically, we can say: the data that flows into and through the US and other countries represents a security risk. The centralization of data in the cloud systems of US companies makes it easier for the US to access data in these systems, for whatever purpose, even outside the US. The most recent example of how this power can be

[60] Google, 'Google terms of use dated 22 January 2019', 22 January 2019. https://policies.google.com/terms/archive/20190122?hl=en

[61] Timm, Trevor. 2015. 'The Snowden Effect: New Privacy Wins Await after Data Transfer Ruling'. The Guardian, 8 October 2015, sec. Opinion. https://www.theguardian.com/commentisfree/2015/oct/08/snowden-effect-new-privacy-wins-data-transfer-ruling

[62] Council of the European Union. 'Report on the findings by the EU Co-chairs of the ad hoc EU-US Working Group on Data Protection', 27 November 2013, 16987/13, https://www.consilium.europa.eu/uedocs/cms_data/docs/pressdata/en/jha/139745.pdf

[63] The New York Times. 'Daily Report: Fallout From Snowden Hurts Tech Company Bottom Lines'. Bits Blog (blog). 21 March 2014. https://bits.blogs.nytimes.com/2014/03/21/daily-report-fallout-from-snowden-hurts-tech-company-bottom-lines/

used is the *US Cloud Act* of 2018,[64] which is intended to give the US government direct access to servers of US companies abroad. The EU is currently negotiating with the US to ensure that such access to servers in Europe can only take place in accordance with European law. Needless to say, the centralization of much of the world's data in the systems of a few mega-corporations creates new security issues. But there are answers to the problems of the current data concentration in US clouds. Because as an alternative to a legal solution, one can also try to build up sufficient cloud capacity in Europe, as is being done with the Gaia X project.[65]

The complex agreements with the US, such as the *EU-US Safe Harbour,*[66] which was repealed by the ECJ, and the later *EU-US Privacy Shield,*[67] now equally repealed by the EJC[68], would not be necessary at all if sufficient cloud capacity were available in Europe for industry and private data, which would then of course also be

---

[64] "Initial legal assessment of the impact of the US CLOUD Act on the EU legal framework for the protection of personal data and the negotiations of an EU-US Agreement on cross-border access to electronic evidence", European Data Protection Board, 10 July 2019 https://edpb.europa.eu/sites/default/files/files/file2/edpb_edps_joint_response_us_cloudact_annex.pdf

[65] Bedingfield, Will. 'Europe Has a Plan to Break Google and Amazon's Cloud Dominance'. Wired UK, 27 January 2020. https://www.wired.co.uk/article/europe-gaia-x-cloud-amazon-google.

[66] Judgment of the Court (Grand Chamber) of 6 October 2015 Maximillian Schrems v Data Protection Commissioner Case C-362/14, ECLI:EU:C:2015:650, https://eur-lex.europa.eu/legal-content/DE/ALL/?uri=CELEX:62014CJ0362

[67] Commission Decision (EU) 2016/1250 of 12 July 2016 pursuant to Directive 95/46/EC of the European Parliament and of the Council on the adequate protection provided by the EU-US data protection shield, C(2016) 4176), OJ L 207, 1 August 2016, p. 1-112, 'http://data.europa.eu/eli/dec_impl/2016/1250/oj'; and 'https://ec.europa'.eu/info/law/law-topic/data-protection/international-dimension-data-protection/eu-us-data-transfers_en' and Reference for a preliminary ruling from the High Court (Ireland), made on 9 November 2009, in the case of 'https://commission.europa.eu/law/law-topic/data-protection/international-dimension-data-protection/eu-us-data-transfers_en'. May 2018, Data Protection Commissioner v Facebook Ireland Limited, Maximilian Schrems, Case C-311/18, https://curia.europa.eu/juris/liste.jsf?num=C-311/18

[68] Judgment of the European Court of Justice of 16 July 2020 in Case C 311/18, Data Protection Commissioner v Facebook Ireland Limited and Maximillian Schrems (Schrems II), ECLI:EU:C:2020:559, https://eur-lex.europa.eu/legal-content/EN/TXT/?uri=CELEX:62018CJ0311;

subject to European law. The processing of European data currently taking place in the US could easily be transferred to Europe, as data and processing programmes are very mobile. The US Cloud giants have already invested billions in Europe to build up Data Centers with sufficient capacity, in a first wave of investment after the Snowdon revelations and a second linked to the entry into force of GDPR.[69]

This example shows that some legal fixes to digital issues are not necessary, and could be remedied by technical solutions.

4.3. The Internet of Things (IoT) expands the ability to control and monitor

The term *Internet of Things (IoT)* describes the connection of movable and non-movable devices with and without sensors, which record and transmit data on their operation, the behaviour of people and the environment, using the Internet. In addition, IoT defines the whole system that results from the interaction of networked objects and the interaction of people with it.[70]

4.3.1. Devices of the Internet of Things, predictive maintenance and workplace monitoring

A small example is a thermostat in your bedroom, which uses a motion sensor or temperature sensor to detect when you get up in the morning. This allows a self-learning algorithm to preheat the car

---

69 https://www.researchandmarkets.com/reports/5550000/europe-data-center-market-industry-outlook-and

70 See EU Commission publication, The Internet of Things, https://digital-strategy.ec.europa.eu/en/policies/internet-things-policy

in front of the door or order the self-propelled car for the learned average time between you getting up and leaving the house. An alternative is parking prohibition tickets being sent directly from the detective's device to the driver's mobile phone. There are also no limits to the uses we can imagine when it comes to maintaining technical equipment. Predictive maintenance is set to become the largest digital business segment after e-commerce.[71] It is made possible by the Internet and AI, which connect sensors in every elevator, vehicle and other technical equipment to control centres that automatically send spare parts and service technicians in advance of technical failures. Over time, it is hoped that AI optimizes through self-learning this type of predictive maintenance.

Problems for the fundamental rights of citizens arise when such devices record data about people, i.e., when the Internet of Things becomes an Internet of Humans. With allegedly up to 67 billion devices connected to the IoT in 2025, the question may be asked whether there will then be any places and activities of people left that are not somehow tracked.[72] A fundamental rights-friendly development of the IoT devices will ensure that the collection of personal data of people in the vicinity of the device is excluded as far as possible.

In America, IoT is also used to constantly monitor the performance of employees. Portable devices on the belt and sensors at the workplace, such as under the desk, on the work chair or in the machine to be operated, collect data on the presence, performance

---

[71] https://www.marketsandmarkets.com/Market-Reports/operational-predictive-maintenance-market-8656856.html

[72] TechJury. '47 Internet Of Things Statistics to Show How Big It Is in 2020', 22 March 2019. https://techjury.net/blog/internet-of-things-statistics/.

and feelings of the employees.[73] In Germany, these technologies and their application in the workplace are subject to strict rules of labour law and co-determination. Constant monitoring by sensors or cameras, for example, is generally not permitted and is also viewed ambivalently by managers,[74] as is reading or listening in on Internet communications, except in special cases, e.g., in the financial markets, to prevent fraud.

4.3.2. Special case of IoT: Video surveillance of public space

A dramatic problem for freedom is the increasing surveillance of public space with cameras. With almost 40,000 of them, Berlin is ranked 93rd in the world ranking of total surveillance of public space, measured by the number of cameras per 1,000 citizens.[75] London has the, perhaps unenviable, title of being the Europe's surveillance champion, ranked sixth in the world. Chinese cities occupy the other spaces between first and ninth. The German Federal Administrative Court has put a stop to the surveillance of public space by private cameras on the basis that GDPR prohibits it. The German video surveillance law previously authorized private operators to decide by themselves whether it is in the public interest for them to monitor public space with cameras. This is no longer possible. It is still conceivable that private operators may carry out surveillance if it is absolutely necessary to avert criminal offences, and the state may do so for security purposes.[76]

---

[73] Holley, Peter. Wearable technology, The Washington Post, 28 June 2019, https://www.washingtonpost.com/technology/2019/06/28/wearable-technology-started-by-tracking-steps-soon-it-may-allow-your-boss-track-your-performance/

[74] Lisa Kuner, Ich sehe was, was du nicht siehst, FAZ 14 March 2019, https://www.faz.net/aktuell/beruf-chance/beruf/ueberwachung-im-buero-wenn-der-chef-die-e-mails-mitliest-16077634.html

[75] Bischoff, Paul, The world's most-surveilled cities, 15 August 2019, https://www.comparitech.com/vpn-privacy/the-worlds-most-surveilled-cities/

[76] Judgment of the German Federal Administrative Court of 27 March 2019 https://www.bverwg.de/270319U6C2.18.0

4.4. 5G extends the connectivity of the mobile Internet

The 5G mobile network will enable the widespread use of IoT devices. This should lead to an industrial transformation, but at the same time create the infrastructure for perfect images and sensor-based monitoring of people. 5G data connections will be able to serve millions of connected devices simultaneously with well over 10 Gigabits per second and latency times of less than 5 milliseconds. Only this network will enable a broad coverage of mobile virtual services in business and society, ranging from virtual reality to remote collaboration, for example in remote operating theatres, online health monitoring or networked cars, to drone delivery and automated driving.

China is leading the way in plans for nationwide coverage with 5G, and Huawei is the technological leader in this field.[77] 5G technology is as good for economic development as it can be for surveillance. This explains the ongoing debate about Europe's possible dependence on Huawei.[78]

4.5. Satellites, drones and autonomous killing robots

As systems of information collection and for the transport of information as well as for physical interventions, we mention here the category of 'everything that flies'. Satellites and drone images

---

77 GSMA. "New GSMA Study: China's Mobile Ecosystem Equivalent to 5.5% of China's GDP," March 20, 2019. https://www.gsma.com/newsroom/press-release/new-gsma-study-chinas-mobile-ecosystem-equivalent-to-5-5-of-chinas-gdp/.

78 Conclusions of the EU Telecommunications Council of Ministers of 3 December 2019 on 5G, document 14517/19, 'https://www.consilium.europa.eu//media/41595/st14517-en19.pdf'. ESA, "ESA - Working towards AI and Earth Observation."

combined with detailed information from the ground open up new frontiers for collecting information on people and places, the natural ecosystem and business. There is also an increasing ability to intervene directly with autonomous or semi-autonomous aircraft for still relatively harmless projects such as the delivery of orders, the control and repair of large installations or the monitoring of the environment. There is no limit to possibilities for observing earth. The earth is already photographed in high resolution several times each day. The analysis of this image by AI can be used to achieve the development goals of the UN.[79] But it can also be put to other purposes including maximizing profits in trade in grain, and strategic analysis of emerging food shortages.[80] The grain trading company Cargill, named the world's worst company in 2018,[81] contributed to a $30 million capitalization round of the emerging AI-based satellite imagery analyst Descartes Labs in California.[82] Earth observation can be used to make profits from hunger and natural catastrophes.

The use of AI for the evaluation of the enormous data treasure of earth observation raises new questions of law and ethics.[83] Satellites and drones controlled by AI can also be used to kill people. Artificial

---

[79] https://www.esa.int/Applications/Observing_the_Earth/Working_towards_AI_and_Earth_observation.

[80] Descartes Lab, 'Creating Actionable Intelligence for the Oil & Gas Sector by Fusing Commercial and Proprietary Data https://medium.com/descarteslabs-team/creating-actionable-intelligence-for-the-oil-gas-sector-by-fusing-commercial-and-proprietary-data-830f09b428ac.

[81] Mighty Earth Report, 'Cargill: The Worst Company in the World'. n.d. http://stories.mightyearth.org/cargill-worst-company-in-the-world/.

[82] Vanian, Jonathan. 'Cargill Satellites Reinvent AI', Fortune, 25 September 2018, https://medium.com/descarteslabs-team/understanding-the-changing-face-of-agriculture-with-global-sensors-f8226443db1'; for Cargill's history in the grain trade and for Earth observation see the documentary "September Wheat", Peter Krieg, 1980, https://www.salzgeber.de/presse/pressehefte/SEPTEMBERWEIZEN_ph.pdf

[83] Parcak, Sarah. "Opinion | Are We Ready for Satellites That See Our Every Move?" The New York Times, 15 October 2019, sec. Opinion. https://www.nytimes.com/2019/10/15/opinion/satellite-image-surveillance-that-could-see-you-and-your-coffee-mug.html

Intelligence and quantum computers in autonomous combat robots awaken the worst predictions of peace researchers and the strategic ambitions of military forces. There are many people who have been or could be detected and killed by a drone through a data network of mobile phone signals, satellite images and Internet use.[84] Mass surveillance, combined with analysis systems in so-called *fusion centres*, which bring together data from all possible sources, is now preparing such killing operations.[85] In doing so, "collateral damage", i.e., the killing of bystanders, is accepted.[86]

Weapons that are beyond human control after exposure, such as biological weapons, i.e., bacteria used in warfare, or landmines, are outlawed by international law. A similar treaty is urgently needed for autonomous, AI-controlled combat robots.[87]

The European Union and Germany are campaigning in the UN for a ban on autonomous combat robots. In the UN negotiations in Geneva, binding rules are aimed for by 2021.[88]

4.6. Blockchain

The term *blockchain* describes a decentralized computer system that allows secure documentation of transactions by exchange and

---

[84] The bureau of investigative journalism: Drone Warfare, https://www.thebureauinvestigates.com/projects/drone-war

[85] Cunningham, Anne C. 2016. Drones, Surveillance, and Targeted Killings. Greenhaven Publishing LLC.

[86] See the documentary film "Drones" by Tonje Hessen Schei, 2014

[87] Gusterson, Hugh. 2017. 'Drone Warfare'. AIP Conference Proceedings 1898 (1): 050005. https://doi.org/10.1063/1.5009234.

[88] European Union, External Action, International Secutity and Lethal Autonomous Weapons, 10.12.2018. https://eeas.europa.eu/headquarters/headquarters-homepage/51679/international-security-and-lethal-autonomous-weapons_en.

decentral storage of fragmented, encrypted data strings. Technically, the system is decentralized. But economically and in terms of control it can be both: decentralized or centralized.[89] At its inception *blockchain* was spurred by a libertarian ideology, without governmental systems and regulation, in order to be able to carry out financial transactions securely in a worldwide system.[90] If *Facebook* had realized its plans for the worldwide blockchain payment system *Libra*, this payment system could have become a new global power factor simply because of the large number of participants in the hands of a GAFAM company. However, the finance ministers and central bank heads of the leading industrialized countries have provisionally rejected the project in October 2019. It must be ensured "that the issuing of a currency remains a matter for states and not for large private companies", said German Federal Finance Minister Olaf Scholz.[91]

Facebook would thus have set up a system capable of undermining monetary policy based on knowledge of financial flows. Payment data could technically very easily be linked to other personal data. What we know from the *WhatsApp-Facebook affair* is that promises by Facebook to refrain from such links are worth nothing. At the time, *Facebook* had told the EU Commission that there would be no linking of data from *Facebook* and *WhatsApp*. Later, however, Facebook did link *WhatsApp* and *Facebook* data. The European Commission punished Facebook for its false statement with a fine of 110 million euros.[92]

---

[89] Hacker Noon, "Blockchain Is Not Trustless or Decentralized | Hacker Noon." https://hackernoon.com/blockchain-is-not-trustless-or-decentralized-nj8i46ji.

[90] Hacker, Philipp, Ioannis Lianos, Georgios Dimitropoulos, and Stefan Eich. 'Regulating Blockchain: Techno-Social and Legal Challenges - An Introduction'. SSRN Scholarly Paper. Rochester, NY: Social Science Research Network, 10 September 2018. https://papers.ssrn.com/abstract=3247150. ; also see https://ec.europa.eu/digital-single-market/en/blockchain-technologies.

[91] Schieritz, Mark. "Libra: Facebook wollte alles – und bekam nichts." Die Zeit. October 25, 2019, sec. economy. https://www.zeit.de/2019/44/facebook-libra-waehrung-soziales-netzwerk-mark-zuckerberg.

[92] Case M.8228 - Facebook/WhatsApp (Art. 14(1) proc.), Document C(2017)3192), 'https://eur-lex.europa.eu/legal-content/EN/TXT/?uri=CELEX:52017M8228(03)'.

It is unclear how data protection can be implemented in the *blockchain*. On the one hand, the systems are transparent, which is often an advantage. On the other hand, it should not actually be possible to delete data. It is precisely this right, however, that the GDPR, Article 17, gives us under European law: it will therefore only be possible to use blockchain legally under European law where there is no right to delete one's own data due to the exemptions to Article 17 GDPR, for example because of a legal obligation to store data, and where the protection of personal data can also be ensured in a transparent system, thus full compliance with GDPR is ensured.[93]

4.7. Artificial Intelligence

The term AI has come to be used in an inflationary way and has become a marketing tool of the first order. Today it is often claimed that a start-up works with AI, or a program contains AI, to drive up the price. But often the program does not go beyond normal automation. Moreover, the term suggests human capabilities for problem solving, which technology cannot offer at present and in all probability will not be able to achieve quickly. To ask intelligent questions, to criticise, to penetrate complex social, political or scientific questions in a thinking way – these features of human intelligence AI is yet to achieve. State of the art AI cannot even master the rules of physics. Therefore, Facebook has launched a test program that can be used to check whether an AI simulates mechanical physics correctly,[94] not

---

[93] See in detail the study by Michele Fink for the European Parliament, Blockchain and GDPR, 2019, http://www.europarl.europa.eu/thinktank/en/document.html?reference=EPRS_STU(2019)634445.

[94] Bakhtin, Anton, van der Maaten, Laurens, Johnson, Justic, Gustafson, Laura, Girshick, Ross. "PHYRE: A New AI Benchmark for Physical Reasoning." https://arxiv.org/abs/1908.05656

to mention causality, time and space relationships, which are still a long way from being handled properly by AI.[95]

Stuart Russell, one of the most important researchers in the field of AI, combines technical knowledge with social responsibility and the ambition to communicate the complex subject to a broad public.[96] He recalls that humans have various forms of intelligence and can use them simultaneously. They can read, write, calculate and speculate, speak and sing, make music, paint, they have physical movement intelligence, social and emotional intelligence and wit, the love, hate and feel compassion. They can deploy these faculties at the same time and blend them together. A program of today's AI like Alpha Go from *DeepMind* can win against humans in the game *GO*. But other than that, AI is currently unable to establish a correlation between the various forms of intelligence.[97]

What AI can currently do is calculate and optimize, starting from a data set to which a learning program is applied. This is called *machine learning (ML)*.

Roughly speaking, ML is being developed and already applied in three areas today: Image recognition and analysis, analysis of unstructured data and human speech and language processing.

---

[95] Marcus, Gary, and Ernest Davis. "Opinion | How to Build Artificial Intelligence We Can Trust." The New York Times, 6 September 2019, sec. Opinion. https://www.nytimes.com/2019/09/06/opinion/ai-explainability.html. Example is the Animal AI Olympics project at Imperial College - where they compare animals (rats or flies) and their capacities for navigation and environmental understanding. The tests there are very rudimentary and yet most AI algorithms (deep reinforcement learning) cannot cope with them.

[96] See also: Smith, Brian Cantwell. The Promise of Artificial Intelligence. Cambridge, MA: The MIT Press, 2019. Russell, Stuart. Human Compatible: AI and the Problem of Control. 1st Edition. Allen Lane, 2019.

[97] Currently, researchers at Washington State University are working on a general IQ test for AI: https://magazine.wsu.edu/2023/02/08/when-will-artificial-intelligence-really-pass-the-test/

Today, *ML* must refer to a very specific area and is narrowly limited, for example, only to image recognition in the diagnosis of breast cancer. It is in this area, as well as in unstructured data, that *ML* is most advanced.[98] It is still weakest in the understanding and creation of language. The fact that language is now properly translated[99] and properly transcribed, that short texts for journalistic purposes, based on figures, in the field of stock market reports and sports, are written by AI, should not expose us to the illusion that AI is capable of free, understanding language creation. Major conceptual breakthroughs would be needed to create AI with human capabilities such as language comprehension, integration of learning and knowledge, and the cumulative collection of concepts and knowledge. In other words: wisdom will not reach AI, even in the long run.[100] What AI can already do, how it develops, how much is invested where and how much AI engineers can earn, is the subject of extensive annual reports, which are themselves part of the hype about AI.[101]

ML always refers to existing data, i.e., empirics. What Kant and Hume taught us, that one cannot conclude from what is to what should be, is a big hurdle for *ML*, which is always based on being. It only reaches beyond optimization processes to a limited extent. It cannot dream and be creative, and certainly cannot determine a "shall". Of

---

[98] Bera, Kaustav, Kurt A. Schalper, David L. Rimm, Vamsidhar Velcheti, and Anant Madabhushi. "Artificial Intelligence in Digital Pathology — New Tools for Diagnosis and Precision Oncology." Nature Reviews. Clinical Oncology 16, no. 11 (November 2019): 703–15. https://doi.org/10.1038/s41571-019-0252-y. See on this: Olk, Julian. "Merantix Healthcare: Erste Zulassung für digitale Krebsdiagnose im Hochrisiko-Bereich." https://www.handelsblatt.com/unternehmen/industrie/merantix-healthcare-erste-zulassung-fuer-digitale-krebsdiagnose-im-hochrisiko-bereich/25149050.html.

[99] The automatic translations of the Cologne-based company DeepL https://www.deepl.com/de/translator seem to us to be better than those of the EU Commission and Google.

[100] AI: an international dialogue, Note of a UK Royal Society – US National Academy of Sciences workshop in Washington, D.C., 23 - 24 May 2019, https://royalsociety.org/-/media/policy/topics/open-science-data/ai-international-dialogue.pdf?la=en-GB&hash=A888136F57440CA9833F3DA79D683320

[101] For example: State of AI Report 2019: https://www.stateof.ai; and the AI Index Report of Stanford University: https://aiindex.org

course, AI can paint pictures in a new way or compose something that sounds quite pleasant and new. But in these examples, rules are applied, then perhaps minor variations around certain elements are inserted. This has nothing to do with creativity and the finely tuned intuition of the human artist. Even if machines were fed with the enormous data about man-made works of art, they could still not create real art because of the lack of corporeality, the lack of feelings like joy and pain and the lack of awareness of their own finiteness.[102]

Please get in touch with us if you discover an AI that composes good songs with interesting lyrics, that creates a new leap in music, such as the development of 12-tone music by Arnold Schönberg or punk rock, that makes innovations in painting such as abstraction, cubism or new realism. In contemporary art, an AI picture may deceive even connoisseurs; but in literature this doesn't work, because real language creation is too complex. Test the new *speech generator GPT-2* and see for yourself how rudimentary it is.[103] It will be a long time before AI easily simulates real literature.

Just as the *creativity* of AI is often overestimated, the importance of knowledge about the subject area in which machine learning is to take place is often underestimated. This is the so-called *domain knowledge,* besides program and data sets, an important third component in the development of *ML*. The idea that one only has to present enough mathematics textbooks to a learning *ML software,* then it will discover Einstein's theory of relativity by itself, is a pipe dream that ML cannot fulfil.[104] This is true even if one acknowledges

---

[102] See Nick Cave's answer to a fan's question about whether AI can surpass humans in writing songs: Cave, Nick. "Artificial intelligence can't write great songs. Only human limitations can do that", "Nick Cave on Why AI Will Never Write a Great Song." January 28 2019, https://www.vice.com/en_us/article/nexbnw/nick-cave-on-why-ai-will-never-write-a-great-song.

[103] Everyone can test the AI here: https://talktotransformer.com and https://chat.openai.com/auth/login

[104] According to Prof. Wahlster, head of DFKI, https://www.dfki.de/web/ in a conversation with Paul Nemitz.

that AI programs already make their own proofs and independently draw conclusions that even surprise mathematicians.

Learning does, however, change the programme and it is conceivable that learning in the workplace may lead to unforeseen results. In order to avoid this, only supervised AI learning is allowed in critical areas. Free, unsupervised learning of AI is risky and not very common in practice.

The theoretical possibility that new causalities are set quasi autonomously, that programs further develop autonomously in a way not foreseeable for the creators of the respective *ML* software, is the reason why ethicists and lawyers are intensively researching *ML* today. This element of autonomy distinguishes *ML* from classical automation, which basically relies on a predictable series of if-then relationships. The autonomous development of the program through learning and the decision-making character of what comes out of these systems together make it necessary to tie them more closely to law and ethics. This is already the case now,[105] and not later when we may have developed a strong or general AI that can transfer learning from one area to another, that gives itself purposes and rules, and that can re-programme itself and thus potentially evade human control. Whether it will then be superior to humans in all areas, as the enthusiasts of AI claim, is a question that is left open. As said, such a general AI is currently not in sight, but it is conceivable. And therefore, its control must be researched and discussed today. In addition, AI - and also algorithms without AI - are already being used billions of times a day. Because of this broad scale of application

[105] See in detail the work of the EU High Level Group on Trustworthy AI, https://digital-strategy.ec.europa.eu/en/policies/expert-group-ai , and the work of the German Government *Data Ethics Commission*, https://www.bmi.bund.de/EN/topics/it-internet-policy/data-ethics-commission/data-ethics-commission-node.html

alone, they must be subject to stricter rules than humans themselves. Humans cannot solve certain tasks a billion times a day, and thus humans as individual decision-makers can cause less damage than an automated AI system; moreover, in making mistakes and biased decisions, humans are less consistent than machines.

This is also seen in America, especially in science:

> *"Even without reaching human-level intelligence, the consequences of AI for individuals, communities and societies are potentially profound. Systems today create convincing fake images, text and video online, which are influencing how people relate to the information they see; they are automating tasks in ways that shape the nature of work; they are enabling the development of automated weapons systems; and they are fostering new forms of dependency on technology.*
>
> *At this stage, research and policy both play important roles in shaping the direction of technology development and in helping to share the benefits of AI technologies across society".*[106]

4.7.1. Artificial Intelligence and human stupidity

Let's come to the problem of the term Artificial Intelligence. The name AI was coined at the Dartmouth Conference in 1956, where participants were focused on raising funds for their research.

The conference participants, all leading mathematicians, computer scientists and computer pioneers, including Marvin Minsky, applied

[106] AI: an international dialogue, Note of a UK Royal Society – US National Academy of Sciences workshop in Washington, D.C., 23 - 24 May 2019, https://royalsociety.org/-/media/policy/topics/open-science-data/ai-international-dialogue.pdf?la=en-GB&hash=A888136F57440CA9833F3DA79D683320 Page 7

for a $13,500 grant from the Rockefeller Foundation. In addition to $1,200 for the graduate students and $700 for the undergraduate students, they also requested money for travel expenses. The origin of the term AI was therefore in some ways marketing. In their grant application, they said:

"The study is to proceed on the basis of the conjecture that every aspect of learning or any other feature of intelligence can in principle be so precisely described that a machine can be made to simulate it. An attempt will be made to find how to make machines use language, form abstractions and concepts, solve kinds of problems now reserved for humans, and improve themselves."[107]

McCarthy suggested the term "Artificial Intelligence." But how exactly did the Dartmouth participants define human intelligence that is to be simulated by machines? Another quote from the grant application provides some insight:

> *"It may be speculated that a large part of human thought consists of manipulating words according to rules of reasoning and rules of conjecture. From this point of view, forming a generalization consists of admitting a new word and some rules whereby sentences containing it imply and are implied by others."*

Intelligence is thus essentially reduced here to the ability to manipulate words. But does this do justice to the manifold aspects of human thinking?

The very meaning of the word intelligence indicates that this is not the case. Intelligence from the Latin *intellegere* means to *recognize,*

---

[107] Quoted from Kaplan, Jerry. Artificial Intelligence. New York, NY, United States of America: Oxford University Press, 2016. p. 27. For the original, see: http://www-formal.stanford.edu/jmc/history/dartmouth/dartmouth.html.

to *see*; to *understand* literally to *choose between* ... from Latin inter (between) as well as *legere* (read).[108]

Legere means, like the Greek *legein*, from which *logos* is derived, reading together, collecting and calculating as well as explaining and giving account.
Can a machine do that?

Recognize? Sure: images, language, data and patterns. But it cannot determine the truth of a statement, or deem if an image is particularly beautiful or arresting. Can machines "choose between"? Yes, but without understanding and without seeing. Patterns "understanding" works where mathematical and logical operations are present. But hardly, if by *understanding* we mean understanding the other person or literary texts.

Hermeneutics is concerned with how to understand texts. It defines the understanding of texts as textual criticism. The aim is to distinguish between the different versions, to check translations and to make sense of the content in context. Such an understanding is not possible without a preceding comprehensive understanding of the world.

Such a comprehensive understanding of the world is a long way off for AI systems, because they can only process data, but not perceive, feel or reflect the world with consciousness. They cannot, like humans, give good reasons, if only because the concept of "good" is impossible for them to ascertain, and even less so an understanding that this "good" for humans can always be something very different

---

[108] From: ‹https://de.wikipedia.org/wiki/Intelligenz XE "Intelligenz" ›. Also: Klotsche, Martin J. quoted in Greenberg, Sidney. A Treasury of the Art of Living. Pickle Partners Publishing, 2018

from different points of view. And instead of the participant's perspective, they are at best capable to simulate an observer's perspective.

Even the young AI stars at Dartmouth couldn't have possibly trusted a machine with all these abilities. Having insight, understanding, accountability and good reasons - a research proposal to teach a machine all this would probably have been rejected, even by the wealthy Rockefeller Foundation.

Minsky and his comrades were clever enough to claim that the machine could only perform specific functions within their research project - and yet still call it intelligence. In doing so, they went beyond pure computation and suggested that computers should also be used in the field of symbolic logic. Especially after the first successes of computers in the Second World War, they hoped to make progress. Jerry Kaplan writes: "*Placed in this historical context, the Dartmouth conference could be seen as an attempt to expand the use of computers beyond crunching numbers and processing data to manipulating symbols.*"[109]

With all the miraculous achievements attributed to Artificial Intelligence today, one should therefore bear in mind that it was already claimed at the beginning that a small excerpt from the possibilities of human thinking would explain the whole of it. Not much has changed in this respect to this day.

The problem of these terms can be described by recalling the classical definition procedure of philosophy: A term is defined by the specification of the next higher genus and then by the designation

---

[109] Kaplan, Jerry. Artificial Intelligence: What Everyone Needs to Know. Oxford University Press, 2016. P. 14

of the specific difference. *Genus proximum et differentia specifica*[110] has been the formula for constructing definitions since scholasticism.

The problem with the concept of Artificial Intelligence is that even the specification of the genus *intelligence* is incorrect because it is incomplete. The difference with human thinking, which is in the word *artificial*, highlights the difference of machine intelligence from human intelligence, but cannot make up for the confusion that both concepts of intelligence include something completely different. Artificial Intelligence is therefore inherently a vaguely defined and thus confusing concept, which should actually be replaced by the phrase 'artificial partial intelligence'. Many people therefore replace it with the more precise term *machine learning*.

However, the term *ML* is also a misleading metaphor: because the processes of learning in a human being have not yet been fully researched. According to general understanding they are an essential part of the formation of the human being into a self-conscious, self-confident and self-determining subject. The statistical and probabilistic calculations of neural networks, on the other hand, lead to the conclusion that their learning is nothing more than optimised probability calculations.

It is already suggested that robots that are operated with *ML* should be regarded as children. Like children, they continue to develop and would certainly one day become independent, which is something parents usually wish for. If this were to happen, as suggested by

---

[110] Genus–differentia is a definition composed of two parts, the genus or family which refers to an existing definition that serves as a portion of the new definition; all definitions with the same genus are considered members of that genus, and the differentia: The portion of the definition that is not provided by the genus. https://en.wikipedia.org/wiki/Genus%E2%80%93differentia_definition#:~:text=A%20genus%E2%80%93differentia%20definition%20is%20considered%20members%20of%20that%20genus

philosopher Janina Loh, for example, we would have to grant such robots human-like rights.[111] Another philosopher, Thomas Metzinger, has pointed out that such an attribution of rights to machines could be a huge obstacle to innovation, because we would then no longer be allowed to treat machines as objects and would need to show consideration for them.[112] The fact that we fundamentally devalue human rights if we grant them to non-human actors is yet another matter of concern.

We will continue to use both the terms Artificial Intelligence and Machine learning in this book despite their blurry boundaries. But by pointing to the fog in the definition, we want to promote a critical use of the terms.

The philosopher Ludwig Wittgenstein views the task of philosophy as giving orientation, thus to "show the fly the way out of the fly-bottle."[113] He probably would have described what is now coined Artificial Intelligence rather as *limited control competence.*

But instead of introducing new terms, we want to be sensitive to the fact that terms should not be adopted and used uncritically, but rather to search for more precise definitions and to ask about the intentions of the word creators in the first place.

---

[111] Loh, Janina. Roboterethik: Eine Einführung. Suhrkamp Verlag AG, 2019.

[112] Metzinger, Thomas, "2015: What do you think about machines that think", https://www.edge.org/response-detail/26091 , and Herwig, Sara, "Will robots suffer in the future just like us? Interview by Sarah Herwig with Thomas Metzinger, SRF, 14 November 2011, 'https://www.srf.ch/kultur/gesellschaft-religion/leiden-roboter-in-zukunft-genau-wie-wir

[113] When asked "what is your goal in philosophy?" Wittgenstein said: "To show the fly the way out of the fly glass" Wittgenstein, Ludwig. Philosophical Investigations. Edited by P. M. S. Hacker and Joachim Schulte. 4 edition. Chichester, West Sussex, U.K. ; Malden, MA: Wiley-Blackwell, 2009. Section 309

4.7.2. Artificial Intelligence or predictability through machines?

Machine learning or Artificial Intelligence can be defined as the ability to solve difficult problems as autonomously as possible, i.e., without clear instructions. The difference between conventional algorithms and machine learning is that in conventional algorithms the model for solving the problem is given but with AI the model is developed automatically from the data. If the latter is the case, it is machine learning, which is made possible by neural networks connected in series.

Machine learning is in parts also *probabilistic learning,* because the programs couple probability calculations with heuristic procedures in order to come faster to a result.

The distance from human learning is still considerable. This doesn't only apply to the number of connections that currently (still) distinguish the human brain from neural networks. Eng Lim Goh, Head of technology at *Hewlett-Packard Enterprise,* who is also responsible for *Artificial Intelligence* at HP, sums up the current gap:

"The human brain has over 100 trillion connections, which are very complex and therefore recognize the context of situations much faster. The currently most complex artificial neural network has only 100 billion connections, so it needs more information to learn."[114]

But let us be wary of inferring qualitative differences from purely quantitative distances. Even if it is true that quantity can turn into

[114] Hewlett Packard Enterprise. 'A Universe Full of Opportunities, for AI and Supercomputing'. n.d. https://www.hpe.com/us/en/newsroom/blog-post/2018/03/a-universe-full-of-opportunities-for-ai-and-supercomputing.html

quality, even a numerical *superiority* of the connections of neural networks would not allow us to draw conclusions about the *superiority* of AI. We still know far too little about the human brain to make such inferences.

4.7.3. AI and automation: Close together, but not the same

Of course, automation gives rise to legal and ethical questions below the level of the learning program. For example, the EU GDPR stipulates that everyone has a right to be the subject of a human decision, i.e., that, in principle, humans may not be the subject of machine decisions, if they object. The GDPR also gives every individual far-reaching information rights in situations of automatic processing of their personal data. Those responsible for processing of the personal data have to give those concerned a deep insight into their algorithms and AI programs.[115] If necessary, this can be done under confidentiality conditions and with appropriate precautions. One example is the in-camera procedure, as lawyers say, in which secrecy is maintained by allowing files or programs to be viewed only in a secure room and without carrying a telephone, camera or personal computer. There is only the pre-installed computer, paper and pencil. GDPR contains an obligation to provide "meaningful" information if the program is too complicated and the person seeking access cannot be reasonably expected to understand it without help.

But such rights must be demanded and exercised! They are not granted automatically! In other words: Those who do not object to automatic processing may receive automatic decisions. And those who never ask for insight and explanation about algorithms and AI should not be surprised if one day they no longer understand the world and what happens to them.

---

[115] Articles 13-15 in conjunction with Art. 22 GDPR.

For the state there are even stricter rules. German administrative law provides that only where there is a specific legal basis, automatic decisions by the state are permitted and possible, and only if the wording of the norm does not contain a margin of appreciation nor the decision does require the exercise of discretion:

> *" § 35a VwVfG - Fully automated adoption of an administrative act*
>
> *An administrative act may be adopted entirely by automatic means, provided that this is authorised by law and that there is neither discretion nor a margin of appreciation."*

In other words, the legislator must first decide on specific rules before the state can decide anything automatically. This is an important fundamental decision that has been taken in our democracy. It ensures that the administration may not autonomously automate entire areas and turn citizens into objects of machine decisions. The democratic legislator has to allow this in a law, that is, a democratic decision on automation, its limits, and the corresponding protective mechanisms are a precondition for automation of state decisions.

If the state, after fulfilling a normative condition, has the possibility to choose between several options for action, thus it has discretion, then this choice cannot be made by a machine. So as soon as a certain level of deliberation is necessary for the decision, then we should hand the decision back to humans. This is a good thing.

The German tax code contains such a special legal basis, which was already accepted in the Bundestag, the German Parliament. It determines for the tax assessment according to § 155 paragraph 4 since 1.1.2017:

> *"The tax authorities may carry out, correct, withdraw, revoke, cancel or amend tax assessments as well as credits for tax deductions and advance payments on the basis of the information available to them and the taxpayer's details, exclusively by automated means, provided that there is no reason for the individual case to be dealt with by public officials".*

An interesting question is whether these regulations also allow the use of special or even general Artificial Intelligence or only mechanistic automation. The exclusion of discretion, i.e., the areas in which learning, experience and weighing play a role, indicates that AI that learns and weighs itself and whose results are not predictable, because it does not make decisions in a mechanistically automated way, must not yet be used on these legal bases. The use of AI for automated state decision making would thus require a new legal basis, which would ensure that the law is observed and that human beings are able to control the processes and that the processes as well as outcomes are verifiable and explainable. In other words, the transition from mechanistic automation with predictable if-then relationships to the use of learning systems in the decision-making process, the results of which are no longer predictable, must be consciously shaped by the legislator in a democracy. This is because the use of Artificial Intelligence transfers more power to machines than simple automation. Depending on the area of application of Artificial Intelligence, the authorization basis will then have to contain special protective provisions to safeguard the rule of law, in particular the explainability of the decision, a precondition for judicial control, democracy and fundamental rights.

4.8. Brain-Internet connectivity and biophysical systems

The dream of the AI scene around GAFAM, in line with the philosophy of total technical feasibility, is to physically connect people directly to the Internet, to promote their health, to optimize their genes. In short: the omnipotence mania of those successful in the digital industry is turning towards biology and the human brain. With a lot of money from Silicon Valley, projects for healing through genetic manipulation, for connecting the Internet and the brain[116] and, in general, for health through AI are being supported. Elon Musk, the boss of the car manufacturer Tesla, is investing 168 million dollars[117] in his project *Neuralink*, which aims to physically connect the human brain to computers.[118]

Projects such as this often rely on the two major global programmes for understanding the brain, the *Obama White House Brain Initiative*[119] and the EU's brain research programme, which has been allocated over 6 billion euros from 2007 to 2020.[120] The still inexplicable of biology and the brain, our thinking and our soul is to be conquered and connected to the global digital analysis and control system.

Much of what sounds great in the vision does not work or only works to a limited extent. Research results are exaggerated in their importance to obtain even more funding.[121] But some perspectives of

---

[116] Martins, Nuno R. B., Amara Angelica, Krishnan Chakravarthy, Yuriy Svidinenko, Frank J. Boehm, Ioan Opris, Mikhail A. Lebedev, et al. "Human Brain/Cloud Interface." Frontiers in Neuroscience 13 (2019). https://doi.org/10.3389/fnins.2019.00112. MIT Technology Review, 'Facebook is funding brain experiments to create a device that reads your mind', 30 July 2019, https://www.technologyreview.com/s/614034/facebook-is-funding-brain-experiments-to-create-a-device-that-reads-your-mind/ ; original Study Moses, David A., Matthew K. Leonard, Joseph G. Makin, and Edward F. Chang. "Real-Time Decoding of Question-and-Answer Speech Dialogue Using Human Cortical Activity." Nature Communications 10, no. 1 (July 30, 2019): 3096. https://doi.org/10.1038/s41467-019-10994-4.

[117] Corbyn, Zoe, "Are brain implants the future of thinking?" The Observer, 22 September 2019, https://www.theguardian.com/science/2019/sep/22/brain-computer-interface-implants-neuralink-braingate-elon-musk

[118] https://neuralink.com/

[119] The Brain Initiative, https://obamawhitehouse.archives.gov/BRAIN.

[120] EU Commission publication: Brain Research, 'https://www.humanbrainproject.eu/en/'

[121] Lewis, Tanya. n.d. 'Elon Musk's Secretive Brain Tech Company Debuts a Sophisticated Neural Implant'. Scientific American. https://www.scientificamerican.com/article/elon-musks-secretive-brain-tech-company-debuts-a-sophisticated-neural-implant1/

technical feasibility no longer seem completely unrealistic. This also raises new ethical questions about the limits of technical penetration into the human mind and soul, as far as this does not serve to cure disease.[122]

4.9. *Virtual Reality* (VR) and *Augmented Reality* (AR)

The technologies of virtual reality immerse us in an artificially created view of the world. Today, smart glasses enable this to happen, controlling what you see, or even offering interactive dialogue with persons or realities represented in the glasses, through speech and your own movement.

Pilots train in cockpit simulators of *virtual reality*: the trainee sees on a screen in front of them, what a pilot sees, the cockpit equipment is running, and the cockpit is in motion, they get to experience how an airplane would move, and hear the sounds of the machines.

Organizations are increasingly relying on VR to onboard and train staff. Before getting a job at *Walmart*, America's largest employer, applicants are subject to a skills test using video glasses. *Amazon* also uses *Virtual Reality* to quickly train employees.[123]

Augmented or *improved/completed reality* adds new elements to the perception of reality. The *Pokémon Go* game is an example of this. The worker in a factory who assembles parts of a machine while

---

[122] iHuman, Blurring lines between mind and machine, The Royal Society, September 2019, ‹https://royalsociety.org/-/media/policy/projects/ihuman/report-neural-interfaces.pdf›.

[123] Holley, Peter, "Walmart Has Added Virtual Reality to Its Assessment of an Employee's Potential." Washington Post. 12 July 2019. https://www.washingtonpost.com/technology/2019/07/12/walmarts-latest-tool-assessing-whether-employees-deserve-promotion-virtual-reality/

looking through a screen at his hands, which appear green when he's assembling correctly and red when he's making mistakes.

Virtual Reality (VR) and *Augmented Reality* (AR) invite manipulation: The fact that the product advertised in VR is not in reality as advertised, is still the most harmless problem around. But this is no longer so harmless when VR is used to re-programme how people think and feel, especially children. The effect of VR on people goes beyond that of television and Internet, as it creates the illusion of real participation and simulates real experience. A code of conduct from scientists warns to be on guard against this potential, which can be used for many purposes.[124]

4.10. Quantum computer and quantum Internet

High-performance computers, right up to quantum computers,[125] process the data collected via the Internet into usable information and perform an important control function. The quantum computer[126] will be joined by the quantum Internet,[127] with its corresponding transmission capacities.

It is generally assumed that operational quantum computers will exceed the computing power of even the strongest of today's conventional computers many times over.

---

[124] Madary, Michael, and Metzinger, Thomas K. "Real Virtuality: A Code of Ethical Conduct. Recommendations for Good Scientific Practice and the Consumers of VR-Technology." Frontiers in Robotics and AI 3 (2016). https://doi.org/10.3389/frobt.2016.00003., https://philpapers.org/archive/MADRVA-3.pdf; find at http://uxmmersive.com/en/meeting-michael-madary-virtual-reality-and-new-technologies-ethics/

[125] EU Commission publication, 'High-Performance Computing', 'https://ec.europa.eu/digital-single-market/en/high-performance-computing

[126] EU Innovation Initiative: Quantum Flagship: 'https://qt.eu/'

[127] https://digital-strategy.ec.europa.eu/en/policies/european-quantum-communication-infrastructure-euroqci

Whoever has the most powerful computers will be able to make the most money out of data collected free of charge for them, said Jaron Lanier even before Shoshana Zuboff. He lamented the fraudulent way in which the large Internet companies keep their accounts, which does not compensate anyone who contributes to the efficiency of GAFAM through data or intellectual performance. For example, those who provide a translation used to train a translation program are not paid. Jaron Lanier received the Peace Prize of the German Book Trade for his book "Who owns the future?".[128] In it, he warns of siren computers that attract all data and processing and give their operators an unassailable lead in the new data world.[129]

Google, one of the richest companies in the world, and the German Research Centre for Quantum Computing in Jülich have decided on a "partnership".[130] *Google* has similar "partnerships" with other research centres working on quantum technology.[131] This is *Google's* tactic - which can also be observed prominently in the mobility and health sector:[132] through these collaborations with smaller partners, Google is acquiring new data and knowledge.

---

128 Microsoft, 'Jaron Lanier Wins German Book Trade's Peace Prize'. 2014. Microsoft Research (blog). 5 June 2014. https://www.microsoft.com/en-us/research/blog/jaron-lanier-wins-german-book-trades-peace-prize/

129 Lanier, Jaron. Who Owns the Future? Export ed. edition. New York: Simon & Schuster, 2014.

130 Forschungszentrum Jülich, 'Forschungszentrum Jülich and Google announce research partnership', 8 July 2019 https://www.fz-juelich.de/SharedDocs/Pressemitteilungen/UK/EN/2019/2019-07-08-quantum-computer-fzj-google.html

131 UCL, 'UCL partners with Google', 25 September 2018. https://www.ucl.ac.uk/news/2018/sep/ucl-partners-google-grow-quantum-software-industry-uk; The Register 'Google, Volkswagen spin up quantum computing partnership', 8 November 2017, https://www.theregister.co.uk/2017/11/08/google_vw_spin_up_quantum_computing_partnership/; Auto Car Po, 'Daimler partners Google to research application of quantum computers', 21 March 2018 https://www.autocarpro.in/news-international/daimler-partners-google-research-application-quantum-computers-28691

132 Google's contracts with hospitals and healthcare services around the world are not consolidated. However, Google's website does contain some research on health: https://ai.googleblog.com/search/label/Health?max-results=11; and on its partnership with the British National Health Service (NHS) https://www.theregister.co.uk/2018/06/15/deepmind_health_independent_panel_report_relationship_google_business_model/

The development of quantum technology and quantum computers, combined with a new quantum Internet, is mainly driven, on the private sector side, by GAFAM, Intel and IBM. *Google* recently announced that it has solved a first previously unsolvable problem by means of a quantum computer.[133] IBM disputed some of *Google's* claims.[134] However, the calculation was generally regarded as a breakthrough.[135] Google itself spoke of it as a way to achieve quantum superiority; that is, to prove the superiority of the quantum computer over all available supercomputers.[136]

---

[133] Arute, Frank, Kunal Arya, Ryan Babbush, Dave Bacon, Joseph C. Bardin, Rami Barends, Rupak Biswas, et al. "Quantum Supremacy Using a Programmable Superconducting Processor." Nature 574, no. 7779 (October 2019): 505–10. https://doi.org/10.1038/s41586-019-1666-5

[134] Novet, Jordan. "IBM and Google Disagree on Quantum Computing Achievement." CNBC, October 23, 2019. https://www.cnbc.com/2019/10/23/google-quantum-computing-supremacy-claim-disputed-by-ibm.html

[135] Tomaso Calargo, Research Centre Jülich in https://podcasts.apple.com/de/podcast/iq-wissenschaft-und-forschung/id259863713?i=1000451929174

[136] Cossins, Daniel. "It's Official: Google Has Achieved Quantum Supremacy." New Scientist. https://www.newscientist.com/article/2220968-its-official-google-has-achieved-quantum-supremacy/

B5 Overview of technologies connected to the Internet and analysis of the power of the GAFAM corporations

The technical system of digital control is based on the Internet as a multi-way medium that connects all people with all people, all people with all devices and all devices with all devices. However, we must also look at the third aspect of the processing of information on the Internet: That everything that is exchanged between people or with machines can and is simultaneously logged and evaluated in central processing systems. Whoever has a hand on these central storage and evaluation systems has the power in the current system of the Internet. At present, it is mainly GAFAM and its sisters in China. States have similar ambitions, not only China. The greater the transmission and processing capacity, the faster, the more attractive and powerful these systems become.[137]

For the philosopher Immanuel Kant there existed

> *"the starry heavens above me and the moral law within me."*[138]

He placed the seeing and thinking human being in the centre of the knowledge of the world, the knowledge of the earth and the universe through science and the knowledge of the right action ("the moral law within me") through human thinking.

The ambition of those who drive the global technical system is to demonstrate that not man, but the technical system is the centre of

---

[137] Lanier, Jaron. Who Owns the Future? Export ed. edition. New York: Simon & Schuster, 2014.

[138] Kant, Immanuel: Critique of Practical Reason. Cambridge University Press, 1997. P. 133

both world knowledge and world control. According to the *Californian ideology*, the Hegelian world spirit[139] thus becomes controllable in the global technical system. It controls all knowledge and all control. It is located in the control centres of the Internet and AI. The moral law that we humans have to determine and that only makes sense if we are free in our decisions, would then lose its meaning. Because in this world view the decisions are made for us by the system.

Hidden behind the term *networking*, concealed by the promise "You can now speak with everyone, everyone will hear you, you can see all information, all possibilities are open to you", is the vision of a total recording system. This should first capture understanding and control of the whole cosmos down to each individual human being and his or her innermost being and the interaction between them. Based on the recording of the state of being and its dynamic changes - both in real time - this system should make possible a total, automated control and manipulation, even unnoticed against the will of the individual or a collective.

It is thus possible for it to impose its will, which is the will of its masters, on all others, whether individuals, collectives, companies or states.

Indeed, the real dependence of central control and functional systems in all areas of government, economy and society on the systems operated by the digital giants is already well advanced. The same applies to the embedding and recording of our own individual lives in these systems, to the feeling of always having to be on our devices, to the addiction to a constant presence on the net and to the helplessness of being at the mercy of the systems that seem to embrace us lovingly and caringly.

---

139 Hegel, Phenomenology of the Spirit, Vol. 3, p. 585: "Before the spirit is not completed in itself, not as world spirit, it cannot reach its completion as self-conscious spirit".

The new thing about digital technologies, AI and the Internet and its connectivity is that they are *general purpose technologies (GPTs)*, which, unlike their predecessors such as the steam engine and electricity, are ideal for use as domination technologies. This is exactly what we are now seeing in China where the Communist Party is trying to update the old dream of achieving communism through technology. Now it is using these new technologies to bring the masses to heel, to offer them comfort, but also to subject them to tangible permanent control through social scoring. Earlier GPTs, such as the steam engine, electricity, or the electric motor, were never developed with the aim of directly influencing people's thoughts and feelings. These earlier technologies have also changed and improved people and their lives. But the new technologies of the Internet and AI have a much deeper and more direct impact on the way an individual lives, communicates, thinks and feels. And so, they also capture and control the public and democracy.

This is exactly what GAFAM is now demanding of all its developers: Whether it is the advertising-driven business models of *Google* and *Facebook*, the iPhone of *Apple* or the applications of *Microsoft*, it is all about creating human habituation and dependence, a feeling of pleasant satisfaction and comfort when the equipment or services are used, a feeling of dissatisfaction and helplessness when they are not used or cannot be used.

All these ambitions of companies and the state, whether in the US or in China, are about this direct control of man and the entire system of man, society and technology. At Stanford University, an institute was founded before the turn of the millennium with the aim of training engineers to use technology to change people's beliefs, the Stanford Persuasive Tech Lab.[140] At the same time, research began on

---

[140] See: https://captology.stanford.edu/ and https://behaviordesign.stanford.edu/

affective computing, i.e. the possibility of using computers to influence people's feelings, to understand them or, vice versa, to equip computers with emotional simulation.[141] The training of engineers in these areas, where attempts are being made to incorporate an emotional function into the machine that simulates compassion and empathy for the human being, shows how far the ambitions for controlling and mastering the human being by machines go - right up to the reports of GAFAM's attempts to find programming that, by stimulating the reward system and by releasing dopamine, puts people into emotional and biological dependence.[142]

Already today, control signals can be transmitted by means of an electrode cap on the head in such a way that a person will stick out the tongue even against its will. Another person connected as a controller by means of another cap gives the command through thinking only. Much more seems possible, such as controlling speech or "playing" videos directly from the brain.[143] *Facebook* was funding research to connect the Internet directly to the human brain and read it.[144] What looks like convenience or even real help, for example when people have no hands and cannot speak, also opens a view directly into the human brain, with the aim of giving the system knowledge of the brain's contents - directly, and not just by observing behaviour.

---

141 Affecting Computer 'https://affect.media.mit.edu/'.

142 Dumaine, Brian. 2020. Bezonomics: How Amazon Is Changing Our Lives, and What the World's Companies Are Learning from It. Simon and Schuster.

143 Brain Computing Interface, 47 Stories, MIT Technology Review ‹https://www.technologyreview.com/humans-and-technology/brain-computer-interface/.

144 Regalado, Antonio, 'Facebook is funding brain experiments to create a device that reads your mind', MIT Technology Review, 30 July 2019, https://www.technologyreview.com/s/614034/facebook-is-funding-brain-experiments-to-create-a-device-that-reads-your-mind/; also see Benjamin, Garfield. "Silicon Valley Wants to Read Your Mind – Here's Why You Should Be Worried." The Conversation. http://theconversation.com/silicon-valley-wants-to-read-your-mind-heres-why-you-should-be-worried-121707.

Convenient services such as entertainment, communication, information, education, mobility, pay, health, trade, energy and security provide access to people and the policy-making system. And this in turn enables GAFAM, combined with the usual means of political influence by large corporations, to control the politically set conditions of their profitability. At least, this is the hope of the magicians of technical and economic total renewal. Whether this calculation works out depends on all of us, not only on our behaviour as consumers of these services of convenience, but above all on our commitment as citizens of a democracy and as entrepreneurs in the development of alternative technologies, technology use and business models.

B6 Concentration of power in the hands of the digital-technological-economic complex

The central players of digital power are the digital corporations like *Google, Amazon, Facebook, Apple* and *Microsoft* from the US and *Tencent, Alibaba* and *Baidu* from China. These companies are surrounded by ecosystems of research and business that depend on them and strengthen their dominance by developing and differentiating technology and business models.

All new technologies, such as Artificial Intelligence, the new mobile phone standard 5G, quantum computers and also the connection of technology and humans by means of biophysical systems[145] will exponentially increase the power of the players who already dominate the Internet, if no specific measures are taken against this concentration of power.

Undoubtedly, one has to distinguish between the Internet and what happens on the Internet, just as one has to distinguish between the theoretical potential of AI in general and the context and objectives for which it is actually developed and used. However, it would be naïve to ignore the fact that for most people today, the reality of how they use the Internet and what the Internet offers them is shaped by a few large corporations.

---

[145] Regalado, Antonio, 'Facebook is funding brain experiments to create a device that reads your mind', MIT Technology Review, 30 July 2019, https://www.technologyreview.com/s/614034/facebook-is-funding-brain-experiments-to-create-a-devie-that-reads-your-mind/; https://www.technologyreview.com/2019/07/30/133986/facebook-is-funding-brain-experiments-to-create-a-device-that-reads-your-mind/ original Study Moses, David A., Matthew K. Leonard, Joseph G. Makin, and Edward F. Chang. "Real-Time Decoding of Question-and-Answer Speech Dialogue Using Human Cortical Activity." Nature Communications 10, no. 1 (July 30, 2019): 3096. https://doi.org/10.1038/s41467-019-10994-4.

The economic power of these corporations allows them to influence politics and democracy with money and to expand their dominant position in the market. We need to see the technology of the Internet and all the technologies connected to it in the context of this enormous concentration of power.

It is true that we need to distinguish between the possibilities of technologies and their use, because in theory they can be used for quite different, namely good purposes, and not only for those that are now mainly imposed on it by powerful corporations and governments. It is precisely these possibilities that we must return to when we ask the question of "what shall we do ?", because they develop from the genuinely human capacity for criticism and dreaming of better, which we want to strive for and let grow against the entrenched powers we see today.

But in the analysis of the problem we must grasp the accumulation of technological and economic power in a few hands, pronounce it and place it at the beginning of the critique.

The use of the Internet as a surveillance medium, especially - but not only - in China,[146] and the development of AI for the purpose of mass behavioural control and manipulation of people must also be part of our analysis.

We must not stop at simple statements about the freedom and great benefits that a free Internet can offer, or at simplified statements about the benefits of AI. The analysis must include the actual and

146 Jiquan, Zhou 'Ten ways China watches its citizens', 4 August 2018 https://www.scmp.com/news/china/society/article/2157883/drones-facial-recognition-and-social-credit-system-10-ways-china.

conceivable effects of the new digital technologies and business activities. At least some of which are already visible today.

The critique of the technology of AI, as already done by Max Tegmark,[147] Sarah Spiekermann[148] and others must be complemented by a critique of the technological-economic power that is exercised by the corporations of the Internet, who are probably the most important developers and users of AI .

---

147 Tegmark, Max. Life 3.0: Being Human in the Age of Artificial Intelligence. 01 edition. Allen Lane, 2017.; also see Youval Harari - Life 3.0 by Max Tegmark review – we are ignoring the AI apocalypse, The Guardian, 22 September 2017, https://www.theguardian.com/books/2017/sep/22/life-30-max-tegmark-review#img-1.

148 Spiekermann, Sarah. Digitale Ethik: Ein Wertesystem für das 21. Jahrhundert. München: Droemer HC, 2019.

B7 Business models of the tech giants

7.1. The strong criticism of GAFAM in the US

The critical analysis of the business models of the Internet giants is very strong in the US itself. There are four reasons for this. First, the GAFAM companies come from the US. They develop their business models and technologies there.

Secondly, there is less legal protection in the US against the abuse of economic and technical power. Much of what happens in the US is unthinkable for Europe because of labour law, consumer protection, data protection, equality law and other protection regulations. The actual and potential abuse of new technologies and business models against ordinary people and small and medium-sized enterprises is more readily discovered and criticised in the US, simply because there is much more abuse of economic and technological power.

The third reason why it is worth looking at America is the constant observation of GAFAM by American journalists, academics and civil society, but also the financial markets. The fierce competition for attention in journalism, the American science market and civil society, which is financed by donations, produces rigorous investigative stories. They are often better researched and produced faster than elsewhere. The New York Times, for example, has invested heavily in *tech accountability journalism* that keeps a close eye on digital companies and brings misconduct to light.

With a $20 million donation from Craig Newmark, the founder of Craigslist, renowned technology journalist Julia Angwin[149] recently established her own independent editorial office for "*Tech Accountability*" journalism, a journalism that seeks to hold GAFAM and other technology companies accountable. *The Markup*[150] is the name of this editorial team, half of which consists of IT engineers and technicians and half of journalists. In addition, there is a plethora of specialized magazines and newsletters. The footnotes in our book provide a cross-section of these sources.

After all, the GAFAM groups are listed on US stock exchanges and occupy the top positions there. A host of analysts observe the companies and make statements about their future prospects and business models. These analysts belong to the public, even if their primary concern is profit. However, it is also possible to believe - and there are quite a few of these analysts who believe this -that the disregard of fundamental rights - such as data protection and privacy - will have a long-term negative impact on the profitability of GAFAM. If such an opinion is formed on the financial markets, more money will flow into business models that are friendly to fundamental rights. *Sustainalytics,*[151] an US stock market analysis firm specializing in sustainability, enumerates every incident in the area of data security and data protection in companies and assesses the risk of every company in this area due to these incidents. Facebook gets a significantly worse rating from Sustainalytics because they see a higher risk in their business model in terms of data protection and data security. In its investor newsletter, *Sustainalytics* called it an own goal of Facebook to have deleted a post by the Democratic presidential candidate Warren, in which she called for the destruction of *Facebook.*[152]

It is therefore worthwhile to look towards the US to understand the business models and technology development as well as for criticism of American tech giants.

---

[150] https://themarkup.org/.

[151] https://www.sustainalytics.com/.

[152] Wennerström, Martin, Governance in brief, 14 March 2019, Sustainalytics

However, there is also clear criticism in the US of the inadequate observation of Silicon Valley by the press and journalism. There are still too few journalists directly on site in Silicon Valley, and there is too little investigative and long-term research compared to political reporting from Washington. This is said, for example, by the author of the best-selling book "Bad Blood", John Carreyrou,[153] a former author of the Wall Street Journal.[154] He showed in his reports and his book on the now infamous *Theranos* start-up, how the lack of control by journalism and naive enthusiasm of the market enables fraud on an enormous scale.[155] The company he portrayed was valued at $9 billion, devouring, among other things, $120 million in investments by Rupert Murdoch - Murdoch is the owner of the Wall Street Journal, for which Carreyrou worked - before *Theranos* almost went bankrupt as a result of his reporting. It finally ceased operations in 2018, with a remaining value of $5 million.[156] The publication of his story in the Wall Street Journal was not blocked in any way, said Carreyrou. In general, he believes that Silicon Valley start-ups are overvalued by at least 30 percent due to the lack of transparency. It's quite possible that GAFAM are equally overrated because of the useful belief that technology will already solve all the world's problems, and perhaps our individual problems, even love, if we think of *Tinder*, for example.[157]

---

[153] Carreyrou, John. Bad Blood: Secrets and Lies in a Silicon Valley Startup. 1 edition. New York: Penguin Random House US, 2018.

[154] At Media en Seine, Paris, 8 October 2019, 'https://www.youtube.com/watch?v=-3VaEuuK3OM', from 17:30 minutes;

[155] Carreyrou, John. "Hot Startup Theranos Has Struggled With Its Blood-Test Technology." Wall Street Journal, October 16, 2015, sec. Business. https://www.wsj.com/articles/theranos-has-struggled-with-blood-tests-1444881901.

[156] Abelson, Reed. "Theranos Is Shutting Down." The New York Times, September 5, 2018, sec. Health. https://www.nytimes.com/2018/09/05/health/theranos-shutting-down.html.

[157] https://tinder.com/?lang=en-GB

7.2. The business models of GAFAM

The business models of each member of GAFAM is different, and we cannot discuss them in detail here. However, a basic understanding of these business models is important to understand whether and how they can collide with public interests, democracy and individual rights.

What unites their business models is that GAFAM benefit from the expansion of the Internet like no other industry. The more people and machines are connected to the Internet, the more people Facebook can connect, the more information can be stored on the cloud systems that are dominated by GAFAM, the more people can use *Google's* search, the more *Amazon* can sell, the more *Microsoft* can enforce its *Software as a Service (SaaS)* strategy. But it is not just about the number of networked people and machines. It is also about moving more and more data and data flows to the Internet. Because the GAFAM companies are the masters of the power to control this data and data flows. They master the technologies to create order in the chaos for their own purposes, to extract information from the data and to use it for control purposes in every conceivable way: from marketing, sales and *IoT* control to the control of political systems, the public and systems for fighting and preventing crime.

From this base, it is then worth considering a simple split in approach: Some GAFAM and some of their Chinese sisters make their money mainly by collecting personal data from people. They use this data for the purpose of advertising their own products or those of third parties. However, they can also be used to monitor people and are therefore of great interest to governments. These companies of the "Stalker Economy" (as Al Gore, the former democratic US vice president, called them as early as 2013) essentially include Google and Facebook in the US and Tencent and Alibaba in China. Shoshana

Zuboff described the business model of these companies in detail in her fundamental work "The Age of Surveillance Capitalism"[158], which some have called the "21st century capital" after "Das Kapital" by Karl Marx.

Apple, on the other hand, has so far made its money mainly with hardware for the Internet and connected services, *Amazon* with trade sales over the Internet and *Microsoft* with software applications. But each company also has some other activities in the business field of its competitors in the other group, which, however, bring in considerably less than their respective main activities.[159] Slovenian philosopher Slavoj Žižek[160] states that, as long as Tech companies do not primarily earn money with targeted advertising on the basis of personality profiles, they all raise a small rent from every mental activity of any person. Whether you are writing something down, designing, organising, communicating, buying or selling, or enjoying it, there is almost always a GAFAM product which you use. The majority of GAFAM companies now no longer exploits its workers, as in classical capitalism, but instead only their users - with exceptions such as *Amazon,* which exploits both users and workers.

---

[158] Naughton, John. "'The Goal Is to Automate Us': Welcome to the Age of Surveillance Capitalism." The Observer, January 20, 2019, sec. Technology. https://www.theguardian.com/technology/2019/jan/20/shoshana-zuboff-age-of-surveillance-capitalism-google-facebook.

[159] Desjardians, Jeff, "Infographic: How the Tech Giants Make Their Billions." 29 March 2019. https://www.visualcapitalist.com/how-tech-giants-make-billions/. Gaudiaut, Tristan, "Comment Les GAFAM Génèrent Des Milliards." https://www.informatiquenews.fr/comment-les-gafam-generent-des-milliards-56199., Statista Infographies. "Infographie: Comment les GAFA génèrent des milliards." 2020.https://fr.statista.com/infographie/13194/source-principale-revenus-google-apple-facebook-amazon/.

[160] Žižek, Slavoj. Like A Thief In Broad Daylight: Power in the Era of Post-Humanity: Power in the Era of Post-Human Capitalism. 01 edition. Allen Lane, 2018.

7.3. Google

Google, now part of the *Alphabet Group,* is, like all GAFAM, a fascinating technology group with a high level of innovative power. Internet search, the Android mobile phone operating system, the *Google Maps* navigation system, the *Google Docs* collaboration system and many other services that work well, are easy to use and require no cash payment, are the basis of *Google's* earning power.[161]

In essence, *Google* makes its money with advertising, based on the exploitation of people and their personal data. *Google's* source of income is the act of manipulating us into buying. Maximizing the consumption of individuals is the goal of the company. With the alleged goal of organizing information on the Internet for us,[162] we are lured into shopping. 84 percent of *Google's* income is advertising income.

Google collects so much data about us that, according to Zuboff, it commands *instrumentarian* power over people, i.e., real power to influence and control. *Google* collects data about people on more and more channels and brings them together to create comprehensive personality profiles: Anyone who searches for information on *Google,* who uses an Android phone, who watches *YouTube,* who writes an e-mail on *Gmail* or a text on *Google Docs,* gives *Google* personal data which it uses for analysis - alongside the contents of the search and communication - through comprehensive personality profiles. Shoshana Zuboff has shown that the mass of data about people collected in this way allows much greater economic returns than are needed to cover the services of Google that are nominally *free.* This

---

[161] For details see https://www.reuters.com/companies/GOOG.O; https://www.statista.com/topics/1001/google/.

[162] Our mission: The information of this world https://about.google/.

is the big lie of the "free" Internet. And it is the basis of a culture of secrecy in *Google* and all GAFAM: much more data about people is collected than is necessary to operate the services on the Internet.

Google also collects personal information from people who do not use the search service, namely through its browser Chrome.[163] The use of Chrome is now strongly discouraged, not only in America. Alternatives are the browsers *Firefox, Brave* and *Safari.*[164] *Google* also collects personal information through cookies that are placed on web pages; and monitors and catalogues the behaviour of users who arrive there, even if they use neither the *Chrome browser* nor *Google search.*[165] For the collection of personal data by *Google* on the Internet it makes no great difference that *Google* is now aiming to phase out cookies by 2022,[166] because most of the personal data it collects comes from the search browser and its own domains. Rather than being misled by such marginal announcements we need to take a holistic view of personal data collection by GAFAM, regardless of the means of collection.

There is now a movement in America that is trying to live completely without *Google*. This seems increasingly impossible.[167]

---

[163] Fowler, Geoffrey A. "Review | Goodbye, Chrome: Google's Web Browser Has Become Spy Software." Washington Post. 21 June 2019. https://www.washingtonpost.com/technology/2019/06/21/google-chrome-has-become-surveillance-software-its-time-switch/.

[164] Fowler, Geoffrey A. "Help Desk: How to Fight the Spies in Your Chrome Browser." Washington Post. 27 June 2019. https://www.washingtonpost.com/technology/2019/06/27/help-desk-how-fight-spies-your-chrome-browser/.

[165] Attention: Almost every website spies out your data. https://www.affiliate-deals.de/magazin/webseiten-ausspaehung/, https://www.sat1.de/ratgeber/sicherheit-im-internet/welche-daten-erhebt-google-analytics.

[166] BBC News, "Cookies Crumbling as Google Phases Them Out." BBC News, 15 January 2020, sec. Technology. https://www.bbc.com/news/technology-51106526.

[167] Bensinger, Greg. "Never-Googlers: Web Users Take the Ultimate Step to Guard Their Data." Washington Post. 23 July 2019. https://www.washingtonpost.com/technology/2019/07/23/never-googlers-web-users-take-ultimate-step-guard-their-data/.

But *Google* also collects and indexes the content that can be found on the Internet. Like personal data, *Google* also takes the contents of the Internet free of charge.

This triple appropriation of data for free, the personal data, the other content on the Internet, and the data in the cloud, all without financial consideration, is central to *Google's* profitability.

This content includes news and reports written by journalists. On *YouTube* you can see not only laymen's content, but also professionally made contributions, including the contributions from public television stations around the world, financed with public money. For example, from Germany you can get content from the main public TV Channels, ARD and ZDF, as well as private broadcasters. The BBC from the UK has more than 8 Million subscribers on YouTube. All this content is available to *Google* free of charge, as is the personal data of citizens, which is collected from the viewers of such content produced with the help of the tax like public broadcasting fees. *YouTube* earns money with advertising placed alongside, and between these contributions. It is astonishing that the public broadcasters don't ask *Google* to give them a share of the revenue.

The idea of the new EU copyright directive[168] was that *Google* pays publishers for reports by journalists that it uses on its services, for example in *Google* News. *Google* has so far refused to do this in Europe.[169] This comes at a time when Facebook has already signed

---

[168] Directive (EU) 2019/790 of the European Parliament and of the Council of 17 April 2019 on copyright and related rights in the digital single market and amending Directives 96/9/EC and 2001/29/EC, OJ L 130, 17.5.2019, p. 92-125, 'http://data.europa.eu/eli/dir/2019/790/oj'.

[169] Politico, 'Plan to Make Google Pay for News Hits Rocks'. 15 February 2017. https://www.politico.eu/article/plan-to-make-google-pay-for-news-hits-rocks-copyright-reform-european-commission/.

a contract with major publishing houses in the US that provides for millions in payments for the use of their content.[170]

In Germany alone, the joint collecting society for press products is demanding licenses from Google of 3.44 billion euros worth of product in 2019, an amount projected to increase to an estimated 8.5 billion euros in 2024.[171] In America, a calculation of the loss of profits for publishers due to *Google* amounts to 4.7 billion dollars in 2018.[172] Against this backdrop, the *Google* News initiative of 300 million euros, which supported technically designed journalism projects, created new dependencies and gave Google new insights into the business models and innovations of the press, is only a drop in the ocean.[173]

Google has paid in the past for access to Twitter content and probably still does. It is said to have paid around 40 million dollars. [174]

Google collects content beyond the Internet. It also scanned the libraries of the world with all their books, when allowed. American courts have considered this practice to be lawful, after lawsuits from

---

[170] Alpert, Lukas I. "Facebook Reaches Deal With Wall Street Journal Publisher, Others for News Section." Wall Street Journal, 18 October 2019, sec. Business. https://www.wsj.com/articles/facebook-reaches-deal-with-wall-street-journal-publisher-others-for-news-section-11571431387.

[171] Speigel, Der. "Google: VG Media fordert Milliardensumme - der Spiegel - Wirtschaft." https://www.spiegel.de/wirtschaft/unternehmen/google-vg-media-fordert-milliardensumme-a-1263608.html.

[172] for criticism of the calculation by the renowned Nieman Lab of Harvard University see 'https://www.niemanlab.org/2019/06/that-4-7-billion-number-for-how-much-money-google-makes-off-the-news-industry-its-imaginary/' Students at the Niemanlab of Harvard University have been financially supported by Google since 2012, see 'https://www.niemanlab.org/tag/google-journalism-fellowship/'.

[173] Dachwitz, Ingo. "Digital News Initiative - Googles Geld und der Journalismus in Europa." netzpolitik.org (blog). https://netzpolitik.org/dni-recherche/..

[174] O'Brien, Chris, "Google's Slippery Slope: If Search Giant Pays Twitter for Content, Should It Pay All Publishers?," VentureBeat, 5 February 2015. https://venturebeat.com/2015/02/05/googles-slippery-slope-if-search-giant-pays-twitter-for-content-should-it-pay-all-publishers/.

book publishers and years of litigation.[175] *Google's* method of making the contents of books available for search, but not for reading as a whole, is considered *fair use* under US law, which is possible free of charge and does not infringe copyright. *Google* wants to enforce this doctrine worldwide for all content, whether text, film or music, in order to obtain a free search and appropriation space for itself. The American *fair-use doctrine* spreads *Google* worldwide as the basis for its own enormous profitability and at the expense of those who produced the content which *Google* than appropriates for free and turns into money.

Google's business model, simply put, is to monetize the free personal information and free content collected from a variety of sources. *Google* does this by selling advertising in addition to the content it collects, and it does so by targeting personal profiles in such a way that as many people as possible buy as quickly as possible. In addition, recently, Google started pushing people to subscribe to own paid services, such as storage on *Google Drive* and *YouTube*.

The model of *Targeted Advertising* has proven to be superior to classical advertising because it has increased the conversion rate from the viewer of an advertisement to the buyer of the product. The basis for this higher conversion rate is the intimate knowledge of the user, to whom the advertisement can then be delivered precisely, addressing the user's interest and at the moment when the emotional disposition to buy is highest. *Google's* remuneration model is in turn structured in such a way that *Google* earns a small amount of money for each visitor directed to a website, i.e., per click, but a significantly higher amount if the visitor then also makes a purchase.

---

[175] Authors Guild, Authors Guild v. Google: Where we stand, https://www.authorsguild.org/where-we-stand/authors-guild-v-google/.

In the US, *Google* thus earns 256 dollars per user per year, worldwide an average of 137 dollars.[176] This means that *Google* collects just under 40 percent of the total expenditure on online advertising in the US.[177] *Google* is by far the world's largest online advertising company, a market that in many western countries will soon account for 50 percent of the entire advertising market[178]. *Google* now earns more than $100 billion a year and, together with *Facebook*, dominates the advertising market[179]. In Germany, according to a subsidiary of the Springer group, a *Google* competitor, the advertising duopoly of *Google* and *Facebook* is said to account for 74.5 percent of the digital advertising market, amounting to 7.79 billion euros.[180]

This lucrative business is obviously not enough for *Google*, especially as it is also burdened with the risks of data protection. The company is pushing into markets with significantly higher revenues per customer, such as healthcare, autonomous driving and home electronics.

Google is constantly striving for better understanding of the structure of language. We have known since Wilhelm von Humboldt that the world is captured through language, and not just through

---

[176] Filloux, Frederic. "The ARPUs of the Big Four Dwarf Everybody Else." Medium, February 11, 2019. https://mondaynote.com/the-arpus-of-the-big-four-dwarf-everybody-else-e5b02a579ed3.

[177] "31 Advertising Statistics to Know in 2018 | WordStream." https://www.wordstream.com/blog/ws/2018/07/19/advertising-statistics.

[178] eMarketer. "Global Digital Ad Spending 2019." https://www.emarketer.com/content/global-digital-ad-spending-2019.

[179] Raconteur. "Google and Facebook: Is the Duopoly Killing Advertising?," March 13, 2019. https://www.raconteur.net/business-innovation/google-facebook-duopoly.

[180] Gehm, Florian. Trotz EU-Strafen – das mächtige Werbe-Duopol im Netz wird immer größer, Die Welt, 17 October 2019, ‹https://www.welt.de/wirtschaft/webwelt/article202021472/Google XE "Google" -und-Facebook XE "Facebook" -Werbe-Duopol-in-Deutschland-waechst-weiter.html Marketing Dive. "Forrester: How Amazon's Ad Growth Will Threaten Google, Facebook, Agencies and Ad-Tech." https://www.marketingdive.com/news/forrester-how-amazons-ad-growth-will-threaten-google-facebook-agencies/562249/

mathematics. Just like *Microsoft*, *Google* invests enormous sums in this area and celebrates every step forward. Recently, for example, it has developed a model for understanding complex search queries.[181] It is also conceivable that with the progress of AI in this field, semantic *targeting* - i.e., targeted advertising based on what we are currently reading on the Internet - will become even more effective than before.

Google has so far systematically abused its position of dominance in search on the Internet. It has a market share of over 90 percent in Europe. In addition, *Google* is developing its own comparison services and products, which are regularly displayed in the search answers before those of competitors. In total, *Google* received fines of 8.3 billion euros for the abuse of its dominant market position following three decisions of the EU Commission.[182] In addition to this were a large number of penalties for violations of data protection and privacy.[183]

One wonders why *Google* is letting these penalties come about and not stopping these practices itself, or at least after initial complaints or indications of possible illegality. One might expect this from

---

[181] Waters, Richard. "Google Uses AI to Boost Search Engine Ranking Efficiency," 25 October 2019. https://www.ft.com/content/d736c314-f6bf-11e9-9ef3-eca8fc8f2d65; also see https://github.com/google-research/bert.

[182] European Commission, 'Commission imposes fine of EUR 2.42 billion on Google for abuse of its dominant position as a search engine by giving undue preferential treatment to its own price comparison service', 27 June 2017, https://ec.europa.eu/commission/presscorner/detail/en/IP_17_1784; Commission imposes fine of 4.34 billion euros against Google for illegal practices in Androidmobile devices to strengthen Google's dominant position as a search engine, 18 July 2018, https://ec.europa.eu/commission/presscorner/detail/en/IP_18_4581 ; Commission imposes a fine of 1.49 billion euros on Google for abuse of a dominant position in the market for online advertising, 20 March 2019, https://ec.europa.eu/commission/presscorner/detail/en/IP_19_1770

[183] See the following articles, each with further references to previous penalties in the US and Europe: Singer, Natasha, and Conger, Kate. "Google Is Fined $170 Million for Violating Children's Privacy on YouTube." The New York Times, September 4, 2019, sec. Technology. https://www.nytimes.com/2019/09/04/technology/google-youtube-fine-ftc.html., Satariano, Adam. "Google Is Fined $57 Million Under Europe's Data Privacy Law." The New York Times, January 21, 2019, sec. Technology. https://www.nytimes.com/2019/01/21/technology/google-europe-gdpr-fine.html.

a company with the former motto "*Don't be evil*", which has now changed to "*Do the right thing*".[184] Either these slogans have always been addressed to the user instead of *Google* itself, or there is not much left of the idealism of the early days in this company. It behaves, where it can, like any other company listed on the stock exchange and maximises profits, if necessary illegally at the expense of its competitors or users or even completely uninvolved parties. The regularity with which legal violations are accepted by the company, at least with approval.

7.4. Apple

Apple is still primarily a manufacturer of hardware, i.e., phones, computers and tablets. But *Apple* is also trying to develop its business model further towards services and streaming. *Apple's* market value is over 1,200 billion, or 1.2 trillion euros. Depending on the stock market situation and the company, this is between 8 and 14 times the value of the most valuable German stock corporations, led by SAP. With a turnover of around 260 billion dollars, *Apple* made a profit of 60 billion dollars in 2018, which is the turnover of *Facebook*.[185] Less than 20 percent of that turnover - but the share is rising - is generated by *Apple's* services.[186]

Apple is trying to break away from the stalker image of GAFAM and is aggressively pursuing data protection. For example, in 2019, it

---

[184] Conger K, Gizmodo, Google removes "Don't be evil" clause from its code of conduct, Gizmodo, 18 May 2018, https://gizmodo.com/google-removes-nearly-all-mentions-of-dont-be-evil-from-1826153393.

[185] https://www.reuters.com/companies/AAPL.OQ; https://de.statista.com/themen/597/apple/.

[186] Statista. "Apple Revenue Breakdown by Product." https://www.statista.com/statistics/382260/segments-share-revenue-of-apple/.

invested in a 54 million dollar advertising campaign and new data protection features.[187] However, the claim that all information on the iPhone remains on the iPhone is untenable, according to Geoffrey Fowler, the Washington Post's Tech Columnist. He has found a total of 5,400 Tracking programs, so-called *trackers*, which transfer personal information from the phone to the app manufacturers, especially at night.[188] There is really something going on. Apple has also made recordings of conversations with Siri without the consent of its customers, which has since been stopped. However, *Apple* writes the dialogues with Siri and keeps the recordings for 6 months in anonymous form, according to *Apple*.[189] An anonymization of extensive dialogues with Siri, for example in combination with several telephone calls and orders, is likely to be difficult. In the US, unfortunately, anonymization is often said, what is actually meant is pseudonymization. The difference is: In the case of anonymization, it must no longer be possible to identify the person who has had the conversation with Siri by means of the written dialogue with Siri or the recording, even with the aid of combinatorics and other information. However, if this is possible, the recorded and written dialogue is not anonymised, but personal data is present, as it is possible to identify the speaker. *Apple* would then have to show that

---

[187] Zeitung, Süddeutsche. "Apple setzt voll auf Datenschutz und riskiert mächtig Ärger." Süddeutsche.de. https://www.sueddeutsche.de/wirtschaft/telekommunikation-apple-setzt-voll-auf-datenschutz-und-riskiert-maechtig-aerger-dpa.urn-newsml-dpa-com-20090101-190605-99-518178. and the same DPA article but with some more links, https://www.welt.de/newsticker/dpa_nt/infoline_nt/netzwelt/article194799861/Apple-setzt-voll-auf-Datenschutz-und-riskiert-maechtig-Aerger.html. Wuerthele, Mike, "Privacy. That's IPhone" Ad Campaign Launches, Highlights Apple's Stance on User Protection'. n.d. AppleInsider. https://appleinsider.com/articles/19/03/14/privacy-thats-iphone-ad-campaign-launches-highlights-apples-stance-on-user-protection

[188] Fowler, Geoffrey A. "Perspective | It's the Middle of the Night. Do You Know Who Your IPhone Is Talking To?" Washington Post. https://www.washingtonpost.com/technology/2019/05/28/its-middle-night-do-you-know-who-your-iphone-is-talking/.

[189] Bloomberg News. "Apple Apologizes over Siri Privacy, Will No Longer Retain Audio Recordings - BNN Bloomberg." BNN, 28 August 2019. https://www.bnnbloomberg.ca/apple-apologizes-over-siri-privacy-and-will-no-longer-retain-audio-recordings-1.1308121

one of the seven reasons exists, of which the user's consent is only one, for the legal processing of personal data under the GDPR.

According to estimates by the bank Goldman Sachs, *Apple* receives an average of over 10 billion dollars a year from *Google* to keep *Google* as default search engine in Safari. With the *Google search* as the default setting, *Apple* earns more than with *Apple Music*. And for that money, Apple feeds its customers, who seek privacy, to the data hungry Google.[190] *Apple's* big words about protecting the privacy and personal data of its customers therefore ring hollow. *Apple* has let *Google* buy its customer's personal information to protect its own reputation on privacy. The default setting of *Google* for searches in Safari, the *Apple* browser, is the opposite of *Privacy by Default*. Geoffrey Fowler argues that if *Apple* were consistent here, it would have to turn off the collection of personal data on the apps sold in the Apple App Store. And we would add it would have to switch to a search engine such as *Duck Duck Go* or *Brave* as a default setting and provide financial and technical support to help them catch up with Google rather than delivering its users to Google. *Apple* could afford that. Every day *Apple* makes 160 million dollars in profit. For the sake of its credibility on privacy and data protection it should make these changes. We wait in hope.

Just by the way: If you obviously don't invest money in *Privacy by default*, i.e. data protection through privacy-friendly pre-settings, and *Privacy by design*, i.e. data protection through technology design, just to cash in 10 billion from *Google* every year and make money in the App Store, you won't be able to hope for the penalty "discount" according to article 83 (2) (d) GDPR when it comes to data protection.

---

[190] Z Leswing, Kif. "Apple Quietly Makes Billions from Google Search Each Year, and It's a Bigger Business than Apple Music." Business Insider. https://www.businessinsider.com/apple-makes-billions-google-search-goldman-sachs-report-2019-2

The maximum fine of 4% of world turn over for non-compliance with GDPR can also hit Apple.

7.5. Facebook

Facebook is number two in the online advertising market after *Google*. *Facebook* attracts people with other people, users' so-called "friends". *WhatsApp* and *Instagram* also now belong to *Facebook*. And, like Facebook, these services focus on networking people.[191] Almost 100 percent of *Facebook's* annual sales of around 60 billion dollars come from the advertising business, the stalker economy.[192]

The economic risk of a business model that is essentially based on the economic exploitation of personal data while neglecting privacy and data protection is most visible on *Facebook*. *Facebook's* shares have plummeted several times due to problems with data protection. In 2018 they lost 40 percent after several scandals, including the Cambridge Analytica scandal.[193] Shareholders are now suing *Facebook* and its management for damages for data protection negligence in connection with the affair. At the time of the Cambridge Analytica incident, *Facebook was* already under an obligation to improve the protection of its users' data following a settlement with the US competition and trade authority, the *Federal Trade Commission (FTC)*.[194] Meanwhile, Facebook was fined $5 billion

---

[191] https://www.reuters.com/markets/companies/META.O/.

[192] Desjardins, Jeff. "How the Tech Giants Make Their Billions." Visual Capitalist, March 29, 2019. https://www.visualcapitalist.com/how-tech-giants-make-billions/; 'https://fr.statista.com/infographie/13194/comment-les-gafam-generent-des-milliards/'.

[193] Rodriguez, Salvador. "Here Are the Scandals and Other Incidents That Have Sent Facebook's Share Price Tanking in 2018." CNBC, November 20, 2018. https://www.cnbc.com/2018/11/20/facebooks-scandals-in-2018-effect-on-stock.html

[194] Doffman Z, 'Facebook Loses Court Battle To Keep Internal Privacy Breach Records Private'. 31 May 2019. https://www.forbes.com/sites/zakdoffman/2019/05/31/facebook-loses-in-court-over-privacy-emails-as-zuckerberg-votes-to-keep-full-control/#15ff0dc3560c

for failing to honor its commitment to privacy as laid down in a prior consent order of the FTC's.[195] Critics emphasized that Facebook took this fine without a significant drop in stock price or profits. Nor did Facebook have to change its business model, which focuses on exploiting personal information, the lives, experiences and feelings, dreams and fears, ambitions and secrets of its users. Interestingly, those members of the FTC who are members of the Democratic Party had called for a significantly higher penalty and published differing assessments of the case. But the *FTC members* from Trump's Republicans overruled them and imposed a fine that was small in terms of Facebook's profit and turnover. However, even under the EU's GDPR, which provides for a maximum penalty of 4 percent of global turnover, the fine that would have been imposed would have amounted to only 2.4 billion dollars.

Here one can already see the embarrassment of the difference in maximum penalties in EU law: In EU competition and antitrust law, maximum penalties of up to 10 percent of world turnover can be imposed, substantially more than the just 4% under GDPR. And there has not yet been any decision to impose this maximum penalty on the basis of the GDPR. The data protection authorities of Europe should view the FTC's decision as an incentive to finally examine the legality of the business model of Facebook and other GAFAM companies in Europe themselves. The *Data Ethics Commission* of the German Federal Government spoke of an enforcement deficit in the area of data protection in Europe.[196] Overall, the penalties for violations of the GDPR in the first year of the regulation, i.e. 2018,

---

[195] Kang, Cecilia. "F.T.C. Approves Facebook Fine of About $5 Billion." The New York Times, July 12, 2019, sec. Technology. https://www.nytimes.com/2019/07/12/technology/facebook-ftc-fine.html.

[196] Section E of the opinion of the *Data Ethics Commission* of 23 October 2019, Para. 65, https://www.bmi.bund.de/EN/topics/it-internet-policy/data-ethics-commission/data-ethics-commission-node.html

amounted to only 56 million euros all across Europe.[197] This means for *Facebook* and its partners: Profits will continue to be made at the expense of citizens of Europe and in disregard of the rules of the GDPR. At the same time, the methods of advertising and tracking as well as the transfer of data to the US are precarious in terms of data protection and have already been the subject of a whole series of investigations and legal proceedings.[198]

Like *Google, Facebook* is pushing into other business areas because it is also fully aware of the present risks of its business model, which is based on the exploitation of personal data. One current example is the attempt to set up a worldwide blockchain-based payment system under the name *LIBRA*. This initiative is now considered a failure following the withdrawal of important consortium partners, especially the major banks and credit card companies, and the unanimous vote blocking the initiative from leading finance ministers. The almost megalomaniacal attempt to push through the project, which calls into question the central currency monopoly of states, without prior clarification of the important regulatory issues with the guardians of the currency, was put in its place by governments and central banks. Mark Zuckerberg's motto "*Move fast and break things*" fortunately did not work here. This motto was and is a basic problem of the corporate philosophy of Facebook, and indeed of Silicon Valley as a whole.

---

[197] The official website of the European Data Protection Board has a heading 'Administrative Fines', but unfortunately at the time of writing this book it had no content, https://edpb.europa.eu/our-work-tools/our-documents/topic/administrative-fines_en'

[198] Hern, Alex. 2018. 'Privacy Policies of Tech Giants "Still Not GDPR-Compliant"'. The Guardian, 4 July 2018, sec. Technology. https://www.theguardian.com/technology/2018/jul/05/privacy-policies-facebook-amazon-google-not-gdpr-compliant

"Facebook Pixel: Ist Das Tracking Aus Datenschutz-Sicht Erlaubt?" https://www.e-recht24.de/artikel/facebook/10585-facebook-pixel.html.; https://noyb.eu/en/search?search=facebook. "Twitter and Facebook Could Be Facing Billions in Fines after Ireland Investigations." CNBC, October 7, 2019. https://www.cnbc.com/2019/10/07/facebook-twitter-investigations-in-ireland-reach-conclusion.html

Neither law nor democracy are exempt from "break[ing] things". And indeed, neither is simple human consideration and the basic principle of human dignity. Does the failure of *LIBRA* mean the end of "*Move fast and break things*" as a business principle?

7.6. Amazon

Facebook connects people, *Google* organizes the contents of the Internet, *Apple* builds the hardware, *Amazon* is the platform for mail order and *Microsoft* provides the programs. At first glance, GAFAM's division of labour looks something like this. And they all offer cloud service, in particular *Amazon*, which controls a third of the cloud market.

The Cloud is the source of 50 percent of Amazon's profits – it manages and processes data from companies in 190 countries, including Netflix, Airbnb, General Electric, Vodafone, the US Navy, and 80 percent of the German stock market listed companies.

Online trading is the other source of profit.[199] *Amazon* makes profits not only in the market, but by owning the market. Amazon is the market and collects percentages for all transactions. In Germany, *Amazon* has a market share of 37 percent in online commerce, according to a study by a consulting firm that counts *Amazon's* competitor *Otto* among its customers.[200]

Amazon learns which products do well and then makes them itself. As the sociologist Philipp Staab writes, *Amazon* is thus a typical

---

[199] https://www.investopedia.com/how-amazon-makes-money-4587523

[200] https://www.handelsblatt.com/unternehmen/handel-konsumgueter/online-handel-amazon-baut-seine-marktmacht-in-der-coronakrise-weiter-aus/25675414.html

representative of digital capitalism, which aims at the development of "proprietary markets".[201] In the long term, Amazon does not only want to dominate the market, but also to be the market. Everyone who wants to trade then has to pay to Amazon as the owner of the market.

In the list of the most valuable companies, *Amazon* is now in third place behind *Microsoft* and *Apple*. *Amazon* is already big in streaming, with its Prime Video channel, and thus part of the opinion-forming cultural industry. Jeff Bezos has plans to make a big move into the advertising business as well.[202]

What is atypical in comparison to the other GAFAM is the high number of employees at *Amazon*. In the US, *Amazon* is the third largest private employer. It has 560,000 employees worldwide.[203] The founder and head of *Amazon* acquired the important American newspaper "Washington Post" and made it profitable again.[204] Another move into the world of political opinion forming.

7.7. Microsoft

Microsoft is trying to build a clean image as a software developer and service provider in terms of data protection and to appear

---

201 Philipp Staab, Digital Capitalism: Market and Rule in the Economy of Scarcity, Berlin 2019.

202 'Amazon's next Ad Move: AmazonFresh "Sponsored Products"'. 2019. Digiday (blog). 1 March 2019. https://digiday.com/marketing/amazons-next-ad-move-amazonfresh-sponsored-products/

203 Hamilton, Isobel Asher. "Thousands of People across Europe Are Protesting and Striking against Amazon on Black Friday." Business Insider. https://www.businessinsider.com/amazon-strikes-and-protests-sweep-across-europe-on-black-friday-2019-11

204 Francesca Guiliani-Hoffman, "How the Washington Post Has Changed under Jeff Bezos - CNN." https://edition.cnn.com/2019/08/16/media/jeff-bezos-donald-graham/index.html

otherwise as a state like representative of public interest. This fits with *Microsoft*, now the most valuable company in the world, selling a lot to the governments of this world. *Microsoft* nevertheless collects considerable amounts of personal data via Office and other services such as LinkedIn and the game Minecraft with 112 million users.[205] In addition *Microsoft* is the leader in the development of facial recognition applications. *Microsoft* researchers have collaborated with the military university in China on facial recognition software in projects that can be used for mass surveillance.[206]

If Microsoft is now calling for face recognition to be regulated, it is also doing so to get off the political defensive in the face of this misstep. Two other reasons are: 1. To keep[207] Chinese competitors, which have been successful in dictatorships, out of the market and 2. To avoid absolute bans on face recognition, such as those that apply to the Californian police since 1.1.2020.[208] And while it calls for regulation of facial recognition, it waters down such regulation in the practical policies of Washington State, where the company's headquarters are registered.[209]

---

[205] Tassi, Paul. “Microsoft Reveals ‘Minecraft’ Has An Astonishing 112 Million Monthly Players.” Forbes. https://www.forbes.com/sites/paultassi/2019/09/14/microsoft-reveals-minecraft-has-an-astonishing-112-million-monthly-players/.

[206] Yang, Yuan, and Madhumita Murgia. “Microsoft Worked with Chinese Military University on Artificial Intelligence,” April 10, 2019. https://www.ft.com/content/9378e7ee-5ae6-11e9-9dde-7aedca0a081a.

[207] Murgia, Madhumita, Yuan Yang, and Anna Gross. “Chinese Tech Groups Shaping UN Facial Recognition Standards,” 1 December 2019. https://www.ft.com/content/c3555a3c-0d3e-11ea-b2d6-9bf4d1957a67 . Murgia, Madhumita, and Yuan Yang. “Facial Recognition: How China Cornered the Surveillance Market,” 6 December 2019. https://www.ft.com/content/6f1a8f48-1813-11ea-9ee4-11f260415385

[208] AB-1215 Law enforcement: facial recognition and other biometric surveillance, http://leginfo.legislature.ca.gov/faces/billNavClient.xhtml?bill_id=201920200AB1215; Rodrigo, Chris Mills. “California Blocks Police Body Cameras from Using Facial Recognition.” Text. TheHill, 9 October 2019. https://thehill.com/policy/technology/464991-california-blocks-police-body-cameras-from-using-facial-recognition.

[209] Simonite, Tom. “Microsoft Wants Rules for Facial Recognition—Just Not These.” Wired. 21 March 2019. https://www.wired.com/story/microsoft-wants-rules-facial-recognition-just-not-these/.

The call for regulation of facial recognition is in the classic, minimalist tradition of the US corporations, whenever regulation is no longer avoidable, to demand only specific sectoral rules, thus averting general rules and leaving large areas unregulated.[210] The right thing to do would be to demand general regulation in the US to protect personal data and privacy for the whole of America, and not just regulation of facial recognition.

In Europe, proper enforcement of the GDPR and the EU directive on data protection in the police and judiciary[211] will also lead to a strict limitation of face recognition.[212] There is no need for special law in Europe for this purpose. Indeed, it is even risky to create a special law in Europe for face recognition, as this in turn raises new questions about the limitations of scope of the expressly broad and technologically neutral GDPR and the Directive of data protection in the area of law enforcement.

Another question, however, is whether the technology of facial recognition itself should be subject to regulation, such as restrictions on production or selling of such technology in the open market.

Each of the GAFAM companies is also active in the other business areas, which are mainly occupied by a competitor. For example,

---

210 Browne, Ryan. "Tech Giants Want Rules on Facial Recognition, but Critics Warn That Won't Be Enough." CNBC, 30 August 2019. https://www.cnbc.com/2019/08/30/facial-recognition-tech-firms-want-regulation-but-critics-want-a-ban.html.

211 Directive (EU) 2016/680 of the European Parliament and of the Council of 27 April 2016 on the protection of individuals with regard to the processing of personal data by the competent authorities for the purpose of the prevention, investigation, detection or prosecution of criminal offences or the execution of criminal penalties and on the free movement of such data and repealing Council Framework Decision 2008/977/JHA. OJ L 119, 4.5.2016, p. 89-131, 'http://data.europa.eu/eli/dir/2016/680/oj.

212 See already the guidelines of the European Data Protection Committee on video surveillance, https://edpb.europa.eu/sites/edpb/files/files/file1/edpb_guidelines_201903_video_devices_en.pdf .

Microsoft has acquired the professional database LinkedIn, which competes in Germany with the *Xing* database of the Burda Media Group. Personal data is also collected here for advertising purposes.

In Europe alone, *Microsoft* earns 2 billion euros a year just from regular license revenues from public administrations. The most important competitor for *Microsoft* was the *open source systems* - until it bought itself into this market.[213] The importance for *Microsoft* of displacing open source in public administration and to appear as the sole, alternative-free provider of package solutions was demonstrated in Munich. The city was given a large new research center by Microsoft, at the same time as it promised not to continue the open source policy in the city, but to introduce *Microsoft's* package for all public services.[214]

The European Commission also uses *Microsoft*. The Commission's contracts with *Microsoft* are currently under intensive review by the European Data Protection Supervisor (EDPS).[215] In any event, *Microsoft's* dominance in European public administration is comparable to IBM's dominance in the 1970s and 1980s.

Microsoft learned early on that lying and failing to keep promises to the authorities can be expensive. In 2009, the company promised the EU Commission that it would unbundle its preinstalled Windows and *Microsoft programs* on computers sold to individual customers.[216]

---

[213] The New Stack. "How Microsoft Finally Came Around to Open Source," 16 February 2017. https://thenewstack.io/microsoft-finally-came-around-open-source/.

[214] Microsoft-Software: Sicher Für Europa?, ARD-Reportage, 19 March 2018, https://www.youtube.com/watch?v=_7583HNrZJs.

[215] EDPS investigation into IT contracts: stronger cooperation to better protect rights of all individuals; https://edps.europa.eu/press-publications/press-news/press-releases/2019/edps-investigation-it-contracts-stronger_en

[216] Unbundling is the obligation to offer the hardware even without the pre-installed software at a lower price. Microsoft had obliged the hardware manufacturers to pre-install Windows on all their devices, without the option of selling the device cheaper without Windows.

However, in 2013 the Commission had to impose a fine of 561 million euros on *Microsoft, the* first company to receive a fine in the history of European competition law for failing to keep a commitment made in a prior settlement.[217] *Microsoft* learned early on that a direct confrontation with the law and the public interest is not a good idea. The company is therefore trying to give the impression that it is acting in the public interest, not least also because of its sales to the public sector. It talks about aligning the company's objectives with the public interests articulated by governments. This is not surprising in view of the high revenues from public-sector contracts. However, in capitalism, the stock market and the profit motive always rule in the end. *Microsoft* did not shy away from weakening data protection in Washington State, where *Microsoft* is based, with considerable lobbying.[218] This, combined with the acquisition of LinkedIn and the game Minecraft, which is also an ideal source for detailed character, intelligence and emotional data, including data about children, indicates that *Microsoft* has not completely written off the exploitative and dishonest business with personal data. The investigations of the European Data Protection Supervisor will show how seriously *Microsoft* and the EU Commission take data protection in their contracts. In any case, Microsoft apparently saw no problem in copying hundreds of thousands of faces of celebrities from the Internet and feeding them into a test database for face recognition.[219]

Microsoft's strategy of standing alongside the public interest and acting as a voice of reason in the development of the digital world is

---

217 European Commission press release of 6 March 2013, Antitrust: Commission fines Microsoft for failing to comply with its obligation to ensure free browser choice, 'http://europa.eu/rapid/press-release_IP-13-196_de.htm'.

218 Scott, Mark. "How lobbyists rewrote Washington State privacy law", Politico.eu, 4 May 2019, https://www.politico.eu/article/how-lobbyists-rewrote-washington-state-privacy-law-microsoft-amazon-regulation/.

219 Murgia, Madhumita. "Microsoft Quietly Deletes Largest Public Face Recognition Data Set," 6 June 2019. https://www.ft.com/content/7d3e0d6a-87a0-11e9-a028-86cea8523dc2

efficiently supported by the foundation activities of Bill and Melinda Gates. The positive image spill over effect from the Foundation to the Company is difficult to measure in Dollars, but certainly not insignificant.

7.8. Google and Microsoft: AI investment focus on health and language

It is possible that the differentiation of business models, where *Google* collects and exploits personal information from people and *Apple* and *Microsoft* try to move away from this model, will continue in the development and use of AI. *Google's* investment in *Nest*, a thermostat and home electronics manufacturer, and its investment in *DeepMind*, combined with commercially acquired health data from the UK *National Health Service* (NHS), has a clear focus on the exploitation of personal information deep into the private sphere. The purchase of *Waze*, a navigator that developed in competition with *Google Maps*, also points in this direction, as navigation data is transaction data, and transaction data is personal data that has a high value for the management and control of mobility and all associated services, and tells a lot about people's habits and interests as well as social relations. This fits with *Google's* core business. However, the question of legality of collections of personal data and profiling of people for profit and whether users will accept this practice in the long term remains unanswered. Therefore, web search and the related advertising business are at risk. The revenues per user are low. *Google's* profitability is essentially based on the huge number of users and its dominant market share, which in many respects is virtually a monopoly.

Revenues per user, for example in the health sector and mobility services, can be considerably higher than those from search and advertising, as user are more willing to pay, in particular for health services. The parent company of Google, Alphabet, is therefore systematically looking for business areas with higher profitability per user than free search and other free services, which are always financed directly or indirectly by advertising.

However, all business developments and purchases by Google always have a dual objective: to tap new sources of personal data and at the same time increase the profitability of the data, including through business services and payment models. This fits in well with the investment in healthcare in the US and the collection of health data from millions of patients[220], partnerships with hospitals around the world[221] and the UK *National Health Service* (*NHS*). There are signs that *Google* is likely to take a nefarious course also in the field of AI to implement the business model described by Zuboff. This is demonstrated by the dissolution of Google's Ethics Council,[222] as well as the Ethics Council at *DeepMind*, the core of AI development bought out by *Google* in London; the release of *DeepMind* co-founder Mustafa Suleyman;[223] and his secondment to *Google's* headquarters, where his activities are better controlled[224]. Also, one has to ask

---

220 Pilkington, Ed. "Google's Secret Cache of Medical Data Includes Names and Full Details of Millions – Whistleblower." The Guardian, November 12, 2019, sec. Technology. https://www.theguardian.com/technology/2019/nov/12/google-medical-data-project-nightingale-secret-transfer-us-health-information. ; CB Insights Research. "Google Healthcare With AI l CB Insights." https://www.cbinsights.com/research/report/google-strategy-healthcare/

221 See for example in the Netherlands :https://clinicaltrials.gov/ct2/show/NCT03295578.

222 MIT Technology Review. "Hey Google, Sorry You Lost Your Ethics Council, so We Made One for You." https://www.technologyreview.com/2019/04/06/65905/google-cancels-ateac-ai-ethics-council-what-next/.

223 Bloomberg, "Google DeepMind Co-Founder Placed on Leave From AI Lab." 21 August 2019. https://www.bloomberg.com/news/articles/2019-08-21/google-deepmind-co-founder-placed-on-leave-from-ai-lab; Murgia, Madhumita. "DeepMind Co-Founder's Leave of Absence Signals Transition," 23 August 2019. https://www.ft.com/content/45bb157a-c4f6-11e9-a8e9-296ca66511c9 .

224 Murgia, Madahumita. "DeepMind founder leaves for policy role at Google" 5 December 2019, https://www.ft.com/content/02757f12-1780-11ea-9ee4-11f260415385.

whether *DeepMind* really developed such great AI, or whether it was bought by *Google* simply to get easier access to NHS health data.[225] *Google* is now the largest collector of health data on the Internet.[226] These data are currently big business and hold great promise. The much-discussed new law in Germany facilitating access and use of health data for profit shows that governments and legislators also believe in this promise.[227] Now we need to watch closely to see how much *Alphabet, Google's* parent company, and its many subsidiaries already involved in health research, benefit from this law.[228] What access to and use of health data will actually benefit patients and the health system? How serious is the protection of health data being taken in Germany at all, now that this new law has apparently been pushed through without taking into account past scandals in the US health sector and worldwide?[229] The reason for the high data protection risk with regard to prescriptions and health data is easy to understand: In the past, more than 80.000 Euros have already been called up for the knowledge of the personal data of diabetes prescriptions, for example.[230] A pharmaceutical company can earn

---

[225] Vaughan, Adam. "Google Is Taking over DeepMind's NHS Contracts – Should We Be Worried?" New Scientist. https://www.newscientist.com/article/2217939-google-is-taking-over-deepminds-nhs-contracts-should-we-be-worried/.

[226] Murgia, Madhumita, and Max Harlow. "How Top Health Websites Are Sharing Sensitive Data with Advertisers," November 13, 2019. https://www.ft.com/content/0fbf4d8e-022b-11ea-be59-e49b2a136b8d.

[227] "Gesundheitsdaten: Den Patienten Fragen." https://www.faz.net/aktuell/politik/inland/gesundheitsdaten-den-patienten-fragen-16466708.html.; Berndt, Christina. "Gesundheitsdaten: Spahn hat die Hacker eingeladen." Süddeutsche.de. https://www.sueddeutsche.de/politik/gesundheitsdaten-schutz-spahn-1.4678619

[228] Kuchler, Hanna. "Can we ever trust Google with our Healthdata?" 20 January 2020, https://www.ft.com/content/4ade8884-1b40-11ea-97df-cc63de1d73f4.

[229] Copeland, Rob. "WSJ News Exclusive | Google's 'Project Nightingale' Gathers Personal Health Data on Millions of Americans." Wall Street Journal, November 11, 2019, sec. Tech. https://www.wsj.com/articles/google-s-secret-project-nightingale-gathers-personal-health-data-on-millions-of-americans-11573496790.

[230] Schmundt, Hilmar. "VSA verkauft Rezeptdaten an Marktforschungsfirmen - Der Spiegel - Backstage." https://www.spiegel.de/spiegel/spiegelblog/vsa-verkauft-rezeptdaten-an-marktforschungsfirmen-a-920845.html

millions per patient with diabetes drugs over a lifetime if it is able to use the prescription data of young patients to replace the previously prescribed drug with its own drug.[231] It is greed for profit that undermines data protection and privacy.

Microsoft is also investing heavily in the profitable and growing healthcare sector, which, according to a country's wealth and demographics, accounts for between 9 and 14 percent of its GDP, and the trend is rising. People are living longer and this, combined with increasing prosperity, is leading to relatively higher health care spending. But *Microsoft's* partnership with Novartis, the Swiss Pharmaceutical giant, to optimize all of its business processes through Artificial Intelligence[232] shows that the focus is on business customers rather than consumers, more on business processes than on individual behavior. However, personal data is also relevant here.

It is quite possible that through this partnership *Microsoft* - as *Google* with the help of *DeepMind* - can open up the most data-intensive areas of application in health. There is much to suggest that the most profitable areas of application for AI and its associated technologies, such as quantum computing and big-data analysis, in the long term will be in the most data-intensive applications, not the advertising that *Facebook* and *Google* have been using to make money. These are areas such as materials research, diagnostics, but also medical research, especially the individualised medicine and remedies beyond pharmaceuticals and chemistry, whose active ingredients

---

231 PMLive. "Sanofi Sets up Diabetes Joint Venture with Google's Verily." Text. PMGroup Worldwide Limited, September 12, 2016. World. http://www.pmlive.com/pharma_news/sanofi_sets_up_diabetes_joint_venture_with_googles_verily_1128741. Einecke, Helga. "Milliardengeschäft Diabetes." Süddeutsche.de. https://www.sueddeutsche.de/wirtschaft/pharmaindustrie-milliardengeschaeft-diabetes-1.1521358.

232 Neville, Sarah, and Richard Waters. "Novartis and Microsoft Join Forces to Develop Drugs Using AI," October 1, 2019. https://www.ft.com/content/93e532ee-e3a5-11e9-b112-9624ec9edc59.

have already been fully discovered. The new field of genomics and genetic medicine, the penetration ever deeper into the nature of cells and their functioning, will open up an enormous treasure trove of data, the Book of Life, which is considered by some to be the most valuable data set of all. There are already initial calls to prevent a monopolization of this data trove and to make it openly available for research.[233] Only recently, the pharmaceutical industry has formed an alliance for AI in Health, with the aim of paving the way for research and applications such as this. Bayer is a German Member of this alliance, which has its Headquarter in the US.[234]

Perhaps some of the GAFAM companies will discover before others what many in German industry, Siemens above all, have been saying for a long time: The most profitable applications of Big Data methodology, AI and Quantum do not lie in the area of fragmented and often imprecise personal data limited by human behaviour, but in research data and economic data that is not personal data, i.e. market data, data on machine operations, physics and technology, biology and earth sciences.

What all GAFAM companies have in common is their work on the next big thing, namely meaningful speech comprehension and the creation of meaningful speech. Will it be possible to move all speech processing to the *edge*, i.e., to the devices using *edge computing*? *Apple Siri, Google Bert, Amazon Alexa* and *Microsoft's* investment in *Open AI* and its speech algorithms, *IBM Project Debater* - all these projects point in a similar direction. The centrality of language for our understanding of the world and for communication in various governance systems makes it the core technology par excellence.

---

[233] Castilla-Rubio, Juan Carlos, "The Book of Life will be the century's most valuable enterprise." Financial Times, 29 October 2019, https://www.ft.com/content/e45fd96a-d3a9-11e9-8d46-8def889b4137

[234] https://www.theaaih.org/

Those who can control and automatically create and understand meaningful language can fully master governance systems, from democracy to law to stock markets to science. *Fake News* will then become the least of our problems. It is therefore extremely important not only to follow the development of AI language technology closely, but also to be right there with it and to regulate it properly. No human being should ever talk or communicate in writing with a machine or AI with him or her believing it is a human on the other side. We need a legal obligation for transparency in the use of this technology to ensure that we always know whether we communicate with a machine.

B8 The eight sources of power of the technological-economic complex

In the following section, we will first describe the core elements of today's economic concentration of power, which, together with their technical potential, poses a threat to both democracy and the functioning of the markets.

The concentration of power was possible because the Internet and software remained largely unregulated. The *infant industry* argument, i.e., the classic neo-liberal argument that any legal regulation would stand in the way of innovation and thus the delicate plant of the new Internet economy, was and is a popular argument. But it no longer makes sense in view of the power of the GAFAM today. It is important to criticise this experience, because with new technological leaps, especially in Artificial Intelligence and quantum computing, we could be faced with new and this time irreversible accumulations of power which will make the GAFAM, which are already powerful today, even more powerful. We must learn from the mistakes of *laissez-faire* with the Internet and software industry.

Specifically, we need to look at eight sources of power from technology companies, individually and collectively, with a particular focus on the activities of the *Terrible Five*[235] that shape our experience with the Internet and digital technologies like AI: *Google, Amazon, Facebook, Apple Microsoft*. These companies, along with two handful of others,

---

[235] Manjoo, Farhad. "Tech's 'Frightful 5' Will Dominate Digital Life for Foreseeable Future." The New York Times, January 20, 2016, sec. Technology. https://www.nytimes.com/2016/01/21/technology/techs-frightful-5-will-dominate-digital-life-for-foreseeable-future.html. Manjoo, Fardah. "Tech's Frightful Five: They've Got Us." The New York Times, May 10, 2017, sec. Technology. https://www.nytimes.com/2017/05/10/technology/techs-frightful-five-theyve-got-us.html

are not only shaping the delivery of Internet-based services to individuals. They are extremely profitable, have the highest value on the stock market and therefore have an economic power that not only allows them to crowd out competitors through innovation, acquisition or imitation, but also gives them disproportionate access to legislators and governments and allows them to distribute, directly or indirectly, financial or in-kind benefits in all areas of society relevant to democratic opinion-forming: governments, legislators, civil society, political parties, schools and education, journalism and journalists in training and, above all, science and research.

The penetration of all areas of social opinion-forming and decision-making by these companies is thus a reality - not only technically, but also in terms of resources and on the ground presence. Politics, civil society, science, journalism, culture and business traditionally try to keep a certain distance from one another, to maintain a certain independence.

But today the Terrible Five are present in all these areas to accumulate knowledge and learn for their own purposes and to promote, diplomatically speaking, sympathy and understanding for their concerns and interests. A critical examination of the relationship of the Internet and AI to human rights, democracy and the rule of law must therefore look at the reality of technology and business models as a whole in their present form. And we need to study how the 'Terrible Five' skip the boundaries of independence of government, parliament, civil society, science, culture and journalism and make "friends" everywhere in order to influence the social climate of discourse and try to call the shots in decision making.

Let us now turn to the eight sources of power of the technology corporations, which must first be considered individually, but then also in their interaction.

8.1. The GAFAM companies are the most valuable and richest in the world

GAFAM have deep pockets. Today, money is the basis of their innovative strength. It is also the classic tool of influence on politics and markets. Google, *Apple, Facebook, Amazon, Microsoft* hold the top positions in global stock market valuations, they are among the largest companies in the world and are also among the most profitable.[236] They have recorded the highest absolute increase in market capitalization since 2009 worldwide. In 2012, *Apple* ousted the US oil company Exxon from the top of the ranking by market capitalization. In 2018 the Chinese companies *Alibaba* and *Tencent* joined the group of the 10 most valuable companies in the world on the stock exchange. *Microsoft* is currently the most valuable company in the world.

Europe and Germany's share of the company value among the top 100 has fallen by 12 percent during this period, to just 11 percent today, while the share of the US has risen by the same percentage to 63 percent today.[237]

The fact that GAFAM have paid almost no taxes to date[238], neither in the US nor in the EU, is certainly not an insignificant contributor of their significant value.

---

[236] https://www.statista.com/statistics/263264/top-companies-in-the-world-by-market-value/ and Fortune 500 Countries by profitability https://fortune.com/fortune500/2019/search/?profits=desc.

[237] see the PWC reports on the top 100 companies in the world, available at https://www.pwc.com/gx/en/audit-services/assets/pdf/global-top-100-companies-2018-report.pdf ; and https://www.pwc.com/gx/en/audit-services/publications/assets/global-top-100-companies-2019.pdf

[238] The following links identify reporting to GAFAM tax avoidance: Facebook https://itep.org/facebook-facing-shareholder-scrutiny-for-its-offshore-tax-avoidance/; Amazon https://itep.org/amazon-in-its-prime-doubles-profits-pays-0-in-federal-income-taxes/; Microsoft https://itep.org/how-to-think-about-the-problem-of-corporate-offshore-cash-lessons-from-microsoft/ Bill Gates and Jeff Bezos https://itep.org/business-insider-theres-a-reason-bill-gates-and-jeff-bezos-both-call-washington-state-home; Apple https://observer.com/2019/09/apple-ireland-tax-lawsuit-european-union-corporate-tax-dodging/ and https://www.nytimes.com/2019/04/30/technology/apple-stock-buyback-quarterly-results.html

8.2. Accumulation of personal data profiles and behavioural predictions about all people

Even when their business models are not based on advertising, the mega-companies are essentially busy collecting personal information where they can and creating a personal profile about each of us based on our *online* and *offline* behavior. *Microsoft, for example,* which has relatively low advertising revenue, collects personal information through LinkedIn and Bing, and Usage patterns through programs such as Windows, Word and Outlook. GAFAM know more about us than we or our friends do[239] - and they use the information for their own business and make this information available for the business of others, for advertising, but also for the state and, under certain conditions, for national election campaigns. They benefit from the claim to strengthen people's autonomy, even though they are centralizing power to an unprecedented extent and thus abolishing autonomy.

To show how powerful *Facebook* computers are, Yann LeCun, then head of AI development at *Facebook,* proudly tweeted in 2018 that *Facebook* was producing 200 trillion behavioral predictions a day. Hello, yes, you did not misread. Two hundred trillion![240]

That's about 1,000 predictions per user per day, not counting the shadow profiles, i.e., the profiles of people who don't use Facebook at all.

---

[239] Egan, Matt. "How to Check What Data Google Holds on You - and How You Can Limit It." Tech Advisor. https://www.techadvisor.co.uk/how-to/security/what-does-google-know-about-you-3592743/ . from Facebook, Quodling, Andrew. "Shadow Profiles - Facebook Knows about You, Even If You're Not on Facebook." The Conversation. http://theconversation.com/shadow-profiles-facebook-knows-about-you-even-if-youre-not-on-facebook-94804. CB Insights Research. "What Are Psychographics?," May 6, 2020. https://www.cbinsights.com/research/what-is-psychographics/ . Hern, Alex. "Facebook Ad Feature Claims to Predict User's Future Behaviour." The Guardian, April 16, 2018, sec. Technology. https://www.theguardian.com/technology/2018/apr/16/facebook-ad-feature-predict-future-behaviour.

[240] Facebook makes about 200 trillion predictions per day, according to Yann LeCun, Chief of AI Development at Facebook, May 3, 2018 at 8:03 on Twitter, https://twitter.com/ylecun/status/991936213249650688

These predictions meet the definition of personal data according to the new EU General Data Protection Regulation.[241] This means that if you live in the EU, you can ask *Facebook* and all other entities that collect data about you to give you these predictions about you, delete them or correct them if they are incorrect. These predictions of behaviour are the most important personal data of the present and the future. Not only do they have the greatest value for companies and the state. They are also the deepest intervention in the informational self-determination of people and therefore enjoy the highest protection. If *Facebook* could not be obliged under the General Data Protection Regulation to release the predictions to those affected, the General Data Protection Regulation would not fulfil its purpose in the age of AI and supercomputers. We would then be at the mercy of control and manipulation based on the behavioral predictions.

8.2.1. Shoshana Zuboff: The digital business model and instrumental power

> *Et qui n'a pas le don de cacher ce qu'il pense*
> *doit faire en ce pays fort peu de résidence*
>
> *And anyone who lacks the gift of hiding*
> *His thoughts should make a brief stay in such a country.*
>
> *Moliere, The Misanthrope*[242]

Shoshana Zuboff's book "The Age of. "The Age of Surveillance Capitalism" has been criticized as misjudging the basic mechanisms of American-style capitalism[243] - a criticism we do not share - it also shows the potential for reform in a still free society like the US. For

[241] Article 4 (1) GDPR and recital 24 at the end.

[242] Molière. Complete Works of Molière. Delphi Classics, 2018.

change begins with criticism, and criticism is still possible in the US. Let us take criticism seriously, because it is no longer possible in some states, such as China. Nor can we rely on technology to do our criticism for us: AI cannot criticise because the programmes do not feel dissatisfaction, nor do they have imagination - the foundations of criticism. There is a risk that people like us, dragged further and further into the comfort trap of the digital paradise, will be too lazy to make fundamental criticisms of the system, leading to a feedback loop of less scrutiny, better and better mechanisms to capture our attention; the slow descent into everyone becoming lazy screen gawkers.

Zuboff has identified and analysed human behaviour as the most important raw material of digital surveillance capitalism. According to the business model developed at *Google* and later adopted by *Facebook* and many others, personal data is merged into behavioral profiles of people who become new tradable investment goods and, if the investment is sufficiently high, this raw commodity can be refined to amount to the lucrative outcome of total behavioral control.

On *behavioral future markets* (Zuboff), the profiles of people are traded, fully automated. *Google's* chief economist, Hal Varian, says: "Because transactions are now computer-mediated we can observe behaviour that was previously unobservable and write contracts on it. This enables transactions that were simply not feasible before."[244]

[243] Morozon, Evgeny, "Capitalism's New Clothes | Evgeny Morozov," The Baffler. 4 February 2019. https://thebaffler.com/latest/capitalisms-new-clothes-morozov.

[244] Zuboff, Shoshana. The Age of Surveillance Capitalism: The Fight for a Human Future at the New Frontier of Power, Main edition (London: Profile Books, 2019). p. 253.

The sequence of value creation according to Zuboff is like this in surveillance capitalism:

Human behavior is transformed by digital capture into data, the collection of which leads to the accumulation of knowledge (in the databases) and thus to an asymmetry of knowledge and power in the relationship between individuals and corporations. Knowledge and power are then used to manipulate people, which ultimately leads to huge profits for the corporations, but can also mean the end of freedom and democracy.

The user is not the *product* of the new economy, but the source of its most important raw material: it provides behavior-based personal data. The raw material of digital business models is therefore the human being himself and his most intimate life!

Initially, the model was also idealistic; it was about sorting the world knowledge available on the Internet and making it accessible. But because this was done for free, Google ran out of money: "*Google's toughest search is for a business model*" wrote Saul Hansell in the *New York Times* on April 8, 2002,[245] when the investors were already waiting impatiently for the search engine's business model and the IPO was imminent. At that time, outsiders were unsure how the search engine would ever make money.

According to Zuboff, *Google* already "cancelled" the reciprocity principle in 2000, according to which only as much personal data should be collected as was necessary to optimise the search.[246] The

---

[245] Hansell, Saul. "Google's Toughest Search Is for a Business Model." The New York Times, April 8, 2002, sec. Business. https://www.nytimes.com/2002/04/08/business/google-s-toughest-search-is-for-a-business-model.html.

[246] Zuboff, Shoshana. The Age of Surveillance Capitalism: The Fight for a Human Future at the New Frontier of Power, Main edition (London: Profile Books, 2019). p. 96.

result was a "new click economy".[247] However, the company kept this top secret fearing competition: "The secrecy dovetailed with Larry Page's hardwired secret-keeping," writes Newsweek editor Steven Levy.[248]

With the introduction of the radicalised, because personalised advertising model, the *fair search* originally promised by the *Google* founders is now revealed as a sham. According to Zuboff, this new *digital capitalism* has in fact created a "perverse amalgam of empowerment inextricably layered with diminishment"[249] that requires a "social response".

From now on, the "behavioral surplus" of the users is skimmed off in order to offer the advertising customer a perfect product, *targeted advertising,* with the promise of the perfect match: The advertising should catch the customer exactly in the mood and at exactly the moment and exactly at the place where the probability that they will buy the product is highest. The aim of this is to find a conversion rate of browser to buyer as close to 100 percent as possible.

What the data reveals about the user and flows into a prediction of his/her future behaviour is confirmed or disproved by the user himself/herself during the subsequent purchase process by the click of the user. With each of these clicks, the data prediction engine is fed and the prediction gets better. However, the user will also first become transparent and then recognized as what he/she is in the end at *Google*: a compliant object of manipulation. This makes him/her the "optimized" user for *Google*. *Google* is therefore not primarily a search

---

[247] Zuboff, Shoshana. The Age of Surveillance Capitalism: The Fight for a Human Future at the New Frontier of Power, Main edition (London: Profile Books, 2019). p. 99.

[248] Levy, Steven. In the Plex: How Google Thinks, Works, and Shapes Our Lives. Simon and Schuster, 2011.

[249] Zuboff, Shoshana. The Age of Surveillance Capitalism: The Fight for a Human Future at the New Frontier of Power, Main edition (London: Profile Books, 2019). p. 43.

engine with which the user searches for information. Rather, *Google* searches its users for data and finds what they are looking for with every search query.

AdWords and *AdSense*, the *Google* programs that bring advertisers and users together, will soon become more important than Search. To achieve this, an analysis of user data and the website content must be carried out in parallel.

Not the data, but the customers' lives and privacy is the oil that is being burned and is fuelling climate change on the Internet: debates are becoming inflammatory, driving a political crisis. The source of profit is an expropriation of personal data, which is called personal because it identifies the person and documents their life and makes it decipherable.

This is where the difference between the US and Europe begins: in Europe, the citizen has a fundamental right to data protection, with the instruments of the GDPR. They have the power to dispose of their data, that is to say, the right to decide whether the data may be collected and whether they should be corrected or deleted.[250] This fundamental right does not exist in the US.

According to Zuboff, the whole thing is now embedded in a policy of fundamental secrecy and lies, which means a relapse into the Middle Ages. Zuboff writes: "The behavioral surplus was necessary for revenue, and secrecy would be necessary for the sustained accumulation of behavioral surplus."[251] Eric Schmidt, the CEO of *Google*, himself spoke of a strategy of "hiding" as baked into the cake of *Google* company policy.[252]

---

[250] GDPR Articles 4 (1), 6, 7, 16 and 17 GDPR.

[251] Zuboff, Shoshana. The Age of Surveillance Capitalism: The Fight for a Human Future at the New Frontier of Power, Main edition (London: Profile Books, 2019). p. 111.

[252] Zuboff, Shoshana. The Age of Surveillance Capitalism: The Fight for a Human Future at the New Frontier of Power, Main edition (London: Profile Books, 2019). p. 111.

The experience of the users becomes behaviour, information about the users becomes *data*, the latter is accumulated and evaluated by the company and thus a knowledge surplus is achieved compared to the access of the individual to his/her own life. We users become *exiles of our own lives*.[253]

The extraction imperative does not stop at any area of life. According to Zuboff, Google proceeds in four phases:[254]

1. assault,
2. habituation,
3. adaptation, and finally
4. reorientation.

The bottom line is that users must pay with their privacy in return for social participation.

We can learn from Zuboff that there are economic imperatives behind the captures, extractions and predictions with which the newly discovered market of "behavioral surpluses", or "data fumes" as the personal data were originally called by *Google*, are managed.

The emerging pressure of competition in this lucrative business, in which - like in no other business - size and scaling are crucial, means that more and more diverse data must be procured. As the amount of data increases, so does the need for powerful algorithms, high-performance computers, data centers, and most importantly, AI that can distill the best forecasts from the sea of data in real time.

---

[253] Zuboff, Shoshana. The Age of Surveillance Capitalism: The Fight for a Human Future at the New Frontier of Power, Main edition (London: Profile Books, 2019). p. 125.

[254] Zuboff, Shoshana. The Age of Surveillance Capitalism: The Fight for a Human Future at the New Frontier of Power, Main edition (London: Profile Books, 2019). p. 168.

AI is therefore the technology needed to drive this highly profitable economy forward, because only intelligent, learning programs can derive patterns and forecasts in this ever-growing sea of data. And quantum computers will soon be able to help with this.

At the same time, the person who has the greatest wealth of data on which to train the AI has an initial advantage in the development of AI. These are mainly *Google* and *Facebook*. In addition to size, speed is the only thing that counts, which is why the state, society and politics with their deliberate, compromise-seeking slowness are the natural enemies of the surveillance capitalists.

A new asymmetry between the platform and the customers is created not only in knowledge but also in the fields of authority and power. For these platforms, the small enemy is the rebellious recalcitrant individual, the bigger enemy is democracy. Both insist on sovereignty, which the surveillance giants claim for themselves. In this conflict there can only be one winner in the long run.

It is also important that the raid for personal data already transcends the inner world of the user to his/her physical environment, which is the next sphere of the ongoing gold rush in the *Internet of Things*.

When talking about the digital oracle that the behaviour manipulators pretend to be, it is important to dampen too high expectations:

The oracle of *Google* & Co. seems to know secretly about its own weaknesses and therefore tries to bring about the state it predicts by its own manipulations. Lenin is alleged to have said that "trust is good, but control is better".[255]

Appropriately, *dataism* provides the image of a compelling course of history and thus also of the future of the individual. It proceeds according to the motto: resistance to our prognosis is futile. Zuboff compares Google's approach with that of the Spanish conquistadors, who read "*El Requerimiento* "[256] or declaration, without translation, to the natives on the newly discovered continent before each fighting action.

The *Google* requirement therefore consists of six declarations:

1. claiming personal experience as raw material
2. the right to their conversion into data
3. the right of possession of such data
4. the right to know what the data reveals
5. the right to decide how this knowledge is used, and
6. the right to determine the conditions of this process

An unprecedented raid aimed at exploiting the most intimate lives of billions of people.

---

[255] What Lenin said is the subject of extensive debate as it is not present in his works and therefore cannot be proven (see History Stack Exchange. "Quotes - Did Lenin Precisely Say 'Trust Is Good, Control Is Better'?" https://history.stackexchange.com/questions/51370/did-lenin-precisely-say-trust-is-good-control-is-better.)

In short, the various phrases attributed to Lenin all mean that one should only rely on what has been previously verified. On the other hand, there is evidence that Lenin very often used the Russian proverb "Trust, but verify" (Russian Доверяй, но проверяй - Dowerjai, no prowerjai). It is assumed that this proverb was slightly modified in some translations of Lenin's texts, since the Russian word proverjai can also be translated as "check" instead of "verify".

[256] The *requerimiento* was a declaration of the Spanish Crown read to the natives to assert the right of the Spanish to conquer the country and subjugate the indigenous population.

The legal scholars and doyens of data protection in Germany, Spiros Simitis[256a] and Paul Schwartz[257], who teaches in California, saw the danger for autonomy and democracy early on. In remembrance of the census decision of the German Federal Constitutional Court,[258] Simitis said: "It is about the structure of our society. A democratic society cannot function without data protection."[259] And: "The right to decide for oneself what happens with one's own data is an elementary prerequisite of a democratic society. It must therefore be understood and practiced in such a way that access to the data is not used to specifically control the behaviour of the individual and to adapt to third party interests".[260]

This practice is not based on evil intentions, but on the logic of data accumulation. After all, data accumulation reveals a picture of a complete life and serves to manipulate behaviour. It thus undermines the freedom to set one's own goals and make decisions.

Google and *Facebook* are also involved in the nightly porn search and the initiation of a page jump. The potential for abuse of such information is illustrated by a headline in the Internet magazine *Readwrite* in the 2012 presidential election campaign about the conservative candidate: "Mitt Romney wants your vote and knows your favourite porn site".[261]

[256a] Simitis died on 18 March 2023 in Königstein im Taunus, near Frankfurt.

[257] Schwartz, Paul M. "Privacy and Democracy in Cyberspace." SSRN Scholarly Paper. Rochester, NY: Social Science Research Network, February 11, 2000. https://doi.org/10.2139/ssrn.205449.

[258] BVerfG, Order of the First Senate of 15 December 1983 - 1 BvR 209/83,http://www.bverfg.de/e/rs19831215_1bvr020983en.html

[259] Spiros Simitis : Kai Bierman, Wider die Staatliche Fürsorge, Die Zeit, 7 November 2008, https://www.zeit.de/zustimmung?url=https%3A%2F%2Fwww.zeit.de%2Fonline%2F2008%2F45%2Fdatenschutz-grundgesetz

[260] Rainer Erd, Data protection threatens to turn out to be fiction, Die Zeit, 18 November 2011 https://www.zeit.de/digital/datenschutz/2011-11/spiros-simitis-datenschutz.

[261] Titlow, John Paul, ReadWrite. "Mitt Romney Knows Your Favorite Porn Sites And Wants Your Vote," ReadWrite. 16 October 2012. https://readwrite.com/2012/10/16/mitt-romney-knows-your-favorite-porn-sites-wants-your-vote/. also see Duhigg, Charles. "Campaigns Mine Personal Lives to Get Out Vote." The New York Times, October 13, 2012, sec. U.S. https://www.nytimes.com/2012/10/14/us/politics/campaigns-mine-personal-lives-to-get-out-vote.html.

The potential for blackmail and the tragedy of such a deep invasion of privacy became clear after the suicide of a priest who was discovered on the website *Ashley Madison* as cheating alongside 33 million other page-turners, after the customer data, probably by hackers, came to the public's attention in 2015.[262]

The *Google* and *Facebook* model is moving away from the passive placement of appropriate advertising, and to an increasing extent towards the shaping and manipulation of desires, which can then be fulfilled by the combination of advertising plus an easy to click purchase option.

To increase the accuracy of advertising hits, not only is user behaviour recorded and future behaviour extrapolated, but active assistance is also provided to arouse specific wishes and then fulfil them immediately.

So, the digital oracle is equipped with a certain cunning. Analogous to the saying "believe only statistics that you yourself have falsified", the most accurate predictions are those whose probability has been increased by the oracle itself, by manipulating the subject of the prediction beforehand. As Zuboff writes, because of the enormous profits that lure, the surveillance capitalists are tempted "to shape the flow of behaviour at the price time and place for maximally efficient influence."[263]

So if the extrapolation of personal data makes it possible to foresee how an individual is *likely to* decide, why not nudging this decision

---

[262] Isiore, Chris and Goldman, David, "Ashley Madison Data Posted by Hackers." https://money.cnn.com/2015/08/18/technology/ashley-madison-data-dump/index.html. Reuters, "Hacker's Ashley Madison Data Dump Threatens Marriages, Reputations." 19 August 2015. https://uk.reuters.com/article/us-ashleymadison-cybersecurity-idUKKCN0QN2BN20150819.

[263] Zuboff, Shoshana. The Age of Surveillance Capitalism: The Fight for a Human Future at the New Frontier of Power, Main edition (London: Profile Books, 2019). p. 235.

a little, i.e., *nudging it,* to help it get started? Especially since the individual in question could experience this manipulation as a benefit: Somehow I have been latently interested in this product all this time, thank you, dear algorithm, for fulfilling my wish, especially now, when it is so good for me, by tricking me into buying it.

So if the first phase of the attack (attack on the data) was connected for the affected persons with the thought "they are welcome to have the data, I have nothing to hide and therefore nothing to fear", then after phase 2 and 3, of acclimatization and adaptation to this situation, phase 4 is now being heralded: the reorientation of the behaviour according to the interests of Google and its business customers.

In this phase it becomes apparent that the thought that one had nothing to fear because one had nothing to hide was just a very pious wish. Because now it is a matter of eliminating the remnants of self-determination in the individual.

But people who no longer have self-determination and autonomous, independent thinking and judgment can no longer keep a democracy alive. For in place of a reorientation in their own thinking, they would have become completely dependent on technology.

Zuboff shows in detail how insights from psychology and the secret services are developed into *life-pattern marketing* to lead this campaign of conquest, which is a campaign of destruction against the freedom of the individual.

According to the inevitability doctrine of Californian ideology, a self-fulfilling prophecy emerges here: with the conquest of the human inner world by psychometric techniques and the conquest of the real outer world by *IoT,* the basic conviction of cybernetics is confirmed that humans can be controlled exactly like machines, since they are also nothing more than information-processing machines.

Quod erat demonstrandum. The destruction of human autonomy becomes proof of the correctness and inevitability of the process that led to its destruction.

Zuboff shows that there are two opposite operations: the recording of human emotions through *affective computing* and the recording of the outside world through sensors in the IoT. It is only through these two operations that social space can be completely virtually reconstructed to control the real from the virtual one.

It is interesting to see how behind the digital veil a recurring pattern is becoming increasingly apparent: Even with *Google* Search, the user is not actually searching, but rather his or her data traces enable the *search* to search behind the user. While the reader reads, he or she is read, while the consumer consumes, he or she is consumed.

The temptations of *personalization* turn out to be a preparation for a perfidious overpowering, in which piece by piece each persons' own abilities and decisions are handed over to the platform. Simultaneously, the dependence on the platform becomes ever greater. An average of 150 daily contacts with the smartphone, sometimes even a thousand with the right *digital natives,* do the rest to promote the habituation effect.

What we are experiencing is the greatest social experiment of all time, carried out on all of us, with means that are more powerful than those of any research laboratory, because they intervene very deeply in our lives. And which, unlike a scientific experiment, is completely without rules. Can you rule out abuse? No, mankind is the target of the entire experiment.

Facebook and *Google* have already been caught several times manipulating without the knowledge of the users in order to test and refine the effectiveness of their instruments. [264] Marc Zuckerberg has justified himself with the sentence: "We decided that would be the social norms now, and we just went for it."[265] A sentence that reveals how immense the self-confidence of the platforms has become, to be able to set the rules themselves, according to which they orientate their practice.

So what will Facebook do in the future? Move fast and destroy things? It looks as if this infantile maxim will also be used in dealing with resistance from democratic institutions. After all, it is all about defending the, currently, most successful business model in the world and exploiting the profit potential it holds. Zuboff writes that in the digital economy, largely heteronomous individuals are already confronted with autonomous capital.[266] This is a huge threat to freedom and democracy.

8.2.2 A European view of Shoshana Zuboff's surveillance capitalism

Zuboff describes the global digital market architecture as "free of democratic obligations, and legal constraints".[267] The actors in the US are not state actors, but purely private corporations, while the same anti-democratic logic of domination is consistently instrumentalized by an autocratic state in China.

---

[264] Goel, Vindu. "Facebook Tinkers With Users' Emotions in News Feed Experiment, Stirring Outcry." The New York Times, June 29, 2014, sec. Technology. https://www.nytimes.com/2014/06/30/technology/facebook-tinkers-with-users-emotions-in-news-feed-experiment-stirring-outcry.html. Australian Associated Press. "Facebook to Hide Number of Likes in Trial Aimed at Improving Users' Wellbeing." The Guardian, September 26, 2019, sec. Technology. https://www.theguardian.com/technology/2019/sep/27/facebook-to-hide-number-of-likes-in-trial-aimed-at-improving-users-wellbeing. Also see https://support.google.com/optimize/answer/6211930?hl=en.

[265] Zuboff, Shoshana. The Age of Surveillance Capitalism: The Fight for a Human Future at the New Frontier of Power, Main edition (London: Profile Books, 2019). p. 314.

[266] Zuboff, Shoshana. The Age of Surveillance Capitalism: The Fight for a Human Future at the New Frontier of Power, Main edition (London: Profile Books, 2019). p. 374.

[267] Zuboff, Shoshana. The Age of Surveillance Capitalism: The Fight for a Human Future at the New Frontier of Power, Main edition (London: Profile Books, 2019). p. 314.

This is what the American writes from a US perspective. In Europe, on the other hand, there is a constitutional obligation to protect privacy and personal data there is the EU General Data Protection Regulation[268] and the *EU-US Privacy Shield*, the later being intended to regulate the EU's data protection relationship with the US.[269]

The Charter of Fundamental Rights of the European Union states

Article 7
Respect for private and family life

Everyone has the right to respect for his private and family life, his home and his correspondence.

Article 8
Protection of personal data

1. Everyone has the right to the protection of personal data concerning him or her.
2. Such data must be processed fairly for specified purposes and on the basis of the consent of the person concerned or some other legitimate basis laid down by law. Everyone has the right of access to data which has been collected concerning him or her, and the right to have it rectified.
3. Compliance with these rules shall be subject to control by an independent authority.[270]

---

[268] Regulation (EU) 2016/679 of the European Parliament and of the Council of 27 April 2016 on the protection of individuals with regard to the processing of personal data, on the free movement of such data and repealing Directive 95/46/EC (General Data Protection Basic Regulation OJ L 119, 4.5.2016, p. 1, 'http://data.europa.eu/eli/reg/2016/679/oj'.

[269] Commission Decision (EU) 2016/1250 of 12 July 2016 pursuant to Directive 95/46/EC of the European Parliament and of the Council on the adequate protection provided by the EU-US data protection shield (notified under document number C(2016) 4176), OJ L 207, 1.8.2016, p. 1, 'http://data.europa.eu/eli/dec_impl/2016/1250/oj'.

[270] EU Charter of Fundamental Rights, Article 7 and Article 8 https://eur-lex.europa.eu/eli/treaty/char_2012/oj

Anyone who reads Zuboff understands better why it is worth fighting for strict compliance with these rights rather than nagging about them.

And let us have no illusions:

Not only GAFAM, but also other companies in the US, Europe and the rest of the world, the government in China, US and other countries, including in the EU, are collecting more and more data for the purpose of instrumentalization. We remember Edward Snowden and the BND Bundesnachrichten Dienst, the German Federal Agency for spying outside Germany investigation committee.[271]

Many want to collect information, and many do. But no one has reached a scale like the American *National Security Agency (NSA)* in combination with GAFAM. No other country in the world, neither China nor Russia, has companies which, like GAFAM, sell technology and services worldwide, constantly work with personal data and confidential information all over the world and centralise it for processing in the US. The pressure on these companies to collaborate with the US government was immense after the attacks of September 11, 2001. Do we really believe that this pressure has decreased after Snowden's revelations? And can the companies in the US without a constitutional requirement, as in the EU, operate data protection autonomously and successfully, even against their own government? At least that is what Brad Smith, the President of *Microsoft*, says in his book Tools and Weapons.[272] Doubts are allowed.

In China, the private giants Tencent, Alibaba and Baidu also collect huge amounts of data. But the reach of these companies outside

---

[271] Biselli, Anna. "NSA-Untersuchungsausschuss." netzpolitik.org (blog). https://netzpolitik.org/nsa-untersuchungsausschuss/.

[272] Smith, Brad, and Carol Ann Browne. Tools and Weapons: The Promise and the Peril of the Digital Age. Penguin Publishing Group, 2019.

China is not comparable to that of GAFAM outside the US. Because their business model does not recognize data protection, and because they depend directly on the Chinese government and must provide it with all data. This means that they will never be able to extend their services to constitutional states with basic rights protection. They are the offspring of dictatorship. Fortunately, the investors have not forgotten this either, who invested 30 percent less in AI in China in 2018 than in 2017. Government, politicians and the public should always be aware of this, especially when signs appear at the cash desks of department stores that *Tencent WeChat Pay*[273] can now be used here. Beware!

8.3. The network and lock-in effects of the platform economy

Platforms are multilateral markets. In the case of Facebook, they bring together those who want to network with their friends. This is one facet of their operation. Another facet is their role in bringing together these networked friends with advertisers and sellers. *Facebook* earns money from these. Because they pay for Facebook to bring them the customers who are most willing to buy or most interested in their advertising, and if possible, in such a way that a purchase is made.[274]

The network effect is that a platform that organises a network becomes more attractive the more users it has. This is just as a market becomes more attractive the more people go there, because that is

[273] South China Morning Post. "China's Tech Giants Struggle with Data Privacy amid Push into US," January 5, 2018. https://www.scmp.com/tech/china-tech/article/2126823/chinese-tech-giants-struggle-data-privacy-they-seek-make-headway.

[274] Kenney, Martin, and Zysman, John. "The Rise of the Platform Economy | Issues in Science and Technology," March 29, 2016. https://issues.org/the-rise-of-the-platform-economy/.

how you meet more friends, and more and better traders come. But the network effect goes a little further: By making the networking of users visible to other users, a new social orientation is created, which itself simplifies networking and provides an incentive for networking. In other words: If I look at the friends of a friend, I meet old friends again. If I see that friends always make new friends, I wonder why I don't do the same. People are usually happy to have friends. And there's more: When I network with others, I see more about their lives. That interests me. You can become addicted to it; you can believe that your own identity depends on being as strongly networked as possible. All this is the network effect, from one side, namely that of the users' side. But because *Facebook* observes all networking and all communication within and beyond it and evaluates it for its own purposes, namely advertising, the network is also becoming increasingly attractive for advertisers and sellers.

The network effect therefore exists on all sides of the market, and the effects on both sides reinforce each other.

Thus, the performance of the platform is improving as more and more players on both sides of the market use it. The gap with the performance of new competitors in the market is increasing simply because *Facebook* has much more data. Thus, the network effect leads to the end of competition, and to a monopoly: *winner takes all*. And once a strong position in the market has been achieved, it also has a huge *lock-in effect*: nobody can leave the network without losing all social contacts - unless there are alternative networks and the data of networking with friends can be transferred there.[275]

In Germany, *Facebook* swept away the student network StudiVZ very early on. Other attempts to create competing networks have

[275] A first step in this direction is the right to data portability, according to Article 20 GDPR.

not been successful so far. Potential competitors such as *WhatsApp* or *Instagram* were simply bought up by Facebook.

The business models of *Apple, Microsoft* and *Amazon* are also based on a mixture of the "*The Winner takes it all*" network effect and lock-in effects in closed systems: You join the network because everyone else is already there, and you don't leave it because that would take too much effort.

The story is similar with search. *Google* has a market share of more than 90 percent in the area of Internet search in all EU member states.[276] This was recently established by the EU Commission in a decision against *Google* for abuse of a dominant position in the market for online advertising. It therefore imposed a fine of 1.49 billion euros on *Google*.

With its approximately 100,000 employees and the acquisition of *DeepMind* in London, *Google* is a powerhouse of Artificial Intelligence. In 2014, only 5 percent of *Google's* employees used *Machine Learning*, or AI. Since then, *Google* has trained about 35,000 employees in *ML*. Google is clearly committed to *ML*. Every day, 15 percent of search queries are new. They are written in all languages and are becoming increasingly complicated. Nobody can write the code for them by hand. That is why *ML* is already being used for programming. *Google* also profits from the network effect, but in a different way: As more and more people use *Google's* search service, it gets better and better. At the same time *Google* gets more and more data from the searchers and can thus better communicate advertising. *Google* also tracks its users on other websites and tries to get more and more data through partnerships with credit card companies like *Mastercard*

---

[276] see the decision of the EU Commission on Google https://ec.europa.eu/commission/presscorner/detail/en/IP_19_1770

and business software companies like *SAP*. *SAP* data is particularly interesting because the Point of Sales (POS) system shows exactly what was purchased. You cannot see this as precisely with credit card data if several items are purchased in one transaction. The data giants are tempted to supplement each other's data. *Google Maps'* new partnership with *Spotify* is another example of this: *Spotify* finds out when its users are where and what music they like to listen to in the car. *Google* learns more about the mood and character of its users through their taste in music.

The network effect also works for business associations, marketing and political lobbying. Dusan Jakovljevic, co-founder of the Platform for Energy Efficiency in Industrial Processes in Europe,[277] has developed a business model with free participation for engineers and technicians working on energy efficiency in industry. Participation is via one of the classic social media channels, from LinkedIn to *Facebook* and Twitter. Paying members of the association are companies that are attracted by the concentration of the relevant audience. If they now want to inform the energy efficiency scene, be it for political mobilization or to sell them something, they can do so via the platform's very detailed lists of the various social media channels. However, these companies can only provide case studies or *best practices*, not hard sales material for products or services. Even paying members must respect the values and neutrality of the network. The model works. Dusan Jakovljevic says he has a total of 137,179 non-paying members and over 50 paying members. The platform is influential and economically successful. It works with a total of 6 employees.

---

[277] https://twitter.com/GreenCogEU

8.4. Dominance in system-integrating innovation in the field of AI

The GAFAM platform companies dominate the development and systems integration of new technologies and services with the Internet. While some of their basic AI research is publicly available, work on systems integration and AI applications for commercial use consumes much larger resources and investments (much larger than the public investment in AI and Internet Technology in many countries of this world) and takes place in a black box. Their previous dominance in the collection, analysis and dissemination of information via the Internet, combined with the new optimization possibilities of AI, will further strengthen their position in the areas in which they are already active and improve it in other areas.

A joint initiative between the EU and its Member States is expected to mobilise €20 billion per year for AI-related research and development over the next ten years.[278] The volume is far below the level of private funding in the US.[279] In China, it is public funds that compensate for the recent decline in private investment in AI.[280]

8.5. Buying up, imitating or displacing competing or subsequent innovation

Where internal innovation reaches its limits or competitors of the own business model appear on the horizon, the method of GAFAM

---

[278] https://digital-strategy.ec.europa.eu/en/policies/european-approach-artificial-intelligence.

[279] O'Meara, Sarah. "Will China Lead the World in AI by 2030?" Nature 572, no. 7770 (August 21, 2019): 427–28. https://doi.org/10.1038/d41586-019-02360-7.

[280] Davenport, Thomas H. "China Is Catching up to the US on Artificial Intelligence Research." The Conversation. http://theconversation.com/china-is-catching-up-to-the-us-on-artificial-intelligence-research-112119.

is buying up, imitating or displacing. They can afford to buy up new ideas and start-ups in an area that is interesting for their business model - and they do so.[281] In the AI sector alone, there were 635 acquisitions between 2010 and mid-2019. The leader in AI acquisitions was *Apple* with 20 purchases, followed by *Google* with 14, *Microsoft* with 10, *Facebook* with 8 and *Amazon* with 7.[282]

Measured against the excitement about acquisitions or investments from China in high-tech companies in Europe and the US, it is astonishing how little criticism the acquisition activities of GAFAM, especially in the field of AI, arouse.

In the core areas of Internet-based business models, competition is being bought up. The most prominent example is *WhatsApp* and *Instagram* by *Facebook*. The Democratic presidential candidate, law professor and long-time US senator Elizabeth Warren planned to create the conditions for reversing these mergers. *Facebook* founder and head Mark Zuckerberg therefore supported opponents of Elizabeth Warren in the election campaign, possibly even Donald Trump.[283] He has announced that he will fight against any attempt to split *Facebook*, using law and justice, precisely the means that *Facebook* has so far undermined, under the slogan "Move fast and break things."[284]

---

[281] Jones, Katie. "The Big Five: Largest Acquisitions by Tech Company." Visual Capitalist, 11 October 2019. https://www.visualcapitalist.com/the-big-five-largest-acquisitions-by-tech-company/.

[282] CB Insights Research. "The Race For AI: Here Are The Tech Giants Rushing To Snap Up Artificial Intelligence Startups," September 17, 2019. https://www.cbinsights.com/research/top-acquirers-ai-startups-ma-timeline/. ; accessed at https://www.statista.com/study/50485/artificial-intelligence/.

[283] Stacey, Kiran. "Mark Zuckerberg Vows to Fight Warren Plan to Break up Big Tech," October 1, 2019. https://www.ft.com/content/29ad3c50-e454-11e9-b112-9624ec9edc59. and Krugman, Paul. "Opinion | Warren Versus the Petty Plutocrats." The New York Times, September 30, 2019, sec. Opinion. https://www.nytimes.com/2019/09/30/opinion/elizabeth-warren-wealth-tax.html.

[284] Newton, Casey. "Read the Full Transcript of Mark Zuckerberg's Leaked Internal Facebook Meetings." The Verge, October 1, 2019. https://www.theverge.com/2019/10/1/20892354/mark-zuckerberg-full-transcript-leaked-facebook-meetings.

Neither the competition authorities in the US nor in Europe have taken action against these mergers. However, the EU Commission subsequently imposed a fine on *Facebook* for making false statements in the proceedings concerning the merger of *Facebook* and *WhatsApp*.

Warren also announced his intention to reverse other mergers, such as *Google's Waze, Nest* and *Double Click* and *Amazon's Whole Foods* and *Zappos*.[285]

8.6. Political influence

In addition to exerting informal influence through money, the GAFAM companies increasingly control the infrastructure of public discourse and the digital environment that is crucial for elections. The digital mega-players can afford to invest massively in political and social influence. GAFAM doubled their lobbying spend between 2017 and 2018 to 55 million dollars in Washington alone. 238 lobbyists work for them there.[286]

In Brussels, the legislative capital of the EU, *Google's* lobbying costs in 2017 amounted to over 8 million euros, with a total of 85 lobbyists [287], *Microsoft's* lobbying costs in 2017 were 5.3 million with 15 lobbyists,[288] *Facebook's* lobbying costs in 2018 were 3.7 million with

[285] Business Insider. "Elizabeth Warren Wants to Force Google to Divest Its Acquisitions of Waze, Nest, and DoubleClick." https://www.businessinsider.in/elizabeth-warren-wants-to-force-google-to-divest-its-acquisitions-of-waze-nest-and-doubleclick/articleshow/68325889.cms.

[286] Kang, Cecilia, and P. Vogel, Kenneth. "Tech Giants Amass a Lobbying Army for an Epic Washington Battle." The New York Times, June 5, 2019, sec. U.S. https://www.nytimes.com/2019/06/05/us/politics/amazon-apple-facebook-google-lobbying.html. Statista Infographics. Richter, Felix, "Infographic: Google Outspends Its Peers on D.C. Lobbying." Statista, 23 January 2019. https://www.statista.com/chart/10393/lobbying-expenditure-of-tech-companies/. ; also see: McCabe, David. "Silicon Valley Learns Washington's Language (and Vice Versa)." The New York Times, December 4, 2019, sec. Technology. https://www.nytimes.com/2019/12/04/technology/personaltech/silicon-valley-washington.html.

[287] Cavallone, Elena. "Tech Giants Are Serious about Lobbying in Brussels." euronews, January 14, 2020. https://www.euronews.com/2020/01/14/tech-giants-are-serious-about-lobbying-in-brussels.

[288] https://www.lobbyfacts.eu/datacard/microsoft-corporation?rid=0801162959-21

20 lobbyists,[289] *Apple's* lobby costs in 2018 were 2.25 million with 7 lobbyists,[290] and *Amazon's* lobbying costs in 2018 were 2 million with 10 lobbyists.[291] No candidate in the democratic process today can afford not to use their services.

It is important to understand that some elements of power reinforce others. Thus, money allows an influence on the formation of public opinion, be it through advertisements and PR campaigns, direct lobbying of politicians, influencing journalists by subsidising their training, project funding and "press partnerships" or influencing scientists through project funding or permanent subsidies.[292]

GAFAM is at the party, with its bags of money, in all areas of public opinion formation and education. It makes friends, seeks understanding and presents itself as a friendly partner with whom goals in the public interest can be achieved. They support election campaigns through training and personnel as well as election appeals. They support journalism schools and journalism projects[293] research projects at universities, incubators, civil society.[294]

For a realistic assessment of the opinion-forming and lobbying power of these companies, it is necessary to include lobbying in the capitals of EU Member States, as it is the governments of EU countries that largely determine the content of EU legislation.

---

289 https://lobbyfacts.eu/representative/64755e0fc2a14e46aa9d8646df6f8f19/facebook-ireland-limited.

290 ‹https://lobbyfacts.eu/representative/1e43aba7ad7041e08fb16b2bdacd5414/apple-inc›.

291 https://lobbyfacts.eu/representative/5615fc9a365b4e0f9e9c0d7929a73f17/amazon-europe-core-sarl

292 Williams, Oscar. "How Big Tech Funds the Debate on AI Ethics." New Statesman, 6 June 2019 https://www.newstatesman.com/science-tech/technology/2019/06/how-big-tech-funds-debate-ai-ethics.

293 Dachwitz, Ingo. "Digital News Initiative - Googles Geld und der Journalismus in Europa." netzpolitik.org (blog). https://netzpolitik.org/dni-recherche/.

294 Previously https://www.googletransparencyproject.org/ and now see https://www.techtransparencyproject.org/

In the US, where laws are now being passed not in Washington, but on the state level, to protect privacy and personal data, but also to maintain net neutrality, lobbying expenditure would also have to be considered consolidated between Washington and all States. The cost of a campaign to water down California's online privacy act alone, which GAFAM have been actively engaged in, has been estimated at $50 million, according to *Microsoft* president Brad Smith. Smith stresses that *Microsoft* has only added $150,000, enough to keep in touch with industry but not enough to really encourage the counter campaign.[295] What he does not mention is that *Microsoft* increased its lobbying budget at the headquarter location of *Microsoft* and *Amazon*, Washington State, and Smith personally lobbied members of the regional parliament to soften the state data protection legislation to which *Microsoft* is bound, according to the newsletter *Politico*.[296]

Even after the California law is adopted, lobbying continues because changes are still possible in 2020. And the Internet Foundation[297] alone is again putting $160,000 on the table for this.[298] But that was only the beginning. According to the initiative supporting the California Online Privacy act, financed by the real estate tycoon Alastair Mactaggart, a new 100 million dollar campaign against the

---

[295] Smith, Brad, and Carol Ann Browne. Tools and Weapons: The Promise and the Peril of the Digital Age. Penguin Publishing Group, 2019, p. 147; access at Fang, Lee. "Google and Facebook Are Quietly Fighting California's Privacy Rights Initiative, Emails Reveal." The Intercept (blog), June 26, 2018. https://theintercept.com/2018/06/26/google-and-facebook-are-quietly-fighting-californias-privacy-rights-initiative-emails-reveal/.

[296] Scott, Mark. "How lobbyists rewrote Washington state's privacy law, Mark Scott", Politico.eu, 26 April 2019, https://www.politico.eu/article/how-lobbyists-rewrote-washington-state-privacy-law-microsoft-amazon-regulation/.

[297] Internet Association. "California Privacy Law Must Be Changed To Avoid Making User Data More Vulnerable • Internet Association." https://internetassociation.org/blog/california-privacy-law-must-be-changed-to-avoid-making-user-data-more-vulnerable/. Callahan, Robert. "California Privacy Law Must Be Changed to Avoid Making User Data More Vulnerable." The Sacramento Bee. https://www.sacbee.com/opinion/op-ed/article223327375.html.

[298] Romm, Tony. "California Adopted the Country's First Major Consumer Privacy Law. Now, Silicon Valley Is Trying to Rewrite It." Washington Post. https://www.washingtonpost.com/technology/2019/09/02/california-adopted-countrys-first-major-consumer-privacy-law-now-silicon-valley-is-trying-rewrite-it/.

law is planned. *Google* alone is willing to spend 40 million dollars on it, Mactaggart said in a letter on the initiative's website.[299]

The figures given here do not include the considerable costs for cultural events thrown by the GAFAM companies, which serve to win allies and to get the ears of political decision-makers, nor do they include expenditure on journalism training, journalism projects, civil society organizations, interest group forums, think tanks and academic research. To truly understand the influence of the tech giants on the public debate on AI and Internet policy, all these expenditures would have to be seen in the aggregate.

It also includes all expenses for membership of associations, such as *Digital Europe,* which appears to represent the interests of the *European* digital economy. In fact, however, *all* GAFAM are members, Microsoft are even on the board. In addition to various associations from EU member states, another 64 multinational corporations are members of Digital Europe. 52 of them have their headquarters outside the EU.

The American Chamber of Commerce *AmCham*[300] and the American government also represent GAFAM interests. They do so not only in Brussels, but in all EU capitals, where they are well represented. They deal with ministers and officials in the individual governments of the Member States as well as members of the European Parliament before they present their own position to the Council of Ministers of the EU and in the European Parliament. Some ambassadors of EU Member States in Brussels invite people to lunch or dinner when a senior GAFAM representative whose European headquarters is located in the ambassador's country visits Brussels and wants to

---

[299] https://www.caprivacy.org/about-us. https://www.ciodive.com/news/how-a-real-estate-developer-gave-california-a-head-start-in-data-privacy-le/555012/.

[300] http://www.amchameu.eu/.

present his interests to the Members of Parliament, the Council and the Commission.

Of course, companies also address Members of Parliament directly if their interests require it. Independently of this, they are engaged in a form of ecosystem management which brings together the community of interested parties and provides social contacts, information and cultural events which people like to go to.

"Informing" the European Parliament and its members is the objective of a whole range of organizations. One should get to know each other in an informal way and make contacts that could be useful if one has special concerns. One such organization is the *European Internet Forum (EIF).* Members of the European Parliament ensure that the association's events can take place in the restaurants of the European Parliament. All costs are borne by the member companies.[301] There are also organizations such as the *Trans-Atlantic Policy Network (TPN)*[302] and the *American European Community Association.*[303] The *Information Technology and Innovation Foundation (ITIF)* plays a special role in the opinion machine of the digital economy. The board of directors of this think tank is full of company representatives, and *Microsoft* is also involved. The studies and statements of the ITIF are often close to industry positions. And the good thing is that the contributions to the ITIF are deductible for tax in the US. Thus, opinion making for industry positions in the US is also tax-deductible.[304] The ITIF has an office in Brussels and maintains a special website with all its positions on digital policy in Europe, tightly aligned with the interests of US industry.[305]

---

[301] https://www.internetforum.eu/about-us/members-new.html.

[302] https://www.tpnonline.org/.

[303] https://aeca.org.

[304] https://www.charitynavigator.org/ein/204403497

[305] https://itif.org/regions/europe/

Digital civil society interests are represented in Brussels by the *European Digital Rights Initiative (EDRI)*. American organizations such as *Access now* and the *Center for Democracy and Technology (CDT)* are also present here. Everyone receives money from GAFAM, although not everyone at the same time and in the same amount.[306] The Washington-based *Center for Democracy and Technology (CDT)*, represented in Brussels by an envoy and with a total annual budget of over USD 6 million, receives 43% of its budget from corporations. GAFAM regularly pay over 200,000 dollars each for membership.[307] *Access Now* also often received funds from GAFAM of $50,000, $100,000 or more per year, with a total annual budget of around $5 million. In comparison to these two American organizations,[308] EDRI's total budget is rather meagre, with less than 1 million euros and GAFAM contributions of 10,000 to 23,000 euros.[309]

The representation of the European digital civil society in Brussels is thus rather weak compared to that of GAFAM - and even more so compared to the digital business interests as a whole.

Among the Brussels-based business associations of the digital economy as well as among the civil society member organizations from the digital scene based in Brussels, there is no single one that has not received GAFAM money in the legislative period 2014 – 2019, in which GDPR and the Copyright Directive was at the core of the debate.

---

[306] See EDRI's 2016 annual report on the contributions of Google and Microsoft, p. 40; following the adoption of the GDPR with the right to be forgotten, which Google is still fighting against today and which EDRI had also opposed, no more contributions from Google are listed in EDRI's annual report from 2017. https://edri.org/about/

[307] Access Now. "FUNDING." https://www.accessnow.org/financials/

[308] https://cdt.org/financials/

[309] https://edri.org/about-us/who-we-are/

8.7. Control of the electronic public and journalism

GAFAM's Internet services are increasingly becoming the sole or main source of political information for citizens, especially the younger generation, to the detriment of the traditional fourth estate journalistic publications, which are so important for democracy and which aim to control the powerful.[310] The targeted online advertising business model robs journalism of its revenue streams and now concentrates more than 80 percent of the new online advertising revenue that has been the basis for the plurality of the privately funded press in the hands of just two companies: Google and *Facebook*.[311] This is not the only cause of the death of newspapers and the journalism profession, but it is the main cause - both in Europe and in the US.

8.8. Ideology of total technical feasibility

Finally, we see the ideology of the total technical feasibility and superiority of machines over humans as the source of GAFAM's power. We have to deal with the ideology of the Internet and computers as alleged freedom-bringers. The claim that individual freedom can be guaranteed by technology and that social problems can be solved by technology alone - this claim of total technical feasibility must be unmasked. This *California ideology,* which prevails in Silicon Valley, combined with that of neoliberalism, is

---

[310] Kleis Nielsen, Rasmus. "Where do people get their news? The British media landscape in 5 charts" Oxford University, Medium 30. May 2017, https://medium.com/oxford-university/where-do-people-get-their-news-8e850a0dea03.

[311] Garrahan, Matthew. "Google and Facebook Dominance Forecast to Rise," December 4, 2017. https://www.ft.com/content/cf362186-d840-11e7-a039-c64b1c09b482; Sluis, Sarah, "Digital Ad Market Soars To $88 Billion, Facebook And Google Contribute 90% Of Growth." AdExchanger, May 10, 2018. https://www.adexchanger.com/online-advertising/digital-ad-market-soars-to-88-billion-facebook-and-google-contribute-90-of-growth/.

incompatible with the basic values of freedom and democracy. It is perhaps the most important instrument of power in the technical-economic complex. Because it enthuses and mobilizes masses of people to support GAFAM interests.

This ideology also contains the promise of comfort and convenience through constant networking, confirmation and wish fulfilment. Combined with elements of popular culture, this amalgam of technical and hedonistic promises of progress contributes significantly to the passive acceptance of GAFAM dominance. This aspect of power can also be described by the currently overused concept of 'the narrative', an overarching story that is supposed to create meaning and gain public acceptance. This concept comes from Jean-François Lyotard's book "The Postmodern Condition" (1979), and through his literary seminars. The problem with narrative is that you can never tell whether it is fiction or fact. Its purpose is to seduce the reader.

As we show, the narrative of GAFAM has the character of an ideology, because it does not only pursue legitimation of their power, but is intentionally-generated false consciousness. GAFAM use significant means to frame their own actions in the public arena to convey not only the superiority of their products but also their underlying technological worldview. This worldview, however, in reality contrasts with promises of progress, which today include universal values such as human rights and democratic structures.

The point is: We must stop letting the narrative of Californian ideology become a fiction that we believe, and instead continue to tell the story of modernity, which is the story of freedom, a freedom of reason and with reason, that must recognize responsibility if it is to have a future in the third millennium.

That is why we now want to devote ourselves to the world view and the human image the economic-technological complex paints in detail. For it is the construction of a convincing *humane* counter-image that the shaping of a human future will depend on to a large extent. We shall not allow the rule of system imperatives and machines to take over.

Part C

The World View and Human Image of the Digital Complex

In its early days, the Internet contained at least two emancipatory promises:

1. Dialogue instead of monologue in public communication: The audience talks back – offering interaction and communication, in contrast to the age of mass communication.
2. Personalization: the tailoring of communication and commerce to the personal needs of the user.

The gains in freedom that the process of individualization has brought, can be combined with the new possibilities of connectivity and participation by an Internet oriented towards the common good, thus fulfilling an essential emancipatory promise of modernity.

The great betrayal of GAFAM is that in the historic moment when, for the first time, a technology was created that opened up direct human networking with extensive new communication possibilities, this tool was used by companies for the primary purpose of profit maximization and to realise plans of economic and technological dominance. These companies have not only appropriated the technology, but they have transformed the ideas and dreams of the idealistic youth movement of the 1960s into the *Californian ideology*. This ideology has become a pillar of their power. And that is why we are here to criticize it. Besides the convenience promised by digital technology, it explains why the new subjugation of people through the GAFAM seems to be completely without coercion.

To explain the tremendous success of capitalism, Max Weber once described what he called the "spirit of capitalism". He saw this spirit as rooted in Protestant ethics and in Calvinist terms of grace. According to Weber, the protestant ethic and in particular the Calvinist ethic formed the ideal prerequisite for capitalism and the rationalization processes of modernity. The demands of hard work, efficiency, and thrift on earth, the daily service in honour of God in the here and now, in order to reap the rewards in eternal life, set the stage for systematization and orderliness that helped lay the foundations for capitalism. According to Weber, *Innerweltliche Askese* "inner-worldly asceticism"[312] meant that a thrifty way of life and a reinvestment of profits formed the basis of capitalist growth, while at the same time it shaped work ethics and adoption of rational rules or laws.

The success of digitization can only be understood if its intellectual foundations are taken into account. These include the promises of comfort and convenience as well as the promotion of rapidly accelerating technological progress culminating in redemption. In place of inner-worldly asceticism comes the promise of a lasting worldly hedonism, crowned by the prospect of virtual immortality. Disinhibition replaces liberation, self-optimization replaces self-determination.

Analysis of the Californian ideology will show that the underlying philosophy considers a world without human dignity, freedom and democracy to be progressive and efficient. The amalgam of the ideas of which it is composed forms the intellectual foundations of the great platforms. It can be called an ideology for various reasons.

The advocates of this ideology - and the business models associated with it - want to shift the boundaries of power between man and machine. However, they themselves want to continue to control the

[312] Weber, Max. 2005. *The Protestant Ethic and the Spirit of Capitalism*. Routledge, 154. We chose the translation “inner-worldly asceticism” following Gordon Wells’ article: Gordon C. Wells, “Issues of Language and Translation in Max Weber’s Protestant Ethic Writings,” Max Weber Studies 2, no. 1 (2001): 33–40.

machine. In the end, humankind may pay the price in terms of loss of humanity.

C1 The Californian ideology

> *Failing to understand the consequences of our inventions while we are in the rapture of discovery and innovation seems to be a common fault of scientists and technologists [...] We are being propelled into this new century with no plan, no control, no brakes. Have we already gone too far down the path to alter course? I don't believe so, but we aren't trying yet, and the last chance to assert control—the fail-safe point—is rapidly approaching.*

William Joy, co-founder of Sun Microsystems[313]

One of the power narratives of digital companies is a comprehensive ideology that uses narrative to promote their technology and business practices aimed at increasing their acceptance by customers and the public. This ideology contributes at least as much to widespread acceptance of the technology as the undoubted benefits that digital offerings bring to everyday life. If acceptance increases, the criticism disappears, according to the calculation.

The enthusiastic claim that GAFAM is trying to make the world a "better place" has been especially effective at securing their position as an indispensable and popular part of our lives. The individual and social benefits, combined with the provision of comprehensive amenities offered by digital services, are deemed so important that the downsides of these technologies have been neglected. The cuts in personal rights and informational self-determination are so great that they have become a major factor in the development of our society, yet the damage inflicted to the public sphere, democracy and law was hardly noticed for a long time.

---

[313] Joy, Bill: 'Why the Future Doesn't Need Us'. Wired, 1 April 2000. https://www.wired.com/2000/04/joy-2/.

The Californian ideology differs from ideologies of earlier epochs in the technical means at its disposal to ensure its own propagation. It draws on human fascination with technologies as a mechanism by which to ensure the adoption of this ideology.

1.1. The concept of ideology

> *"Criticism becomes ideology critique when it wants to show that the validity of the theory has not sufficiently detached itself from the context of its origin, that behind the back of the theory an inadmissible mixture of power and validity is hidden and that it owes its reputation to this."*

Jürgen Habermas[314]

By ideology we mean: first, a certain world view, a system of basic attitudes and values, and a system of political ideas that are intended to achieve political goals. Second, following Marx, we understand ideology as "false consciousness", which means that the self-image disseminated in ideology differs from the objectively correct image. As a rule, this deviation is deliberately created to distract from power relations or interests.

In this second sense, ideology is a self-understanding based on faulty or incomplete basic assumptions and it is consciously generated to conceal power relations.

When talking about ideology, we do not follow Marx in thinking that it is "socially necessary" that humans must develop false

[314] Habermas, Jürgen, 'The Devouring of Myth and reconnaissance. Remarks on the "Dialectic of Enlightenment" - after a second reading'. in Karl-Heinz Bohrer, Mythos und Moderne, p. 415.

consciousness. The postulation of alleged necessities is itself a weakness of Marx's thinking, which it ultimately shares with the Californian ideology, which also contains an inevitability doctrine. According to this deterministic thinking, the "scientifically recognized" laws of historical materialism or the "exponential growth" of technological potential, causes history to necessarily follow a certain logic of development.[315]

This assertion of inevitability serves to eliminate the power of action of the individual and of society. It is an essential component of anti-democratic ideologies. Historically, critique of ideology has its origins in the Enlightenment. It is directed against dominant worldviews that spread superstition, prejudices and errors that people are supposed to hold on to, even though they contradict their own potential as rational beings. For such a critique one needs a position outside the ideology, which for its part must be reasonably justifiable and show that the ideology criticized contradicts reasonable claims. Alternatively, ideology critique can also be practiced immanent. In this procedure the ideal postulated in ideology is measured against reality.

As a rule, ideologies instrumentalize concepts in order to legitimize rule. In ideology critique, the aim is to grasp this context and to show the wrong basis of legitimation. Thus, it is at the same time a critique of power based on ideology, and as a critique of concepts, it is by its very nature always also a critique of language.

The Enlightenment subjected the world views of Christian origin to a radical process of criticism. It led to an extensive secularization of society and a "disenchantment of the world" (Max Weber) by

---

[315] On inevitabilism with Marx and the Californian Ideology see Shoshana Zuboff; The Age of Surveillance Capitalism, p. 256.

the advancing natural sciences. According to Georg Lukács, this disenchantment of the world with science and technology leads to a "transcendental homelessness"[316] of modern man. The Californian ideology now promises to lead people out of this homelessness through the magic of digitalization into a zone of well-being, where they exchange the efforts of self-determination for the promises of comfort and self-similarity.

Unlike the algorithms and source codes on which the power of the platforms is based, the ideology is by no means secret. It is openly disseminated in books, is distributed at conferences and with each step wins more and more followers. In a certain sense, it exudes the charm of success that can be measured in stock market value and the market shares of its technologies on the world markets, and the top positions on the stock exchange that GAFAM occupy.

The Californian ideology was founded in the 1960s by activists like Stewart Brand in the environment of hippie counterculture. With the advent of the Internet in the 1990s, this ideology was then developed in a second step into a defensive reflex against any political and legal interference in cyberspace, culminating in John Berry Barlow's Declaration of Cyberspace Independence of 1996.[317]

After the founding and triumphant advance of the major platforms such as Google and Facebook in the new millennium, finally that meta-theory of the Californian ideology was formed, for which Noah Yuval Harari coined the term *dataism*.[318] This was at the time that *Big*

---

[316] Lukács, Georg, and György Lukács. 1971. *The Theory of the Novel: A Historico-Philosophical Essay on the Forms of Great Epic Literature*. MIT Press. p.185

[317] John Perry Barlow, 'A Declaration of the Independence of Cyberspace', Electronic Frontier Foundation, 8. February 1996, https://www.eff.org/de/cyberspace-independence

[318] Y. N. Harari, Homo Deus. Although Harari may not have invented this phrase, that was probably David Brooks in 2013: David Brooks, "Opinion | The Philosophy of Data," *The New York Times*, February 4, 2013, https://www.nytimes.com/2013/02/05/opinion/brooks-the-philosophy-of-data.html

Data was being considered for the first time and powerful programs were becoming available to generate forecasts for the future from the analysis of data. However, *dataism* goes far beyond the description of these possibilities and, according to Harari, becomes a new religion, which contains an independent ontology ("All Being is Information") and its own ethics ("Good is what processes as much data as possible"). Authors such as Ray Kurzweil also construct a cosmological meta-narration based on these elements, making predictions about future technological developments and the role of mankind in the future.

Just as Shoshana Zuboff uses Google's chief economist, Hal Varian, as the "*Adam Smith of Googleonomics*"[319] for her analysis, so we analyse and criticise in detail the work of the creator of the Singularity religion, Ray Kurzweil, who is also Google's influential head of development. Like Varian, Kurzweil's creative power reaches far into Google. It is clear from our analysis that not only the "digital dream darkens" (Zuboff) as a result of surveillance capitalism, but that reason itself also is put under a shadow (Horkheimer) as a result of the totality claim of the associated thinking in conjunction with instrumental power.

1.2. Stewart Brand and the Whole Earth Catalog - technology instead of politics

In the first years and decades of its existence, computer technology and the Internet were able to develop without any legal framework. To understand why it could develop without guard rails and why this went wrong in the end, let us look at the beginnings of digital

[319] Zuboff, op. cit. p. 86.

culture in California and that stupendous amalgam of ideas that became the ideology of the same name.[320]

The rejection of law in the original culture of Silicon Valley has its roots in the youth movement of the 1960s. This movement, politicized by the Vietnam War and Martin Luther King, among others, split into two camps after the 1967 Summer of Love ended in an orgy of violence and crime. The hippies split up: on one side was the new left wing, those who continued the fight against the war in Vietnam on the other dropouts who withdrew to rural communities and the deserts of California to avoid being drafted to Vietnam, Washington-dominated politics and the technical dominance of IBM-*mainframe systems*.[321] This second wing of the youth movement was guided by a strong impulse for personal freedom and individual empowerment through decentralization of decision-making and technology. "Freedom must now be won by liberating us from conventions. (...) Many icons of counterculture develop a contempt for politics in general."[322] Between 1965 and 1973, 750,000 young Americans left home in this way to join the new communities.[323] This movement became increasingly distanced from the New Left in the US. In contrast to the political activists, who continued to try to change society with manifestos and political actions, this new communalism, as Fred Turner calls it, "turned away from the politics of struggle and embraced the politics of consciousness."[324]

---

[320] The term Californian Ideology was introduced by: Barbrook, Richard, Cameron Andy: The Californian Ideology, 1995: http://www.imaginaryfutures.net/2007/04/17/the-californian-ideology-2/.

[321] Here we follow the detailed analyses of Fred Turner in: The Deceptive Promise and Fred Turner, From Counterculture to Cyberculture: Stewart Brand, the Whole Earth Network and the rise of Digital Utopianism, Chicago 2006.

[322] Virtual Essay in: The Whole Earth, California and the the disappearance of the Outside, Berlin 2013, The Whole Earth Catalog, 1968. https://monoskop.org/images/0/09/Brand_Stewart_Whole_Earth_Catalog_Fall_1968.pdf p. 75

[323] Turner, Fred. 2010. From Counterculture to Cyberculture: Stewart Brand, the Whole Earth Network, and the Rise of Digital Utopianism. University of Chicago Press, p.64

[324] Turner, Fred. 2010. From Counterculture to Cyberculture, p. 64

In 2005 the House of World Cultures in Berlin presented an exhibition entitled: *The Whole Earth Catalogue*[325]. It dealt with the influence of the 1968 movement and the American counterculture on the development of the tech scene in Silicon Valley. The curators examined the counterculture's geographical and cultural anchoring in the American dream of the settlers who once had to stop on their way west along the Pacific Ocean, where American expansion was driven by the slogan Go West! in the mid-19th century.[326] After the land grab on the new continent was completed, the conquest could, it seemed at first, only be continued in cyberspace.

In the discovery and reclamation of cyberspace, one man stands out: Stewart Brand. He came from the counterculture and was set to have an enormous influence. This activist was one of the founders of the hippie scene in Haight-Asbury, experimented with LSD and organized festivals featuring bands like the *Grateful Dead*. In 1966, Brand launched a campaign demanding that NASA publish photos of the Earth. He sold buttons with the caption *Why Haven't We Seen a Photograph of The Whole Earth Yet?* After the first photo was finally published, he published the *Whole Earth Catalogue* in 1968, in the middle of the worldwide year of the student and pop-cultural revolution. The subtitle of the catalogue was: Access to tools, with the longed-for image of the whole earth seen from space as cover. Next to the picture of the earth was written: "*The flow of energy through a system acts to organize that system.*"

The motif is in a way the first selfie on earth and of earth. One can speak of a selfie of the earth, because the photo owes its existence to the intelligence that the earth itself has produced, the human being.

---

325 The Whole Earth, California and the Disappearance of the Outside, exhibition catalogue, House of World Cultures, Berlin 2013, see Footnote 334

326 This is the title of a book by Jean-François Lyotard, The Wall of the Pacific, 2006.

The first picture of the "whole" earth thus unites in a peculiar way a double perspective: The view of the earth from outside makes us all simultaneously viewers and observed.

Is this photo the beginning of a new peaceful epoch of mankind, a new chapter of enlightenment, like Brand believed? Or does a direct path lead to the selfish-Selfie-Society as we know it today?

Until then, the iconic photograph of the 20th century had been the photograph of an atomic explosion mushroom cloud, symbolizing the destructive power of modern large-scale technologies. The image of the *Blue Marble,* on the other hand, showed people for the first time the beauty and wholeness of their planet, but at the same time its vulnerability. It therefore marked in Brand's eyes the beginning of a new epoch, in which through the expansion of consciousness a new holistic view of the world and thus a new harmonious age could dawn.

Brand at least was convinced of the possibility of utopia, because the *Whole Earth Catalogue* euphorically promotes the idea that through technology all human beings are connected as free beings while simultaneously feeling themselves to be part of the whole. The first picture of the whole earth was connected with the hope that the threats posed by large-scale military technologies and the dominant central *mainframes* of IBM would be replaced by a peaceful alternative through the networking of individual *personal computers,* a concept that Brand immediately co-invented. Through free access to the knowledge of the world for everyone and through decentralized networking, a new libertarian utopia was to be realized, according to Brands vision. With the view from outside on the whole of the earth, the old dream of mankind to be able to take God's view on the world had finally come closer.

Brand's utopia of a new, holistic, harmony between humans and technology led him to create the first local network: *WELL, The Whole Earth 'Lectronic Link,* which connected the users of PCs with each other. Through the network, a local forerunner of the *World Wide Web,* on which the *Dead Heads,* the fans of the Grateful Dead, exchanged ideas, the liberated individual was to connect himself or herself harmoniously with the greater whole - without any middlemen or institutions. The devout Catholic and media theorist Marshall McLuhan enthusiastically celebrated the opportunity to "instantly translate any code into any other code and any language into any other language. In other words, the computer offers by technology a Pentecost festival of universal understanding and unity."[327]

In view of the impending nuclear apocalypse at the time, for Brand and his companions, the alternative was a "cybernetic vision of the world, one in which material reality could be imagined as an information system. To a generation that had grown up in a world beset by massive armies and by the threat of nuclear holocaust, the cybernetic notion of the globe as a single, interlinked pattern of information was deeply comforting: in the invisible play of information, many thought they could see the possibility of global harmony."[328]

The *Whole Earth Catalog,* a magazine with a mix of essays, book recommendations and lists of computer accessories soon develops into the bible of the Californian hippie and alternative scene. Steve Jobs later said that the catalogue was the analogue forerunner of the search engines: " It was sort of like Google in paperback form, 35 years before Google came along."[329]

---

[327] Marshall McLuhan, Understanding Media: The extensions of man, London and New York, p.80. Access at https://designopendata.files.wordpress.com/2014/05/understanding-media-mcluhan.pdf.

[328] 'From Counterculture to Cyberculture: Stewart Brand, the Whole Earth Network, and the Rise of Digital Utopianism by Fred Turner, Introduction'. https://press.uchicago.edu/Misc/Chicago/817415.html.

[329] Steve Jobs, Stanford Commencement Speech, 2005, quoted in Dirksen, Kirsten, 'Steve Jobs' Google of the 60s: Whole Earth Catalog Revisited'. HuffPost, https://www.huffpost.com/entry/whole-earth-catalog_b_1075638

A large part of this effect was down to the pathos of Stewart Brand, which was also clever marketing for the magazine. It contained articles on books, including Norbert Wiener's *Cybernetics*, but also descriptions of technical components, which Brand then sold directly to readers. He took a bus to the drop-outs from politics who had committed themselves to the project of self-discovery in the valleys of California, after the political hopes of the counterculture had failed for the time being.

The American counterculture was based on the myth and libertarian ethos of the frontier: a key concept of American self-image.[330] The historian Frederick J. Turner used the term "*frontier*" to describe the feeling and aspiration of the American settlers at the end of the 19th century: being on their own and able to survive and become successful not by the guidance of an authority but only by their own abilities.[331] The term has since become a metaphor for freedom, individualism and independence from state authorities. The ideology of the frontier brings the anarcho-libertarian touch to the Californian ideology, which is still leading the way today.

In space, borders do not play a role. The circle of the earth's entirety replaces with the iconographic first image of the earth the horizon that the settlers had had in mind until then and which had shown them the direction in their expansion.

The beginnings of the Californian ideology thus lie between enthusiasm for technology and hippie revolt. They still fall into the

---

[330] John F. Kennedy seized upon this myth after the Sputnik shock and in 1960 announced the New Frontier's government program, with which the US, as the technologically most highly developed nation in the world, was to take up the fight for supremacy in space. As part of this New Frontier program, Kennedy also announced the first moon landing, which became a reality in 1969.

[331] Turner, Frederick Jackson. 2008. *The Significance of the Frontier in American History*. Penguin UK.

time when the horizon was geographically bounded. What then began to happen was the beginning of the colonization of a space that no longer knows any boundaries. This space, the Internet, becomes the projection surface of a cosmic feeling of unity, in which the individual can seemingly effortlessly escape the lowlands of individuation and, without any rules and laws, find harmony with others and with creation.

1.3. Excursus: Freud and the boundless feeling of omnipotence

Freud had, in 1927 written about "Uneasiness in Civilization"[332] and described the "oceanic feeling" of primary narcissism, in which the individual feels a sensation of infinity, without a boundary between the individual and the outside world. Freud develops the adult's sense of ego from the infant's pleasure ego, which is all-embracing and boundless. In a certain sense, both the immediate fantasies of unity and the godlike feelings of omnipotence of the discoverers of cyberspace correspond to this narcissism as described by Freud. Already the founder of psychoanalysis had described the "oceanic feeling" by the feeling of "limitless extension and oneness with the universe".[333]

The image of the *Blue Marble* not only became an icon of the environmental movement. It also became a symbol of universalization through technology without borders. The end of the expansion towards the Pacific Ocean steered it in a new direction: to a discovery and then conquest of space, which is at the same time a discovery of the subjective interior.[334] The man of modernity, who according to

[333] Freud, Sigmund, 1929. Civilization and Its Discontents. World Heritage Encyclopaedia.

[334] Diederichsen, Diedrich, and Anselm Franke. 2013. The Whole Earth: California and the Disappearance of the Outside. Sternberg Press.

Max Weber and Georg Lukács is condemned to a "transcendental homelessness" by the scientific "disenchantment of the world", can apparently find a new home in this absolute immanence of the "planetary interior".

For this, Brand suggested that all you need is a computer and a modem for networking. From this a new holism was to emerge, which enabled an intimate connection between self and earth, because man and earth were equally conceived as organisms. The analogy of human organism and organism of nature is based on an idea that was developed in the 1970s in the form of the "Gaia Hypothesis" by the biophysicist James Lovelock.[335] According to him, all life is part of the evolution of the super-organism Earth. Politics, it was concluded, is unnecessary, even obstructive, for the self-organization of this organism, just like other forms of institutionalization. In its place is a "spontaneous harmony" that will begin to exist among the subjects in the greater whole. Since, according to the Gaia hypothesis, the inner world and the outer world are characterized by the same structures, an immediate experience of unity between the subjects and Mother Nature is possible.

Spectacular photos from the Apollo missions, such as the image of the rising earth taken from the moon, especially the so-called "Earthrise" image, further increased the enthusiasm for space. The use of new tools and the networking of the users with each other - both should serve the transformation of the countercultural consciousness to a cosmic sense of unity. "The Whole Earth Catalogue [lists] as a map the tools for transforming consciousness".[336] It was intended to bring about an order in which no rules were required to create harmony.

---

[335] Lovelock, James. 1995. Gaia: A New Look at Life on Earth. Oxford University Press.

[336] Turner, Fred "The politics of Gaia around 1968 - and today", in: The Whole World, Catalogue, p. 45

The perspective from outer space on the planet was until then reserved for God. The first feelings of Homo Deus[337] who, with his own deification, begin to develop with fantasies of omnipotence and immortality. Stewart Brand wrote euphorically at the sight of the new photo icon: "We are as gods and might as well get good at it."[338]

The big picture can be seen for the first time with the photographs from space. Systems theory and cybernetics provide the terms with which this wholeness can first be grasped and finally mastered. In the process, the opposites of technology, nature and culture seem to dissolve. For in the concept of cybernetics, both technical and natural systems are equally controlled by information. Then they must also be able to be programmed in the same way. But who writes the code?

The Cambridge physicist C.P. Snow described in 1959 in the essay *The Two Cultures*[340] the contrast between the scientific-technical and the humanities-literary world view in such a way that mutual understanding between these two intellectual approaches seemed impossible. Since then, the gulf between the "explanatory" natural sciences and the "understanding" humanities has been covered by the concept of the two cultures. Many researchers subsequently felt the urge to overcome this contrast with a super theory of systemic order. A quotation from Claude Lévi-Strauss, the eminent anthropologist and founder of structuralism, is an example: "We have had to wait until the middle of this century for the crossing of long separated paths. That which arrives at the physical world by the detour of communication, and that which as we have recently come to know, arrives at the world of communication by the detour of the

---

[337] Harari, Yuval Noah. 2016. Homo Deus: A Brief History of Tomorrow. Random House.

[338] Whole Earth Catalog 1968, The Whole Earth, California and the Disappearance of the Outside, exhibition catalogue, House of World Cultures, Berlin 2013 p. 43.

[340] Snow, C. P. 1993. The Two Cultures. Cambridge University Press.

physical. The entire process of human knowledge thus assumes the character of a closed system"."[341] Lévi Strauss argued that the error of the humanities and social sciences in the past was that they did not pay enough attention to the relationships between the elements of the respective systems: "The error of traditional sociology, like traditional linguistics, is to have considered the terms and not the relationships between the terms".[342]

Structuralism and cybernetics wanted instead to develop a grand unified theory of nature and society.

Cybernetics accelerates the change in Californian ideology from its utopian beginnings to the increasingly anti-politics theories in which technology increasingly takes on the role of the saviour: "The new communalists [...] bought the central faith of the military research world: that experimentation and the proper deployment of the right technologies could save the world,"[343] writes Fred Turner about the new technology scene. Changing the world through technology and solving all its problems, this is the denominator of the new faith of the technology-mad New Communalists in California in the 1970s. Politics can only slow down this mission through regulation and laws. Against such attempts by politicians, targeted lobbying was used early on, which has long been a natural part of the practice of the technology companies.

The role of Stewart Brand as the computer industry's first "storyteller" makes him so important for an analysis of Californian ideology even today. "Silicon Valley engineers revered the man because he showed

---

[341] Lévi-Strauss, Claude. The Savage Mind, p.130

[342] Lévi-Strauss, Claude. Structural Anthropology, p.44.

[343] Turner, Fred. 2010. From Counterculture to Cyberculture: Stewart Brand, the Whole Earth Network, and the Rise of Digital Utopianism. University of Chicago Press. P.244

them the importance of their work in a way that they themselves could not."[344] Brand provided the tech scene with a narrative in which computers were given the status of saviours of humanity, according to Franklin Foer: "If politics couldn't save humanity, maybe computers could."[345]

With this, Brand set a dangerous exaggeration and misconception of the world, which is still spread by leading companies in Silicon Valley today. It has become their first commandment that all problems can be solved by technology or computers, which in turn make no mistakes. In doing so, they overlook the fact that the reduction of human action to technical action itself again produces a mythological and non-rational belief system, which they actually want to overcome in the name of technical rationality.

The Californian ideology thus laid the foundation for the current attitude of the GAFAM companies early on: technology solves problems better than politics. An attitude which, in its extreme form, concludes that politics can therefore be replaced by technology.[346] In the end, it creates a new technocratic form of politics that believes it can do without human, democratic institutions. It believes itself to be free of ideology, and yet it is precisely for this reason that it is ideological in its foundations.

Also Harvard sociologist Daniel Bell postulated that information technologies would replace ideologies and value orientations in the information society. As he put it: "The old ideologies of the West are exhausted. The intellectual and political groundlessness

---

[344] Foer, Franklin. World Without Mind: The Existential Threat of Big Tech (New York: Penguin Press, 2017). p.26

[345] Foer, Franklin. World Without Mind: The Existential Threat of Big Tech (New York: Penguin Press, 2017). p. 26.

[346] See below our remarks on Transhumanism.

is obvious."[347] A scenario that leads to a society determined by technology with an information order that gets by without any moral values beyond that. The consequence of these changes, and overarching aims of the Californian ideology, is to replace politics entirely with technology.

1.4. John Perry Barlow and the unbroken freedom of the Internet

John Perry Barlow, lyricist for the Grateful Dead and co-founder of the Electronic Frontier Foundation (EFF),[348] was one of the most influential activists in the early days of cyberspace. To this day, the EFF, financed by the digital giants, remains one of the most important lobby machines for an unregulated Internet. In 1996, Barlow summarized the credo of the Californian ideology in his "Declaration on the Independence of Cyberspace", as it has been handed down in the tech scene ever since: No law passed by parliaments was suitable for the Internet, traditional forms of government based on the rule of law could "have no sovereignty where we (the actors of cyberspace) gather".[349] It was more by chance that this declaration was presented at the World Economic Forum in Davos in 1996.[350] But this multiplied its impact, because since then it has been considered a quasi-official document of the tech scene.

The ideology of self-regulation was also reflected in the discourse on American jurisprudence: law professors David Johnson and David

[347] Bell, Daniel. 2000. The End of Ideology: On the Exhaustion of Political Ideas in the Fifties : With 'The Resumption of History in the New Century'. Harvard University Press. P.402

[348] Barlow, John Perry. 'John Perry Barlow Library'. 2018. Electronic Frontier Foundation. 9 February 2018. https://www.eff.org/john-perry-barlow.

[349] Barlow, John Perry. 2016. 'A Declaration of the Independence of Cyberspace'. Electronic Frontier Foundation. 20 January 2016. https://www.eff.org/cyberspace-independence.

[350] A video of John Perry Barlow reading the text of his 1996 declaration can be found in the "Internet Hall of Fame", https://internethalloffame.org/news/in-their-own-words/declaration-independencecyberspace.

Post declared that if the users of a particular space on the Internet wanted to establish a set of rules that did not violate the vital interests of non-users, then "the right of sovereigns in the physical world should be established for this new form of self-government".[351] In other words, parliaments should determine that the Internet and cyberspace should be self-regulating.

Even today, the strong, poetic language of John Perry Barlow continues to fascinate.[352] Offering ethics in the form of self-regulation over valid democratic law, Barlow has established a method of argumentation, a *topos,* that continues to shape the debate on regulation on the Internet today. Barlow rejected the binding law that emerged from and was legitimized by a democratic process.[353]

With this in mind, Barlow invokes the Internet as a paradise of natural, childlike fantasy. The impact of Barlow's manifesto on the issue of Internet control to date cannot be overestimated. Drawing on the linguistic style of the American Declaration of Independence and the sound of pop culture, he calls on governments not to interfere with laws or rules on the Internet. The world of governments is "of flesh and steel", he says, while cyberspace is "the new home of Mind."[354] The Internet is a sphere without a body. "I ask you of the past to leave us alone, you are not welcome among us."[355]

---

351 Johnson, David R., and David G. Post. 'Law and Borders - the Rise of Law in Cyberspace'. SSRN Scholarly Paper. Rochester, NY: Social Science Research Network, 1 February 1997. https://doi.org/10.2139/ssrn.535.

352 Barlow, John Perry. 2016. 'A Declaration of the Independence of Cyberspace'. Electronic Frontier Foundation. 20 January 2016. https://www.eff.org/cyberspace-independence.

353 A readable obituary of Barlow was published on the UC Berkley alumni website in March 2006: "John Perry Barlow: Is Cyberspace Still Anti-Sovereign?," Cal Alumni Association, 12 February 2018, https://alumni.berkeley.edu/california-magazine/march-april-2006-can-we-know-everything/jp-barlow-cyberspace-still-anti.

354 All quotes from Barlow, John Perry. 2016. 'A Declaration of the Independence of Cyberspace'. Electronic Frontier Foundation. 20 January 2016. https://www.eff.org/cyberspace-independence

355 The lyricist of the Grateful Dead alludes to Bob Dylan's The Times they are a changing, where it says to adults: "Don't criticise what you can't understand".

The calls for independence come from many who benefit from the platforms: "You are afraid of your own children because they are natives of a world in which you will always remain immigrants". It goes on to say: "You don't know our culture or our rules." This motif still runs through the debates today, most recently when it came to the regulation of EU copyright law when angry YouTubers repeatedly emphasised that adults could not say anything because they were not YouTubers. Only those who are YouTubers can therefore regulate YouTube, and no YouTuber wants that, so it is best to just leave it alone. In debates like this, Google, whose parent company Alphabet YouTube is part of, has no difficulty in using a broad-based campaign with PR videos in the run-up to the copyright reform of the European Parliament to arouse fear among users of the desire for regulation, for example by threatening to shut down the popular video platform.

Barlow's second assumption shows a naïve utopianism and the author's basic techno-liberal attitude: "Cyberspace [...] is an act of nature and it grows through our collective actions". In his view, harmony results from the concrete actions of the individual - controlled only by an invisible hand." Where Adam Smith saw the market as an anonymous actor at work, joining the self-interests of the economic subjects into a whole that provides all participants with optimal results, Barlow explains the technical infrastructure itself as the ultimate regulatory principle.

Barlow confidently declared, "We are forming our own social contract." Analogous to Rousseau's *Contrat Social*, his construction of cyberspace draws its normative legitimation from natural law, whereas for him technology takes the place of nature as the source of ideals. Norms are ultimately reduced to technical rules or derived from them. In Barlow's case, however, there is a special feature that the Internet is a pure realm of the mind. "Our world is everywhere and nowhere, and it is not where bodies live." Full of "back to nature", the motto is, so to speak, "back to the spirit", to pure understanding, organized solely by technology.

This radically idealistic position is supposed to legitimize a radically anarchist state: "Your legal concepts of property, expression, identity, movement, and context do not apply to us. They are all based on matter, and there is no matter here."

Radically idealistic is then also the only standard that Barlow applies: "The only law that all our constituent cultures would generally recognize is the Golden Rule."

The golden rule, an ethical formula such as "Treat others as you would like to be treated yourself" is considered the lowest common denominator when it comes to ethical standards. At first sight it sounds like a nice and convincing rule, especially as it basically expresses a fairness that consists of putting oneself in another's position. However, the rule has also been criticised many times in the philosophical tradition, since it is precisely from its simplicity that a one-sidedness can follow, which in extreme cases contradicts the principle of fairness.

Kant therefore introduced the categorical imperative as the supreme moral maxim; he explains the difference between the two moral rules in this way:

> *"Now all imperatives command either hypothetically or categorically. The former represent the practical necessity of a possible action as a means to something else that is willed (or at least which one might possibly will). The categorical imperative would be that which represented an action as necessary of itself without reference to any other purpose i.e. as objectively necessary."*[356]

[356] Kant, Immanuel. 1879. Kant's Critique of Practical Reason and Other Works on the Theory of Ethics. Longmans, Green, p. 38

In contrast to the categorical imperative Kant refers to the golden rule as a hypothetical rule:

> *"If now the action is good only as a means to something else, then the imperative is hypothetical, if it is conceived as good in itself and consequently as being necessarily the principle of a will which of itself conforms to reason, then it is categorical"*[357].

With an example from Kant the difference can be explained like this: According to the golden rule, I can make a faulty promise to someone else because I can bet that the motives behind my reasoning will ultimately excuse my lie. What I am not thinking about is that by lying I am damaging a social law. If lies are accepted, even if only in emergency situations, for Kant they damage the basic social order. According to his reasoning rules that allow such things are therefore not suitable as moral law.

For Kant the hypothetical rule may be reasonable, but it is a pure rule of wisdom, which can be used as a guide in everyday life. However, it is merely a subjective, personal principle that becomes problematic when it is generalised. Only a rule that can be generalized without contradiction can justify moral actions in the eyes of Kant.

In itself, the golden rule according to Kant, cannot be a general law, since it could be used to deny the benefits to others or to refuse to be punished by a judge.[358] Namely a criminal could argue against his judge, objecting that the judge would not wish to be imprisoned by

---

[357] Kant, Immanuel. 1879. Kant's Critique of Practical Reason and Other Works on the Theory of Ethics. Longmans, Green, p. 38

[358] Kant, Immanuel. Grounding for the Metaphysics of Morals, trans. J.W. Ellington, 1981 originally published 1785 p.37, or for a more recent translation see Sally Sedgwick, ed., "Section II: Transition from Popular Moral Philosophy to Metaphysics of Morals," in Kant's Groundwork of the Metaphysics of Morals: An Introduction, Cambridge Introductions to Key Philosophical Texts (Cambridge: Cambridge University Press, 2008), 83–167, https://doi.org/10.1017/CBO9780511809538.005

a criminal, with the help of the Golden Rule,[359] since he would not like to be convicted in his place.

The hypothetical imperative has the form: If you want A, do B. For Kant such rules are only "mere rules of skill"[360] and therefore cannot be sufficient to establish moral principles.

This regularity corresponds in a certain way to the structure of algorithms. "If you want B, do A" can be understood as algorithm in its simplest form. This basic structure of algorithms makes them unable to draw "moral" conclusions because they cannot decide whether they should want B at all or not. This is a practical reality that is of great importance for attempts to teach ethics to Artificial Intelligence.

What is important is Kant's insight: that the mere insistence on such rules as the only principle of morality would violate the most important provision of ethics, namely, to treat other people never as a mere means, but always as ends in themselves. Kant can be used to show that moral action guided solely by hypothetical imperatives or algorithms, will always miss the importance of human dignity. Dignity in Kant's understanding requires considering each human subject as an end themselves, because each person is self-determining. "In other words, what applies to the moral law also applies to this legislative subject, being nothing but humanity in person."[361]

For Kant, Barlow's reference to the golden rule would not be sufficient

---

[359] Immanuel Kant, Grounding for the Metaphysics of Morals, trans. J.W. Ellington, 1981 originally published 1785 p.37

[360] Kant Section II, Volume VII, p. 126. G 4:417

[361] Hawamura, Katsutoshi. Kant's criticism of the Golden Rule// Kant between West and East. To commemorate the 200th anniversary of Kant's death and 280th birthday. Ed. by Prof. Dr. Vladimir Bryuschinkin. Vol.2. Kaliningrad, 2005. pp. 179 - 186.

to form a moral law because, like any hypothetical imperative, it describes only an expediency-means relation. It is a rule that indicates the means to be used to achieve a certain end. It cannot justify why one should respect the other person in principle and as an end in itself.[362] But it is precisely in being an end in itself that Kant ultimately sees the dignity of every individual. Only man sets purposes and is an end in himself. Technology can never be this.

It should also be mentioned that the golden rule was criticized as insufficient in the 20th century, among others by the philosopher and psychoanalyst Erich Fromm and the jurist Hans Kelsen. For Erich Fromm the persuasive power of the golden rule is derived from the capitalist barter law "I give you as much as you give me". According to this logic, one respects the rights of others without feeling responsible for them. Fromm criticizes the rule for only excluding fraud, but not leading to a genuine connection and community between people.[363]

For the jurist Hans Kelsen the golden rule was nothing but a meaningless formula of justice. If a criminal could not be punished because nobody would like to be punished, the rule could override any legal system. It is therefore inappropriate as the sole moral principle.[364] Only through the insight that the other human being is not only an end of my will but, like every human being, an end in themselves, can the mere rule of prudence also become a moral rule of action for Kelsen.[365]

---

[362] "In the Grounding for the Metaphysics of Morals (1785)[3] Kant criticizes the Golden Rule that it cannot be a general law of moral action because, among other things, it does not contain any justification of "duties of love to others" , Kawamura, Katsutoshi. Kant's criticism of the Golden Rule// Kant between West and East. In commemoration of Kant's 200th anniversary of death and 280th birthday. Ed. by Prof. Dr. Vladimir Bryuschinkin. Vol.2 Kaliningrad, 2005. pp. 179 - 186.

[363] Fromm, Erich. The Art of Loving: The Centennial Edition (A&C Black, 2000). P.117

[364] Kelsen, Hans. What Is Justice?: Justice, Law, and Politics in the Mirror of Science: Collected Essays (The Lawbook Exchange, Ltd., 2000).

[365] See on this: Hawamura, Katsutoshi. Kant's Critique of the Golden Rule// Kant between West and East. To commemorate the 200th anniversary of Kant's death and 280th birthday. Edited by Prof. Dr. Vladimir Bryuschinkin. Vol.2. Kaliningrad, 2005. pp. 179 - 186.

Barlow in cyberspace recognized the golden rule as the only remaining moral law. He in this way reinforced the anarchist tendency of the World Wide Web, which at the same time increased in opposition to any legal system. His utopia culminated in the vision of a "civilization of the mind in cyberspace", fuelled by the "virus of freedom" which, according to Barlow, would defy all legal regulations and restrictions. "Morally", this space is controlled at best by the golden rule, which, entirely in the sense of this argumentation, also has the advantage that it can take the form of algorithms.

Barlow very clearly separates cyberspace from public space, which for him is ruled by classical mass media, for which laws and rules are necessary because they are products of the "information industries". Barlow justifies this division with the mercantile form of the media industry, in which its membership in the material world is rooted. Because of their commercial character, the classic media fall within the jurisdiction of the legislator for him, but cyberspace does not. An interesting reasoning, as applied to today's Internet, it means that because of its commercial structure, it must also now be regulated by law. In Barlow's vision of cyberspace, on the other hand, ideas should oscillate freely, completely detached from economic and legal rules: "In our world, whatever the human mind may create can be reproduced and distributed infinitely at no cost. The global conveyance of thought no longer requires your factories to accomplish." Property rights and copyrights are rejected on the grounds that earthly laws have no validity in the realm of the absolute spirit.

With this plea, Barlow prevented US legislators from acting on their intentions to treat and regulate the new communication structure of the Internet like a media structure. It is largely thanks to his efforts that the *Communications Decency Act (CDA)*, which was part of the *Telecommunication Act* in 1996, introduced the so-called provider privilege according to which an Internet service provider is not responsible for the content distributed by third parties through its service. This provider privilege was copied in Article 14 of the EU Directive on electronic commerce.[366]

What was hailed as a victory for freedom of speech at the time was a pyrrhic victory, since the Internet is at least as "mercantilist" as the media for which Barlow reserved his privileged position in 1996. It has taken over the supply of information for more and more people.[367] The platform privilege prevents fair competition between platforms and classic media houses. It treats the Internet platforms like mere providers of a communication network, without any responsibility for the content, neither in terms of the truth of content nor in terms of their effect while leaving them exempt, unlike the classic media, from editorial duties and the need to moderate ex ante and to answer for the consequences in society. Thus, the privilege that Barlow enforced caused the division of the public sphere into one which was professionally regulated and aspired for truth, the media, and another freely manipulatable area, the electronic platforms and social networks, – and thus ultimately the destruction of the functioning public, so necessary for democracy.

But Barlow did not simply want to fend off law, he was concerned with the right to freedom of opinion, which he wanted to enforce for the Internet in absolute form. Like Richard Barbrook and Andy Cameron as early as 1995, Barlow thus stands for the "left" expression of the Californian ideology. This political theory is based exclusively on the negative freedom of action, i.e. "freedom from" which

---

[366] See: 'https://eur-lex.europa.eu/eli/dir/2000/31/oj'.

[367] According to Reuters Institute, Digital News Report 2019 p. 42/43 42% of the population in the US and Europe use the Internet and social media as a source of political information, access at "Digital News Report 2019," Reuters Institute for the Study of Journalism. https://reutersinstitute.politics.ox.ac.uk/our-research/digital-news-report-2019

should be distinguished from positive freedom ("freedom to") the former being very popular in the US given its libertarian, anti-state, tradition.[368] According to Barbrook and Cameron, in addition to the free flow of information in cyberspace, the Internet simultaneously propagates the pure form of the free market: "They want information technologies to be used to create a new 'Jeffersonian' democracy' where all individuals will be able to express themselves freely within in cyberspace."[369] They believe that the California ideology offers a fatalistic vision of the natural and inevitable victory of the high-tech free market.But they are not aware of the side effects they bring with them: "However, by championing this seemingly admirable ideal, these techno-boosters are at the same time reproducing some of the most atavistic features of American society, especially those derived from the bitter legacy of slavery."[370]

With his ideas, Barlow follows on from McLuhan's utopia of the "electronic agora", which he created through the fusion of information technology, media and telecommunications. In this utopia, McLuhan hopes that the power of the individual will overcome both *big business* and *big government*. A development that McLuhan more or less inevitably links with the increasing computerization of society. He could not foresee that the utopia of the freedom-giving electronic agora would one day become the electronic public sphere we know today, dominated by GAFAM.

Barbrook and Cameron showed in their essay that the utopia of boundless freedom began to turn dark when the first followers of the Californian Utopia became entrepreneurs, and their rapidly growing

---

368 The Liberal Movement was founded in the 1950s in sharp rejection of Roosevelt's New Deal and represents radical market attitudes following John Locke's idea of self-ownership.

369 Jordan, Tim. 2002. Cyberpower: An Introduction to the Politics of Cyberspace. Routledge.

370 Barbrook, Richard, and Cameron, Andy. 1996. 'The Californian Ideology'. Science As Culture 6 (January): para.41. https://doi.org/10.1080/09505439609526455

companies discovered the commercial possibilities of the net and profited from it. To these authors' eyes the mixture of conservative economics and left-wing libertarianism leads to a systematic devaluation of the political debate: "The prophets of the Californian Ideology argue that only the cybernetic flows and chaotic eddies of free markets and global communications will determine the future. Political debate therefore, is a waste of breath."[371]

In the developing cyberspace, the free flow of thoughts became more and more the free flow of targeted advertising, goods and services - the agora of opinions became the agora of capital and trade. In contrast to antiquity, or even the more recent past however, the online marketplace is not limited to the polis, but spans the globe, transcending borders. Despite this accurate analysis and criticism by Barbrook and Cameron as early as 1995, the tech scene and politicians more or less ignored these conclusions until 2015.

Rather, the Californian ideology shows itself capable of following the teachings of the neoliberals, for whom the right of the individual is based above all on the right of possession and the right to engage in free trade. When markets are unregulated, the first movers, the first conquerors, can profit most. The prospect of profits leads to the postponement of the original goals and visions of the world-wide web. This libertarian reduction to economic *free flow* has disavowed the original impulse of the "new home of the mind" in cyberspace. A good part of cyberspace today consists of e-commerce, and the "realm of the mind" has become a swamp of Hate Speech, Fake News, troll factories and chatbots up to no good. The Internet is also to a large extend "broken" when it comes to freedom of speech, when it is misused as an excuse for lies and propaganda.

---

[371] Barbrook, Cameron. "The Californian Ideology." Text. Mute. Mute Publishing Limited, September 1995. https://www.metamute.org/editorial/articles/californian-ideology.

It is now possible to see, twenty years later, why things went wrong. First, the propagandists of the free Internet pushed a reduced concept of freedom, involuntarily helping to transform the Net from a utopian place to a romping hotbed of dark forces that know how to exploit the absence of rules for their own purposes. It is the concept of freedom, which was shortened from the outset to negative freedom ("freedom from stately interference") and to rights of defence against state institutions in the tradition of liberalism, which has taken revenge.

For good reasons, German Constitutional provides for the natural law elements of fundamental rights, which embody rights of defence against state arbitrariness, to be complemented by a developed positive law, which ensures the implementation of fundamental rights in social and economic, including among private parties, practice through democratic procedures and legal control.

Freedom does not automatically lead to the boundless freedom of the individual freed from all external. It ends at the place where the freedom of the other begins. This "insight into necessity"[372] (Hegel) reflects the need to institutionally secure freedom in the community.[373]

---

[372] Even if the famous phrase "Freedom is insight into necessity" comes from Friedrich Engels, Hegel's basic idea is that freedom does not mean unrestricted action, but one that is bound to reason. see Vieweg, Klaus, Hegel, The Philosopher of Freedom, 2020 Hegel, Georg Wilhelm Fredrich. 1991. Hegel: Elements of the Philosophy of Right. New Ed edition. Cambridge: Cambridge University Press. At para. 467A

[373] Later, ITIF (the think tank that lobbies for the Californian Ideology) claimed that Barlow had reneged on his manifesto, but WIRED magazine refuted this claim in 2016. Quoting Barlow: "The main thing I was declaring was that cyberspace is naturally immune to sovereignty and always would be....I believed that was true then, and I believe it's true now." ‹https://www.wired.com/2016/02/its-been-20-years-since-this-man-declared-cyberspace-independence/› see also: see also Andrew Keen, John Perry Barlow and the struggle for the independence of cyberspace, https://engelsbergideas.com/essays/john-perrybarlow-and-the-struggle-for-the-independence-of-cyberspace/ Greenberg, Andy. 2016. 'It's Been 20 Years since John Perry Barlow Declared Cyberspace Independence'. Wired, 8 February 2016. https://www.wired.com/2016/02/its-been-20-years-since-this-man-declared-cyberspace-independence/. See also Andrew Keen, John Perry Barlow and the struggle for the independence of cyberspace, https://engelsbergideas.com/essays/john-perry-barlow-and-the-struggle-for-the-independence-of-cyberspace/

The then CEO of Google, Eric Schmidt, was still wondering in 2013 about the paradisiacal freedom that the Internet offers companies: "The Internet is the first thing that humanity has built that humanity doesn't understand, the largest experiment in anarchy that we have ever had."[374] Google and Schmidt have earned billions from this experiment in anarchy and the people and societies around the globe are increasingly paying the price.

In the discussion about the ethical and regulatory handling of emerging AI today, the mistakes that opened up this anarchist space must not be repeated. AI has the potential to make technical processes independent of human input and non-transparent, thus posing an even greater threat to freedom than the "old" Internet 2.0. AI systems can help to put an end to anarchism once and for all, and create an electronic iron cage (*stahlhartes Gehäuse*)[375] in which there is no room for autonomy and democracy - not even for the libertarian freedom that Barlow had in mind.

Even Barlow argued unilaterally and radically in his manifesto: by rejecting any regulation by the legislator, he also prevents fundamental rights from being safeguarded in cyberspace. The very freedom that he wants to secure in this way is endangered by his rejection of any rules. For cyberspace is not a power-free space either. Power is also exercised in digital space. The only thing is that power does not lie with the state alone, except in China and other dictatorships, it lies with powerful global corporations.

The protection of basic and human rights in cyberspace must, as we are realising today, include defensive rights against state and corporate surveillance and manipulation techniques.

---

[374] Schmitt, Eric. Internet World Trade Show, New York, 18 November 1999

[375] An Iron Cage was Talcott Parson's Translation of Max Weber's stahlhartes Gehäuse (literally steel-hard casing) that reigned in modern times: "...from the coat the doom made a steel-hard case. Today the spirit of ascenticism has fled from this casing, whether with finality, who knows? Victorious capitalism, in any case, ever since it came to rest on a mechanical foundation, no longer need asceticism as a supporting pillar." Quote from Weber, Max. 2013. The Protestant Ethic and the Spirit of Capitalism. Routledge. P.124

Today, more than 20 years after Barlow's declaration, questions of freedom and regulation are once again arising, this time with a view to the AI. Will we this time find and create the rules that are needed to secure freedom? In doing so, we must respect the correct hierarchy of democracy and digitalization. German President Frank-Walter Steinmeier emphasises this point: "It is not the digitalization of democracy that we must concern ourselves with first and foremost, but the democratization of the digital!"[376]

20 years of the Wild West on the World Wide Web emerged because the ability for corporations to build unregulated cyberspace into a model that enabled them to accrue economic and political power was underestimated. This process is ongoing, new parts of Silicon Valley are still drawing energy from the old ideology, and have only recently begun open combat against democracy.

1.5. Dare more technology! Peter Thiel's opposition to democracy

The conflict between democracy and digital technology is encapsulated in the views of influential Silicon Valley investor Peter Thiel, who has German roots and who has long rejected democracy: "Above all, I no longer believe that freedom and democracy are compatible."[377] Thiel makes a clear plea for freedom, by which he means primarily the freedom of the entrepreneur and with this statement sets himself up in opposition to democracy. Thiel is one of the initial investors in Facebook and co-founder and shareholder of *Palantir Technologies*, a company that provides data products and surveillance software primarily for

---

[376] Source: 'http://www.bundespraesident.de/SharedDocs/Reden/DE/Frank-Walter-Steinmeier/Reden/2019/06/190620-Kirchentag-Podiumdiskussion.html'. translated by the authors

[377] Peter Thiel in Cato Unbound, "The Education of a Libertarian," 13 April 2009, https://www.cato-unbound.org/2009/04/13/peter-thiel/education-libertarian/

the police and security services. With the software provider and data dealer *Palantir*, Thiel operates one of the darkest surveillance companies grown out of Silicon Valley.

Thiel calls for the renunciation of politics and demands the freedom of the entrepreneur to establish monopolies, which he considers necessary because they arise from the logic of the network effect. He rejects competition: "Competition is an ideology—the ideology—that pervades our society and distorts our thinking."[378] This is a surprising statement for someone who would generally be classified as a neoliberal. Thiel's struggle for market supremacy has turned into a struggle for global power with Palantir as a tool. This provocation shows the radicalization of liberal ideas by libertarians, who consistently think through to the end the unleashed freedom of the individual. The aim of the monopoly, as Thiel openly explains, is to crown technology as the leader of world affairs: "In view of these realities, one would despair if one were to limit one's horizon to the world of politics. I do not despair because I no longer believe that politics encompasses all possible future prospects for our world. In our time, the great task for libertarians is to escape politics in all its forms – from the totalitarian and fundamentalist disasters to the mindless demos that guide so-called 'social democracy'(…) A better metaphor is that we are in a deadly race between politics and technology".[379]

These lines can be read as a declaration of war on democracy by a man who belonged to Donald Trump's circle of supporters and advisors. It is a radical expression of solutionism inherent in the Californian Ideology. His mantra could be "dare more technology".[380]

---

[378] Masters, Blake and Thiel, Peter. Zero to One: Notes on Start Ups, or How to Build the Future (Random House, 2014).

[379] Quoted in Mitchell, Liam. 2018. Ludopolitics: Videogames against Control. John Hunt Publishing.

[380] This contrasts to "Dare more Democracy", the election campaign slogan of Willy Brandt, the first Social Democratic Chancellor of Germany.

In view of such statements, economic liberalism will finally turn into cryptoanarchism, an ideology that seeks to create an information infrastructure which is immune to government or other external actors' attempts to gain control over activities. But the "deadly race" between politics and technology has not yet been decided. The concept of "deadly race", which makes an irreconcilable contradiction out of the differences in the system logic of politics and technology, overlooks the fact that both spheres could instead be placed in symbiotic exchange.

Thiel criticises, above all, politics' tendency to slow down technical development through rules and laws. Technological development and democratic decision-making take place at different speeds. Democracy, especially in its deliberative form, depends on the laborious weighing of reasons and negotiation of compromises, because this is the only way to determine what is in the interest of the majority. In contrast, technology develops rapidly and at an increasing pace. This does not, as *Moore's Law* suggests, take place at an inevitable speed. Instead, it is determined by individual entrepreneurs, the resources available to them, and government intervention. However, one does not have to accept Kurzweil's law of exponential growth as the natural law of technology in order to recognize the increase in knowledge production and the speed of technological development. Even without exponential growth, the acceleration of scientific and technological progress already poses problems for democracy.

Thiel attempts to set entrepreneurial freedom and the development of technology against democracy. This is evidence, on the one hand, of the aberrations in which the Californian ideology has become entangled in the meantime. On the other hand, it also demonstrates the blatant claim to power with which certain circles around GAFAM are now formulating their technology induced dominance in all areas of society and politics.

From Thiel's thesis can ultimately follow a sequence in which first you replace politics with technology, then people. It is therefore not surprising that a so-called *World Transhumanistic Party* has already been founded, which aims to hand over the affairs of government to an AI. Its founder Zoltan Istvan published a manifesto in 2014[381] which calls, among other things, for the creation of a "cultural mindset" in the United States, based on the assumption that "embracing and producing radical technology" is in the best interests of the United States and humanity as a "species". Istvan ran as a candidate in the 2016 US elections, was smiled at but enjoyed great popularity in the media as an object of curiosity. The Watson Foundation also promoted IBM's electronic mastermind Watson, known as Jeopardy Winner, as a nominee for the presidency. And wouldn't a predictable machine at first glance be a better choice than an unpredictable Trump?[382]

A few years ago, the statement that a controversial reality-TV celebrity, who understood the machinations of the entertainment business, would become president of the US would have been dismissed. Even for Trump, who has mastered the laws of the medium, the TV-affine staging as an entertainer in popular reality shows was a decisive criterion for conquering first broadcast time and attention, then the White House. To smile at political activities of transhumanists like Istvan as clownery can later turn out to be a fatal error, especially when its views are reaching influential and powerful figures like Thiel. The demand to put algorithms in the place of politicians will probably soon no longer be an exotic outsider position.[383] The business community is also leading the

[381] Istvan, Zoltan, "Should a Transhumanist Run for US President? | HuffPost," December 08 2014 https://www.huffpost.com/entry/should-a-transhumanist-be_b_5949688.

[382] Adrian Lobe has made these considerations in a German-language article that is worth reading: Adrian Lobe, "Künstliche Intelligenz: Ein Bot im Weißen Haus," Die Zeit, January 23, 2017, sec. Kultur, https://www.zeit.de/kultur/2017-01/kuenstliche-intelligenz-computer-politik-ersatz-demokratie-maschinenraum.

[383] Joshua Davis, "Hear me out: Let's elect an AI as President", Wired, 18 May 2017, https://www.wired.com/2017/05/hear-lets-elect-ai-president/.

way here: in 2014, the Hong Kong-based venture capital firm Deep Knowledge Ventures "appointed" an AI to its board of directors, to make recommendations to the other board members on investment decisions.[384] As a result, the AI attracted attention primarily through "nepotism" - it wanted to invest in companies that advocated giving algorithms greater power.[385]

The road to a surveillance state doesn't really worry Kurzweil, the other main representative of today's Californian ideology. Rather, he also fears that the accelerated technological progress could be slowed down by law and politics: he has used the concept of singularity not only to coin one of the most dazzling buzzwords of Californian ideology, he also wrote a grand narrative on the concept that hardly seems capable of improvement.

1.6. Ray Kurzweil: Technology in place of man or Écrasez GAFAM!

In Ray Kurzweil's Book "Humanity 2.0. The Singularity approaching,"[386] Californian ideology appears in the form of a mathematical-scientific ontology linked to a radical idea of technical progress which serves as a source of a naturalistic ethic.

The basic thesis of the book is that technological growth in the age of information technology is exponential and not linear. As a result, the growth of information is constantly accelerating. For Kurzweil,

---

[384] Found at Harari, Yuval Noah. Homo Deus: A Brief History of Tomorrow (Random House, 2016). P.322

[385] Found at Harari, Yuval Noah. Homo Deus: A Brief History of Tomorrow (Random House, 2016). P.322

[386] Kurzweil, Raymond. 2006. The Singularity Is Near. London: Duckworth. Original edition: Ray Kurzweil, The Singularity Is Near: When Humans Transcend Biology, New York, Penguin 2005

this process inevitably steers us towards the state of what is called the singularity. A term used in mathematics, systems theory and astrophysics, singularity has a different meaning in each subject, which makes it a good source for Kurzweil, as he can borrow definitions from all of these areas.

For Kurzweil, singularity denotes a point in time in which technological growth turns irreversible and uncontrollable, in such a way that no further predictions can be made beyond that point, in many ways similar to a *black hole* in astrophysics. In view of this comparison, Kurzweil, dares to make a surprising number of predictions for the time after the singularity. In his vision, singularity occurs when the increase in knowledge on an exponential curve becomes so steep that the knowledge explodes at a stroke. He predicts that this event will take place in the year 2045, with serious consequences: Ray Kurzweil transforms the singularity into a breakthrough of technology, at which point the technology can improve and generate itself, heralding man's salvation from biological existence.

We mention Singularity here, not because it is a spectacular prediction, that lends itself sometimes to parody, but because it belongs to an analysis of the Californian ideology. Kurzweil's theory has earned him the important position of *Alphabet's* head of development. His ideas are largely shared by Larry Page.[387] If Page, like Max Tegmark writes, is the man who is currently changing the world the most, Kurzweil is the brain behind him. In his position as *Director of Engineering,* Kurzweil also directly and significantly controls Google's investments in new technologies. Through his strategic leadership at Google, he influences countless programmers in their work on AI, as well as other investors who invest in Google

---

[387] Tegmark, Max. Life 3.0: Being Human in the Age of Artificial Intelligence. 1st Edition. Allen Lane, 2017

and companies similar to Google/Alphabet. This is reason enough to deal with the theses of Kurzweil's *The Singularity is near*.

Kurzweil decides to become an inventor at the age of 5 and he starts programming computers at an early age[388]. One thing fascinates him about his computer: "Unlike us, it makes no mistakes."[389] This naïve conviction of the 20-year-old Kurzweil still runs through Californian ideology as a basic assumption today.

According to Kurzweil, the singularity will emerge from three key technologies in particular: genetics, nanotechnology and robotics. Genetics and nanotechnology should help people to improve their genetic make-up. Robotics, finally, should achieve strong Artificial Intelligence at the singularity stage, which surpasses human intelligence in all areas. The strong Artificial Intelligence will then give information technology such a boost that unimaginable inventions and developments will become possible.

At this stage of development, humans can at best keep up in some areas by radically revitalizing their biological existence through nanotechnology and genetic design, or by uploading their consciousness to an AI hard drive to participate in super intelligence and become immortal. This idea of an "upload" of the human mind has often been quoted It is mostly dismissed as humbug by serious AI researchers. However, Google has long been researching interfaces that are supposed to connect the human brain with hard drives or the Internet. Google founder Sergey Brin is convinced of the purpose of this research: "Certainly, if you had all the world's information directly attached to your brain, or an artificial brain that

---

[388] Kurzweil, Raymond. 2006. The Singularity Is Near. London: Duckworth. p.1

[389] Kurzweil, Raymond. 2006. The Singularity Is Near. London: Duckworth p.1

was smarter than your brain, you'd be better off".[390] In this world view, "good" and "better" mean having maximum access to and the fastest possible ability to process information. These are purely quantitative values.

Many of these visions sound like the continuation of the Sci-Fi children's books from the Tom Swift Jr. series, which Kurzweil read as a child and which, according to him, strongly influenced his life. Above all, while reading the volumes, Kurzweil developed what he calls his personal philosophy: "To this day, I continue to be convinced of this basic philosophy: no matter what quandaries we face—business problems, health issues, relationship difficulties, as well as the great social and cultural challenges of our time—there exists an idea that will enable us to prevail. Furthermore, we can find that idea. And when we find it, we need to implement it. The power of an idea—this is itself an idea, one that has inspired me since a young age."[391] This is basically a sympathetic life motto, but Kurzweil's worldview is that only technology offers these solutions to all problems. On his journey through space, Kurzweil's hero Tom Swift solves problems with technical means, at different points in the story ingeniously using a flying lab, a pedal-powered plane, an electric hydrolung and an underwater robot.[392] All these objects could theoretically be developed in *Google Labs* today. In any case, they are suitable inspirations for Google's *'moonshot'* projects, with which the company regularly tries to make the impossible possible. Kurzweil's theory pursues the same goal. The super-intelligence he imagines is intended to solve the great scientific, social or cultural challenges of our time, with technology alone being the key. In

---

[390] Quoted in Newsweek Staff On 3/28/04 at 7:00 PM EST, "All Eyes on Google," Newsweek, March 28, 2004, https://www.newsweek.com/all-eyes-google-124041

[391] Kurzweil, Raymond, 'The Power of an Idea', https://www.kurzweilai.net/the-power-of-an-idea-2

[392] See note in Kurzweil, Raymond. 2006. The Singularity Is Near. London: Duckworth , p. 515.

The Singularity is Near, Kurzweil tries to argue for this approach to problem-solving drawing strictly from science in order to make his prognosis as credible as possible. For him, the findings of the natural sciences, which he draws from in his theory, are not simply findings. But instead they represent the sacred data and patterns of nature that reveal themselves to the prophet of mankind's redemption by technology.

For Kurzweil, the question of whether humans should conform to technology or vice-versa is clear: it is the human to the "superior" machine. Kurzweil uses in his book the device of a fictional conversation between people at different points in time to answer this question.[393] We want to reproduce this briefly below.

Besides "Ray" (that is Mr. Kurzweil himself) the conversation includes "Molly" speaking in 2004, "George" in 2045 (when, according to Kurzweil's calculations, the singularity occurs) and, on the sidelines, is Charles Darwin (no precise time is given). First, Molly asks Ray when "it", by which she means the singularity, will occur. She says she needs some time to prepare for this. When Ray asks her what for, she says: "For starters, I'll want to fine-tune my resume, I'll want to make a good impression on the powers that be."

Freud might call this statement "identification with the aggressor". Because with Molly, the prospect of the emerging singularity gives her a queasy feeling: "I might also want to erase a few documents-you know, where I'm a little insulting to a few machines I know." So Molly is thinking about getting incriminating material to safety before the machines take over and possibly take revenge for their "sincerity". But George, the voice from the singularity, can reassure her: "Oh, the machines will find them anyway – but don't worry,

[393] Kurzweil, Raymond. 2006. The Singularity Is Near. London: Duckworth, p. 31

they are very understanding."[394] The superiority of the machines postulated by Kurzweil thus includes both total knowledge and control of personal data and a moral goodness of the machines, which also show themselves superior to humans in this field: "They are extremely understanding" - It is clear that if these machines have been defined beforehand in such a way that they are superior to humans in all areas, they must ultimately be so in moral terms and in social behaviour.

Despite all the superiority and trustworthiness of the machines: Molly has an uneasy feeling about having to change her identity in the singularity to give up completely. "And should I wish to remain as I am?" she asks anxiously.[395] George replied: "I'll be devoted to you in any event. But I can be more than just your transcendent servant."

When Molly expresses her fear that the singularity AI could destroy her biological existence, Ray comforts her with the words: "You represent such a tiny fraction of the available matter and energy. Keeping you biological won't appreciably change the order of magnitude of matter and energy available to the Singularity. It will be well worth it to maintain the biological heritage." This makes it clear: the transcendent servant has long since been the new master who decides on life and death. It would not be the first servant to rise to become the lord of the original sovereign.[396]

Man will therefore survive in the Kurzweil's fantasy only because he is so insignificant to cosmic intelligence that its biological heritage has become preserved for a museum. What a lucky escape. Perhaps we should all make ourselves as small as possible, to maximise the chance that we will be spared as rule by AI approaches.

---

[394] Kurzweil, Raymond. 2006. The Singularity Is Near. London: Duckworth, p. 30

[395] Kurzweil, Raymond. 2006. The Singularity Is Near. London: Duckworth p. 32.

[396] See Hegel Works Vol. 3, Phenomenology of the Mind, Chap. Dominion and Servitude.

The dialogue ends when the doubting Molly is finally converted: Molly: "So, basically, I'll have everything I need at my fingertips? George 2045: "Indeed". Molly 2004: 'Sounds like King Midas. You know, everything he touched turned to gold. Ray: 'Yes, and as you will recall he died of starvation as a result. Molly 2004: "Well, if I do end up going over to the other side, with all of that vast expanse of subjective time, I think I'll die of boredom". George 2045: "Oh, that could never happen. I will make sure of it."[397] What sounds like a final threat of Artificial Intelligence is revealing in several ways: it is not death by starvation that represents the great fear of the technology believers, but boredom in the techno paradise. This poses no problem for Kurzweil, his imaginary super AI promises to provide entertainment.

If one looks at these fantasies from an ideology-critical point of view, it is obvious that they are intended to train people to adapt to the machine. The particular attraction is that adaptation to the computer is not celebrated as subjugation, but as the final step of human "overcoming oneself". This *enhancement* leads to one's own demise, which experiences its apotheosis as "immortality". Man survives, if at all, only as a kind of human being. As a mere biological human being, according to the vision, he will disappear. This is in keeping with the fact that human intelligence in Silicon Valley is often discredited as "wet ware", a term drawn from the computer-related idea of hardware or software, but applied to biological life forms, which is supposedly inferior to so-called silicon-based intelligence.

According to Kurzweil, the replacement of man by a "superior" intelligence is taking place with relentless inevitability, it is "without

[397] Kurzweil, Raymond. 2006. The Singularity Is Near. London: Duckworth p. 34 .

[398] Turing, Alan, On Computable Numbers, with an Application to the "Decision Problem" May 28, 1936.

alternative". Kurzweil quotes theories according to which the entire universe is software. In his words, it consists of *Computation Rules*, i.e., algorithmic processes, wherever you look. This statement has consequences. Alan Turing had described the *Turing machine* as capable of solving any conceivable mathematical problem, provided that it can be solved by an algorithm.[398] But if now the whole universe is nothing other than a big calculation, why shouldn't his superintelligent algorithm be able to decode it first and then master it?

Philosophically speaking, Kurzweil's theses represent an absolute mathematical idealism. There are parallels of this idea of the universe as a huge computer with Hegel's metaphysics of absolute knowledge but also a significant difference. Hegel pointed to a structural equality that came from recognizing subject and recognizing object, which has its pivotal point in his thesis that "substance is subject". Nature is thought of by Hegel in a structure that corresponds to the structure of human self-consciousness, which enables man to fully recognize nature. When Hegel, in the foreword to the *Science of Logic,* claims that "It can therefore be said that this content is the exposition of God as he is in his eternal essence before the creation of nature and of a finite spirit."[399] what he is saying seems to be very close to California ideology. Hegel did not want to be satisfied with the answer given by Augustine to the question of what God was doing before the creation of the world (namely to prepare hell for people who pry into mysteries).[400] But as said, with him the "substance is subject", i.e., the universe ultimately has the same structure as human thought and self-consciousness. With Hegel, nature is completely subjectivized. In the Californian ideology, the subject is completely naturalized, more precisely mathematized.

---

[399] Hegel, Georg Wilhelm Fredrich. The Science of Logic (Cambridge University Press, 2010). p.29.

[400] Saint Augustine and John St Augustine, The Confessions (Hendrickson Publishers, 2004).

Kurzweil designs human self-consciousness in analogy with a computer program. He also assumes that the universe has the structure of a highly complicated quantum calculation program. His idea of structural equality starts from mathematics and computer science, not from the subject. Kurzweil thinks of the subject as a mathematical substance or a mathematical pattern. Hegel see the substance as subject. By placing the subject as centre Hegel can develop a theory of freedom, whereas in Kurzweil's theory no space for human freedom remains. Kurzweil, like many representatives of Californian ideology, has no problem with a deterministic world view. After all, he himself dared to make predictions such as the one that in 2045 the singularity will occur. He regards all deviations from the strict path of determinism as random phenomena that could be calculated by randomized programs.[401]

The connection to the theory of the functional equality of organic and technical systems by Norbert Wiener, the founder of cybernetics, is obvious.[402] Kurzweil totalizes this structure by first connecting it to the timeline of evolution and thus historicizing it, and then basing it on the development of the universe. In the end he assigns the role of a divine creator to the programmer: programmes recommended in his book *How to create a mind: the secret of human thought revealed set* "God Parameters"[403] that draw from evolutionary algorithms to create "learning and recognising systems". The chief task of the programme is to find usable rules for these parameters (...)" The programmer sits in heaven like God at the great computer of creation, for which he is to write suitable software to control its further development. If Stewart Brand's success consisted in providing the developers with a great

[401] Kurzweil, Raymond. 2006. The Singularity Is Near. London: Duckworth

[402] See chapter D 2.1 below.

[403] Kurzweil, Ray. 2013. How to Create a Mind: The Secret of Human Thought Revealed. Prelude Books.

narrative of their actions with his enthusiasm and vision,[404] Kurzweil further bolstered the narrative of programming with this image. The programmers are called upon to crack the code of the universe in order to program the course of cosmic development. If hopeful young programmers are still looking for a motive for hiring at Alphabet, a job description can hardly offer more than the promise of deciphering God's plan and possibly later gaining his power at work.

But on closer inspection, not everything in this edifice of ideas turns out to be directed towards the future: By defining living beings as "cellular automata", Kurzweil shows himself to be a backward-looking heir to Descartes' mechanistic worldview. For Descartes, man was a machine created by God, similar to the animals he defined as reductively explainable automatons. The corresponding world view is described as "mechanistic". Kurzweil follows this up with his concept of "cellular automata" and mentions that democracy would be improved by his idea of progress - even if he avoids any deep discussion of this question and admits a few pages later that not all people are subject to the blessings of the singularity.

Also in passing, Kurzweil mentions the interlocking of technological development with economic processes. The enormous market that computers have newly created has increased investment, which in turn drive the acceleration effects of development, he writes. But he is far from critical of this connection between the laws of the financial markets and technology. Kurzweil quite naturally relies on the affinity of the neoliberal economic form with the cosmologically interpreted acceleration of information processing. He believes in exponential growth in both areas, and attributes a natural logic

---

[404] "The Complicated Legacy of Stewart Brand's 'Whole Earth Catalog' | The New Yorker," https://www.newyorker.com/news/letter-from-silicon-valley/the-complicated-legacy-of-stewart-brands-whole-earth-catalog

to both processes. Like Norbert Wiener's view of organisms and technical systems as functionally identical, Kurzweil describes economy and technology as systems subject to the same control mechanisms. From the theory of exponential growth finally follows the *law of increasing yield growth*.[405] The very formulation suggests that the whole theory is driven at least as much by an economic as by a technological growth fantasy.

Kurzweil also bluntly pays homage to an economic Darwinism that does not yet sound as self-evident to our ears as in neoliberal Silicon Valley. "*Economic imperative* of the free market is the driving force behind technical development (...) Growth pressure is the counterpart of natural selection in biological evolution."[406] Capitalism amplifies evolution. And both united work on the *optimization* of the human factor. This makes man more and more of a disturbing factor and must finally lead to its overcoming. In this view, man as a biological being appears as a defective mutant, whose possibilities of further developing biologically within the evolutionary process to date are limited compared to the speed of *exponential growth* of technology. They are also further limited by the obstacles that emanate from politics and regulation, which to the displeasure of Kurzweil, are hindering development. This is an idea Kurzweil shares with other transhumanists. Digital mastermind Peter Diamandis, co-founder of *Singularity University*, writes: "The changes to the global environment were so rapid and so extreme that the dinosaurs – the dominant form of life that the time – were unable to adapt. Instead, they went extinct. [...] In fact, this tale of colossal impact, radical transformation, and spectacular rebirth has exceptional relevance today – especially for business."[407]

---

[405] Kurzweil, Raymond. 2006. The Singularity Is Near. London: Duckworth, pp. 35-42.

[406] Kurzweil, Raymond. 2006. The Singularity Is Near. London: Duckworth p. 94.

[407] Diamandis, Peter H., and Kotler, Steven. 2016. Bold: How to Go Big, Create Wealth and Impact the World. Simon and Schuster. (from introduction p. x)

The enthusiasm for the power of ideas, which Kurzweil cites as his motif at the beginning of the book, is thus exhausted in his enthusiasm for the idea of a boundless growth in economy and technology, which is accelerating more and more and thus inevitably heading for a point where it transforms itself into a new dimension: into the singularity. There is no part for the individual, biologically-constrained human, in this world view, except as a biological intermediate carrier of information that is in the process of making its way through it to a more powerful silicon-based carrier medium. Man is the means of genetic information, which is the new end in itself. There is no more *dignity* for the human being.

Evgeny Morozov has called this ideology "solutionism".[408] He asks how the conviction that technology can, in principle, solve every problem in the world affects society. Can it really be the case that deeply entrenched complicated political, moral and interpersonal problems can be solved with the right app? If only someone could just program it. Kurzweil openly advocates this technological absolutism; one could also say technical *totalitarianism*: "Technology alone can solve the problems that human societies have had to contend with for generations."[409]

When it comes to difficult political problems, however, this absolute solutionism seems to crack even for him. When his book considers *The Threat from Fundamentalism* and contemplates the question of how best to fight terrorists, Kurzweil's mind starts to spin: "How can authorities now dismantle a worldwide network of terrorist cells if they always have to wait until a crime has been committed? (...) But I see no technical miracle cure to solve this dilemma."[410] Apart

---

[408] Morozov, Evgeny. To Save Everything, Click Here: The Folly of Technological Solutionism. PublicAffairs. 2013.

[409] Kurzweil, Raymond. 2006. The Singularity Is Near. London: Duckworth p. 381.

[410] Kurzweil, Raymond. 2006. The Singularity Is Near. London: Duckworth p. 435.

from the fact that Kurzweil here limits the problem of counter-terrorism entirely to the prediction of terrorist acts and that a deeper investigation of causes seems to be alien to him, such self-critical passages in his book remain the exception.

Any thinking in political categories is alien to Kurzweil, since in his eyes the greatest enemies of technical progress are politics and regulation: "Excessive regulation in the introduction of new therapies, for example, costs many lives."[411]

Kurzweil holds the "fundamentalist humanism", which he sees as the driving force behind the regulation of technology, as one of the great enemies of progress. In his list, "fundamentalist humanism" comes immediately after "radical Islamist terrorism"[412] as a blocker to technological progress, even if it is in the name of human rights. For example, he accused the Council of Europe of fundamentalist humanism after it adopted a directive prohibiting changes to human genetics. Kurzweil is indignant that the Council of Europe believes that "human rights include the right to a genetic pattern not artificially altered".[413] Kurzweil cannot understand such ethical positions: They seduce him into the cynical rhetoric that "the Council also endorses the right not to be cured of natural diseases artificially."[414]

Here we encounter a common argumentation pattern of Californian ideology: First, future techniques are imagined, and their blessings are described in detail. Whoever then stands in the way of the implementation of these plans for salvation bears the responsibility for human suffering, which could be prevented very quickly without the inhibitions caused by democratic concerns. Singularity first, concerns and democracy second.

---

[411] Kurzweil, Raymond. 2006. The Singularity Is Near. London: Duckworth p. 384

[412] Kurzweil, Raymond. 2006. The Singularity Is Near. London: Duckworth p. 429.

[413] Kurzweil, Raymond. 2006. The Singularity Is Near. London: Duckworth p. 488.

[414] Kurzweil, Raymond. 2006. The Singularity Is Near. London: Duckworth p. 488.

A technological moral cudgel, which works on the model that those who want to check their risks before using a new technology are held responsible for the suffering and death of people who, due to these allegedly unnecessary delays, can only later enjoy the technological blessings. Kurzweil speaks of the "unbearable slowness of social institutions"[415] and, of course, he sees civil servants and institutions, as well as political regulations, as a single "brick on the leg of progress".[416] This moral cudgel is the outbidding of the common argument that those who advocate impact assessment or legal regulation of new technologies prevent economic growth and thus prosperity.

"Excessive regulation is economically counterproductive and ethically unsustainable, given the possibilities of curing disease."[417] Behind it is the reversal of the principle of responsibility by Hans Jonas. This is the basis for the responsibility that man bears for the remote effects of the technology that he himself develops, and on which he should therefore examine the consequences before using it. Jonas' philosophical principle has found its way into European constitutional law as the "precautionary principle" in Art. 191 (2) ot the Treaty on the Functioning of the European Union.

Kurzweil comforts himself in view of the current misery with the political brakemen with the fact that in the age of a general Artificial Intelligence it will finally be over with them: "And as far as the debate about whether such extensions are desirable is concerned, it is easy to predict who will win, because people with extended intelligence will be much better at debating." The representatives of technical progress will be doped with technology, using "advanced

[415] Kurzweil, Raymond. 2006. The Singularity Is Near. London: Duckworth p. 489.
[416] Kurzweil, Raymond. 2006. The Singularity Is Near. London: Duckworth p. 490.
[417] Kurzweil, Raymond. 2006. The Singularity Is Near. London: Duckworth p. 426.

intelligence" in order to beat politicians and citizens with arguments. This scenario has also become reality in some rudimentary form, as the development of the *Debater project* by IBM which, with the help of AI, supposedly 'beat' human players in debates, just like AI software in chess, go or poker.[418]

Kurzweil's argument reveals a second important pattern of Californian ideology: it is recognized that not all people will benefit from technical optimization. If only because some "fundamental humanists" will refuse to insert implants into the body or brain. Kurzweil's vision therefore, at its core, requires the creation of a new human race, the cyborgs, which will be superior to purely "biological" humans.

Surprisingly, Kurzweil believes that despite everything, "the human being remains at the centre". For man has been able to create the technology that has triggered the "accelerated progress". But: "Man 2.0", the title of the German edition of Kurzweil's book, will lose his human face after this "redemption". Whether a fully developed cosmic intelligence, as Kurzweil imagines it to be, will still be indebted to its creator, the human being, who fades out, is unclear. As too is the answer to the question of what values such an AI would pursue and whether humans can influence these values.

Max Tegmark describes this position as one of *digital utopianism*, whose best proponent he identifies as Larry Page. The main accusation against this position, according to Max Tegmark[419] is that while digital

---

418 "It's Interesting How IBM's AI Debater Sways Opinions, But Who Wins - Man Or Machine?," https://www.forbes.com/sites/cognitiveworld/2018/12/31/its-interesting-how-ibms-ai-debater-sways-opinions-but-who-wins-man-or-machine/#6da07e043528.

419 Tegmark, Max. Life 3.0: Being Human in the Age of Artificial Intelligence. 1st Edition. Allen Lane, 2017.

utopians realise the fundamental ambivalence of all technology, Kurzweil draws the wrong conclusion from it. For if technology can be used for good or bad, the question arises as to who defines the good and the bad. Instead of bringing human values or even the human being into play here, Kurzweil suggests that one simply lets technology answer this question itself, by not letting a human being get in the way of its uninhibited exponential development.

Since Kurzweil sees problems primarily as mathematical problems, he is also convinced that they can all be solved by improving computer code. In the end he stylizes technology into a self-controlling system. Progress is consistently reduced to technical progress; cultural or social progress does not occur. All problems of mankind can then be solved by technical means. But progress cannot be thought of in terms of itself, it is always progress on the way to something else, be it health, happiness, peace, justice or other values. Technology does not contain these values in itself; they have to be given to it by man. Even in the age of AI, technology is only the realm of means, the ends must be given by humans. In Kurzweil's thinking, technology pursues a given purpose - the subjugation of the universe: "We will gain power over our fates. Our mortality will be at our fingertips".[420] But will these fingertips still belong to humans or an Artificial Intelligence?

For Francis Bacon, the Renaissance thinker, who is considered the founder of modern philosophy of science, the aim of science was the *Regnum Hominis*, the restoration of the paradisiacal state when man gave names to animals and living beings and gained power over them by capturing their essence. Like Bacon, Kurzweil's Mission, which can be broadly described as Google or Alphabet's mission, is nothing more than an incomparable program of domination through

---

[420] Kurzweil, Raymond. 2006. The Singularity Is Near. London: Duckworth

technology, which can easily turn into domination by technology. Those who want to get the cosmos "into their own hands" are not afraid of manipulation. Emancipation, which in the Latin sense of the word refers to stepping out of the hand (Latin: *manus*) of authority, is answered in Kurzweil's picture in the manner of Google: When the whole universe is in their hands, emancipation is over. The dialectic of this program of domination is that at the end of the process "we", who are supposed to "hold" the domination of the universe in our hands, no longer exist, because technology has replaced humans. The book thus ends in a true and delusional dialectic of digitalization.

So, to sum up:

- The Ideology of Singularity is misanthropic in a literal sense: technology takes the place of the human being, who is first *optimized* by technology and finally uploaded.
- The theory is hostile to freedom: instead of emancipating man, Artificial Intelligence is supposed to have the entire cosmos "at its fingertips".
- The theory is hostile to democracy and ethically irresponsible: the principle of responsibility, i.e., the impact assessment of technology, is seen merely as an enemy of progress. In the age of AI, political and ethical debates are to be decided in favour of an uninhibited use of technology.
- The theory endorses treating humans unequally, setting the stage for new forms of discrimination: cyborgs are intended to breed a technically tuned superior race of humans.

Kurzweils transhumanism is anything but the harmless enthusiasm for technology of a nerd who has remained childish. It is the ideology of a leading representative of one of the most powerful corporations the world has ever seen. The low public awareness of these sometimes crude theories of technological superhumans and the conquest of space by Artificial Intelligence is in stark contrast to the impact these theories are already having on people's lives today.

Kurzweils' theories do not remain just theories as they are put into practice. Google runs a company that does research on abolishing death called Calico.[421] This new daughter of Google aims to combat ageing and age-associated diseases. It wants to overcome death, which, as Kurzweil says, is the interface between artificial and human intelligence.[422] The name of the company was both a kind of acronym of the California Life Company, and also a nod to cats and their nine lives.[423] Company founder Levinson wrote: "Calico is an abbreviation for the "*California Life Company*," but if you're thinking about cats, we like the old saying that they have nine lives".[424].[425] By 2018 more than 2.5 billion dollars had been invested in Calico.[426] For a long time now, *Calico has* been conducting real experiments and trials to achieve the goals of prolonging life or even victory over death.

From fun weird future scenarios, it can be learned that the transition from humans to computers was already taking place long before the "singularity", namely where people believe they have to adapt to the "superior computer". The computer pioneer Konrad Zuse, who is quoted by Kurzweil because he wasone of the first to argue that all of space is a single computer,[427] said very wisely: "The danger of

---

[421] https://calicolabs.com/

[422] Wolf, Gary. 'Futurist Ray Kurzweil Pulls Out All the Stops (and Pills) to Live to Witness the Singularity'. Wired, 24 March 2008. https://www.wired.com/2008/03/ff-kurzweil/.

[423] Walter, Chip. Immortality, Inc, 2020. Simon and Schuster.

[424] Dunn, James. "Former Genentech CEO Art Levinson to Head Google's Calico Health Spinoff." San Francisco Business Times. 18 September 2013. https://www.bizjournals.com/sanfrancisco/blog/2013/09/apple-art-levinson-google-calico-health.html.

[425] Dove, Jackie. "A Fond Farewell to Grumpy Cat, the Internet's Most Famous Feline," Digital Trends, May 18, 2019, https://www.digitaltrends.com/social-media/a-fond-farewell-to-grumpy-cat-the-internets-most-famous-feline/.

[426] LaVito, Sara Salinas, Angelica. 2018. 'Google Sister Company and Drug Giant Chip in Another $1 Billion to Cure Age-Related Diseases'. CNBC. 26 June 2018. https://www.cnbc.com/2018/06/26/alphabet-backed-calico-and-abbvie-chip-in-1-billion-to-cure-aging.html.

[427] Zuse, Konrad. The Computer - My Life (Springer Science & Business Media, 2013). quoted following Kurzweil p. 83.

computers becoming like humans is not as great as the danger of humans becoming like computers."[428]

Interesting in this context is an interview that Google director and singularity inventor Kurzweil conducted with MIT researcher Marvin Minsky under the title "Is singularity near?" which can be viewed on YouTube.[429]

Minksy, who in 1956 co-founded the *Dartmouth Conference*, where the concept of AI was introduced, was undoubtedly one of the outstanding experts in this field. Throughout his life he was full of confidence that AI would soon surpass human intelligence. He also founded the Laboratory for Artificial Intelligence at MIT and received numerous scientific awards.

Kurzweil and Minsky talk in the video about the *high-level management of* the brain, which is responsible for switching between different regions of the brain and keeping track of things. Kurzweil wants to know: "What is this master management program? Philosophers would say that they are in search of self-confidence. Minsky, however, does not see a "self" in humans: "I don't see there is any self". (8:45)

Minsky explains that one has to imagine the brain like 400 computers connected in series, but he complains that all attempts to get to the bottom of human consciousness by connecting 400 computers in series have unfortunately failed. The hope of finding the high-level management agent that plays a key role in the development of a general AI nevertheless rests on AI for both Kurzweil and Minsky. In any case, Minsky seems to be sure that AI will develop a self-confidence or a *High Level Management System*: "Well, they (the computers) may have

---

428 Zuse, Konrad. quoted in Youngman, Paul A. We Are the Machine: The Computer, the Internet, and Information in Contemporary German Literature (Camden House, 2009).

429 Kurzweil Interviews Minsky: Is Singularity Near?, 2014, https://www.youtube.com/watch?v=RZ3ahBm3dCk

recompiled themselves.[430] Suddenly the AI has a "self", that ominous something that was previously denied to human consciousness.

They can even imagine that AI will have a "heart". Minsky has said: "We better take control, because it is not clear what interests they have at heart." And later he says, "Someone has to decide." By "someone" what he seems to mean is the people who develop AI. So, in the end, despite all the enthusiasm for AI, Minsky still demands human decisions to control these systems.

In any case, it is difficult to imagine what interests the AI would have at its heart. How could an AI have these values: their dead view of the world and humanity lacks the necessary concepts. In their concept of consciousness and their anthropological assumptions, they do not, like humans, have the fundamental characteristic of behaving consciously towards themselves. This self-relationship has been described by philosophy as *daimonia* (Socrates), as self-confidence (from Descartes to Hegel), as care (Sorge) or as the ability to create the meaning of language through its use (Wittgenstein). The nature of self-relationship which is the basis for consciousness are utterly alien to machines.

Philosophers have always tried to grasp this specific self-relationship of man as a social being endowed with reason and language, and thus to determine his special "place in the cosmos".[431] With the question of whether machines are fundamentally capable of attaining or imitating human consciousness, these concepts must be taken into account.

"If you had Scanners in the brain, that are able to find a structure that you can upload, than you have a backup, would you consider

---

[430] Recompiling means that something is translated from a higher programming language into the machine language of a particular computer.

[431] This is the title of a work by Scheler, Max, Eugene Kelly, and Anthony J. Steinbock. The Human Place in the Cosmos. Translated by Manfred S. Frings. Evanston, Ill: Northwestern University Press, 2009.

it a copy of you?" asks Kurzweil, alluding to his idea of "uploading" consciousness to a hard disk as a victory over mortality. Minsky reacts approvingly but does not yet want to accept the idea of eternal life in cyberspace.

In the interview Kurzweil makes it clear time and again that he expects a "quantum leap" in technology to bring redemption from human suffering. And again, the point is that AI should first of all perfectly copy human intelligence, but technology only imitates human intelligence, and only in some areas. It can hardly copy it – probably not even as general AI. It lacks the will, corpus and imagination of human consciousness, as well as the ability to be simultaneously subject and object. It also lacks the ability in the unique structure of self-consciousness to be simultaneously subject, object and the activity of this self-assurance, a structure of the human mind that is a prerequisite for freedom and self-determination.

So with AI we will only ever have to deal with an imitation that is based on different, because technical, basic principles, regardless of what storage capacities and computing speeds are achieved. Strictly speaking, machines are not even capable of complete imitation: human feelings and intuition, good judgement and common sense too, are only extrapolated using the method of statistical induction. The term imitation assumes someone who consciously imitates something and must therefore be put in quotation marks for AI. So, if such systems give the impression that they understand, can sympathise and empathise, this is only mathematical illusion and is miles away from real empathy. This does not make the vision of a general AI any less threatening: For such systems can be faster than humans in many areas and give the impression of a "divine" omniscience, from which divine omnipotence is derived for their admirers. And they can exercise this power in a very real way if we allow them to make essential decisions without their arithmetic steps being made transparent and without a human being having to make the final decision in the end.

We dedicate ourselves to Kurzweil's theories not least because of his enormous power as Google's chief developer and his equally great influence on Larry Page, whom his admirers consider to be the person who can change the world the most. Franklin Foer writes that Larry Page's vision of the future of a world reminds him of Voltaire's satirical novel *Candide or the optimism*, especially the character of Pangloss: thanks to technology there will be no need, no sickness and no death, in short, an "optimal" world is within reach.[432] Pangloss was the teacher of Candide. He accompanied him on his adventurous journey, holding steadfastly to the conviction that, despite all the suffering they encountered, the world was the best of all possible worlds. He was not alone among philosophers; Leibniz was also convinced that everything on earth would inevitably develop to the best possible state.

But this is an attitude that leads to human inactivity, because in a world where everything is pre-determined, man does not matter. At the same time, determinism has the effect of a self-fulfilling prophecy, which also characterizes the fatalism in Voltaire's biting satire: if we run everything by technical criteria because we believe we have no choice, the world will actually run by technical criteria. Against the idea of the world as a process of optimization according to mathematical formulas, which inevitably leads to the best of all possible worlds, Candide's sentence helps: "If this is the best of all possible worlds, what are the others like?" This is a question we should never tire of asking.

If, like Kurzweil (and in a certain sense Leibniz), we define the criteria for "optimization" mathematically, the best of all possible worlds

---

[432] Foer, Franklin. 2017. World Without Mind. Random House

can consequently also be controlled mathematically only. Leibniz, however, unlike his Californian successors, wanted to program only one machine as he did with the Leibniz *machina arithmetica* in 1671, the first calculating machine for the mechanical solution of all four basic operations.

Voltaire strongly cautioned against the Leibniz-style mathematically derived optimism. His battle cry against those who blandish the evil state of the world was: *Écrasez l'infâme*! Destroy the vile, namely the infamous lies of priests and rulers, the alliance of throne and altar, who believe themselves to be in possession of the last truths of revelation and thus lead to fanaticism, intolerance and conformism. The battle cry of the great Enlightenment philosopher could today be transformed into Écrasez *GAFAM*!

For the alliance of throne and altar in the 18th century has been replaced today by the alliance of economic and informational technological power. The altar of our time is the religion of optimization. The new slaughtering block on which the subject is to be sacrificed is the database.

Kurzweil explains right away in the preface to his book that he had become aware of the profound consequences of the "meta-idea" early on, that "the power of ideas to change the world is growing ever greater."[433]

But the idea of relying exclusively on technical solutions to solve future problems, as Kurzweil does, does not lead to the desired results.

---

[433] Kurzweil, Raymond. 2006. The Singularity Is Near. London: Duckworth, p. 3.

C2 The three sources of Californian ideology

In the following we will look at the main sources that make up the Californian ideology:

- Cybernetics/computer science
- Darwinism
- Neoliberalism

2.1. Cybernetics

> *"The surest general characterization of the European philosophical tradition is that it consists of a series of footnotes to Plato." –*
>
> *Whithead, Process and Reality, S. 91*

Shortly after the end of the Second World War, two books were published, each of which in a very different way summarised a very individual contribution to the war effort, and each of which directly or indirectly referred to Plato. Both had a significant impact on how we think about the world today.

They are Karl Popper's *The Open Society and its Enemies*[434], published in 1945, and Norbert Wiener's *Cybernetics: or Control and Communication in the Animal and the Machine*[435] from 1948. Both books reflect in their own way the totalitarian systems of Stalinism and National Socialism, but draw very different lessons from the experience of totalitarianism.

---

[434] https://press.princeton.edu/books/paperback/9780691210841/the-open-society-and-its-enemies

[435] https://direct.mit.edu/books/book/4581/Cybernetics-or-Control-and-Communication-in-the-Animal-and-the-Machine

The mathematician and philosopher Norbert Wiener coined the term *cybernetics* which is a made-up word comprised of the Greek κυβερνητικός (kybernetikos, governmental), κυβερνήτης (cybernete, helmsman) and κυβέρνησις (cybernesis, leadership, rule).

The innovation of cybernetics marked a fulfilment of an ambition that Wiener had already expressed in the title of his 1948 book: *Control and Communication in the Animal and the Machine.*[436] Cybernetics is intended to unite technology and the life sciences. In this way, Wiener also wanted to solve some of the problems that he himself had worked on as an engineer during the Second World War, when his job was to calculate the trajectories of aircraft and anti-aircraft missiles. The aim at that time was to develop an "anti-aircraft predictor" which would enable the prediction of the future position of aircraft. Wiener believed that by modelling the feedback mechanism he had discovered, the ability for machines and humans to "modify their patterns of behaviour on the basis of past experience" could be enhanced.[437] In other words, he believed to be able to control future behaviour by analysing past performance.

Because Wiener in his research during World War II also had to take the behaviour of aircraft pilots into account, the cradle of cybernetics was the machine-human interface. Cybernetics describes a principle of action that can be demonstrated in organisms and technical systems alike. It focuses on the concepts of communication and regulation. The regulation between a measured actual state and a given target state is achieved by negative feedback. This has the aim of automatically compensating for any deviation from a functional optimum.

---

[436] Wiener, Norbert. Cybernetics: Or Control and Communication in the Animal and the Machine. Second edition. edition. Cambridge, Mass: MIT Press, 1961.

[437] Wiener, Norbert. The Human Use Of Human Beings: Cybernetics and Society. New edition. edition. New York, N.Y: DaCapo Press, 1988.

Wiener's fundamental insight was the benefit of a process in which a controlling force exerts control over the controlled flows of energy or information through feedback. Systems, whether biological or technical, are thus completely controlled. "Certain kinds of machines and some living organisms [...] modify their patterns of behaviour [...] to achieve anti-entropic ends, [...] both exemplify locally antientropic processes, which perhaps may also be exemplified in many other ways which we should naturally term neither biological nor mechanical".[438] Cybernetics cannot justify human and social control principles such as moral or ethical standards. Wiener did not even consider such values to be necessary, but firmly believed that totalitarianism could best be fought through technical information control. "For Wiener, the solution to the problem of totalitarianism was to see the world as a system of shared communication on an equal footing, which can be modelled and managed by computers. Implicitly, this means that the right way to realise the committee's vision and democratise society is to take power away from politicians and put it in the hands of technicians".[439]

Norbert Wiener introduced the concept of cybernetics with reference to James Maxwell. The eminent physicist Maxwell had described centrifugal regulators through feedback as *governors* in his work *On Governors* in 1867. For Wiener, the control of a ship is the archetype for a feedback mechanism. "We have decided to call the entire field of control and communication theory, whether in the machine or in the animal by the name Cybernetics which we form from the Greek for *steersman*. In choosing this term, we wish to recognise that the first significant paper on feedback mechanisms is an article on governors, which was published by Clerk Maxwell in 1868, and that governor

---

[438] Wiener. The Human Use Of Human Beings: Cybernetics and Society. New Ed edition. New York, N.Y: DaCapo Press, 1988.

[439] Turner, Fred., '[Essay] Machine Politics by Fred Turner'. 2019. Harper's Magazine. 1 January 2019. https://harpers.org/archive/2019/01/machine-politics-facebook-political-polarization/.

is a Latin corruption of steersman. We also wish to refer to the fact that the steering engines of a ship are indeed one of the earliest and best-developed forms of feedback mechanisms."[440]

Plato had used the image of the steersman in one of his famous parables, in which he compared the state to a ship. The neologism *cybernetics* thus also refers to the parable of the ship in Plato's *Politeia*.

In Plato's parable the ship is a metaphor for the state, the owner of the ship is the people. The people are strong, but deaf and unable to steer the ship through wind and waves due to a lack of nautical knowledge. The question of who is best equipped to steer, is a source of dispute between the crew members. But they too are all characterized by half-knowledge at best. But in Plato's eyes the crowd prefers half-knowledge to wisdom. In the parable of the ship, he wanted to explain why philosophers, despite their love of truth, are not well regarded by the people.[441] For Plato the answer is clear: only those who have the greatest expertise in the form of the best nautical knowledge can become steersmen, whether or not they meet with the approval of the people. Who else should keep course in wind and waves and bring the crew safely to their destination?

Plato's metaphor can certainly be understood as a critique of democracy. He was influenced by political struggles in Athenian democracy, which increasingly led to instability. Even his beloved teacher Socrates had been accused of godlessness and sentenced to death. Plato's theory of the state must therefore be understood against the background of a crisis in the society and culture of Athenian democracy. As always, when traditions and inherited

---

440 Wiener. Cybernetics: Or Control and Communication in the Animal and the Machine. Second edition edition. Cambridge, Mass: MIT Press, 1961. p.11-12

441 Plato, Protagoras, 319d3 quoted in Hunter, Richard. 2012. Plato and the Traditions of Ancient Literature: The Silent Stream. Cambridge University Press.

values lose their binding force, the need for something new to take their place grows. For Plato this new can only come about as a result of reasonable insight and planning. Plato sees a problem of democracy in the fluctuating opinions of the people and the constant danger that demagogues could influence them and persuade them to make wrong decisions. Salvation for him can only come from the philosopher ruler who presides over a hierarchically structured community in which everyone fulfils a function for the whole.

This anti-democratic tendency of Plato's theory of the state has rarely been criticised as sharply as by Karl Popper with the proverbial title of his main work: *The open society and its enemies*. Written during the Second World War, it was to be a "contribution to the war effort"[442]: a philosophical-theoretical turn against all totalitarian ideas that had led to the systems of rule of Stalinism and National Socialism. Popper identifies Plato as having paved the way for the totalitarian ideologies of the 20th century with the design of a closed society.

The steersman, who sets the course for all others on the basis of his "highest knowledge", is in his eyes the archetype of the dictator, or at least of the technocrat, who legitimizes an absolute claim to power with his elitist knowledge. Popper separates the ideas of Plato and Socrates; and identifies Plato as having taken the fateful step of defining the philosopher as capable of seeing and knowing eternal knowledge, whereas Socrates had always spoken of the search for and love of truth. "Plato's ideal philosopher approaches both to omniscience and to omnipotence. [...] He is the philosopher kind. It is hard, I think, to conceive a greater contrast than between the Socratic and Platonic ideal of a philosopher."[443] The omniscient

---

[442] Popper, Karl. The Open Society and its Enemies, Volume 1 The Magic of Plato, Tübingen 2003, Preface to the 7th German Edition, p.IX.

[443] Popper, Karl. The Open Society and Its Enemies (Routledge, 2012). P.125

and omnipotent ruler presides over a closed tribal society, which in Popper's eyes represented Plato's ideal. Popper sees this as the antithesis of the open society, which forms a system capable of learning, in which its members always take into account their own fallibility and are thus capable of correcting their course. As in his theory of society, his theory of science focuses on the ability for self-correction. Openness and the ability to criticise are the genesis of the scientific method; with the ability of scientific assumptions to be falsified is an essential prerequisite to scientific progress. At the same time, they lead to a humane and open society that does not succumb to the terror of closed systems such as Stalinism and Nazism. Popper's criticism of the Platonic state myth aims at the absolute form in which the king of philosophers should rule according to Plato, rather than at Plato's actual rationalism, which wanted to establish the new rule through rational insight.

Wiener's cybernetic steersman, on the other hand, claims the helm for himself because 'he knows more'. In this context it is interesting to note Wiener's double derivation of the term cybernetics, which comes from the words for helmsman and management/rule. The helmsman not only uses the right means to steer the ship safely, but also has control over the objectives and course of the ship and thus the purpose of the voyage. If the "control of message transmission in the living being and the machine regulation of the transmission of information"[444] could be united under one theory, societies and living beings could ultimately be controlled and thus dominated like machines, because human behaviour could be represented in computational processes. The information flow secures the processes in the system, in which errors and deviations appear as pure communication disturbances. It is also crucial that Wiener identifies the "steering engine of a ship", which for him is the archetype of

[444] Claus Pais, Cybernetics for Introduction, in Media Theories p. 427.

the feedback mechanism, without further ado with the captain who sets the course. With Plato, the helmsman was still the one who says where to go and then takes responsibility for setting the course. "Cybernetics thus replaces concepts such as consciousness, life and soul with message, control and feedback."[445]

Wiener himself knew about the importance of technology for the self-image of an epoch. "The thinking of each age is reflected in its technology," he says the chapter on regulation and message transmission in the living being and in the machine.[446] He explains the importance of connection to the current epoch a little later in the book: "If the seventeenth and early eighteenth centuries were the age of clocks and the later eighteenth and the nineteenth centuries constitute the age of steam engines, the present time is the age of communication and control." The same regulation in machines, living organisms and social organizations, brought together in a single great supersystem, that is the cybernetic dream. Which can also become a nightmare.

Weiner believed that technological innovation could spur a new era of rational control of social processes. For him, the advance over totalitarian systems consisted in the fact that his information theory put an end to the power of myths on which rule had previously been based. But this rationalization, in turn, could lead to the information machines themselves becoming enchanted. And humans then adapt to the machines they admire for their superior information processing. Thus, information theory itself threatens to become a myth.

According to Plato, the person at the helm of government should be the one who has the most insight and knowledge, even if he is

---

[445] ibid. p. 428.

[446] Wiener, Norbert. 1965. Cybernetics or Control and Communication in the Animal and the Machine. MIT Press. P.39

not elected. In Wiener's system, the one with the most insight and knowledge may ultimately be a machine, given the superiority in terms of information processing capabilities of machines to man. The exercise of power in highly complex societies by means of cybernetic processes of steering and control was supposed to solve the problem of totalitarianism in Wiener's imagination, but can also lead to totalitarian structures, just like the myths that Wiener originally criticized. Popper's analysis of how the "all-knowing and all-powerful" expert rules shows that a society in which communication serves exclusively to control processes and in which the individual is reduced to his or her function in a system is the archetype of a closed society. His distinction between an open system, which corresponds to an open, humane society, and a closed system, which resembles an authoritarian tribal society, is helpful for our analysis. Also useful is Popper's concept of "historicism" which refers to the mistaken belief that historical processes can be scientifically recognized and predicted. In Popper's eyes, these closed worldviews form the basis for "prophecies", i.e., unconditional predictions that lead either to fatalism or to activism, which calls for the acceleration and bringing about of that which is inevitable anyway. Both images of history serve the attempt to manipulate society. It is interesting that Popper's analysis is active today: A cybernetic-technical control of information flows, experienced through non-human actors in digital communication, leads back to the tribal society controlled by single individuals and single interests, rather than bringing new resources to the open society. A pluralistic society, which relies on free individuals who are self-determined because they are free to inform themselves, must be governed by procedures and instruments that determine the helmsman while simultaneously limiting his power, rather than simply submitting to a machine.

2.1.1.Cybernetics and systems theory

The theory of feedback through information in systems relates to the media theory of Marshall McLuhan - on media as "extensions of the body" - as well as functionalist social system theories from Parsons to Niklas Luhmann. In these theories the basic principles of the still nascent computer science were progressively elaborated to the "basic principle of a new political vision".[447] Some assumptions of both theories are reflected in the current manifestations of Californian ideology.

McLuhan describes in his media theory the replacement of the Gutenberg Galaxy by the television and computer. In this historic upheaval of media, the media appear to him as the technical central nervous system of mankind. McLuhan is interested in how technical artefacts are used for increasing the performance of the body. He argues that men have sought to become *Prosthetic Gods,* seeking "extensions of man"[448] following Freud in *Civilization and its Discontents* that "with every tool man is perfecting his own organs, whether motor or sensory, or is removing the limits to their functioning". This *prosthesis theory* of technology, from Freud[449], states that: "All media are extensions of certain human dispositions -

[447] Fred Turner, From Counterculture to Cyberculture: Stewart Brand, the Whole Earth Network, and the Rise of Digital Utopianism (University of Chicago Press, 2010). P.53.

[448] McLuhan, Marshall. Understanding Media: 2 edition. London: Routledge, 2001.

[449] Freud has described man as "prosthetic God", who enthusiastically develops fantasies of omnipotence through the magical possibilities that technology gives him: "It does not only sound like a fairy tale, it is the direct fulfilment of all - no, most - fairy tale wishes, what man has done on this earth through his science and technology, in which he first appeared as a weak animal being and into which every individual of his kind must enter as a helpless infant (...). (...) For a long time he has formed an ideal of omnipotence and omniscience, which he embodied in his gods. He attributed to them everything that seemed unattainable to his wishes or was forbidden to him. So one may say that these gods were cultural ideals. Now he has come very close to achieving this ideal, has almost become a god himself. (...) Not completely, in some pieces not at all, in others only halfway. Man has become a kind of prosthetic god, so to speak, quite magnificent, if he puts on all his auxiliary organs, but they are not fused with him, and occasionally still cause him a great deal of trouble." Freud, The Discomfort in Culture, op-ed, p. 222.

be they mental or physical."[450] The theory was also further developed by Arnold Gehlen.[451] This gives technology a double character: on the one hand, it can constantly expand the fields of human action and at the same time change humans. According to McLuhan, media first influence people's cognition and then their social organization. But the question remains whether and, if so, how this control could be self-regulated. So: Who controls? And: who controls, who controls?[452]

Talcott Parsons had introduced the concept of *symbolically generalized communication media* in the 1950s, which for him played an important role in the social systems of action. According to Parsons, modern societies differentiate themselves into systems of action, each of which functionally deals with specific problem situations. The famous *four-function schema* includes:

1. *adaptation* as a function of the economic system, which deals with the requirements of the goods system.
2. *goal attainment* controls the political system in which social power is at stake.
3. social systems are controlled by *integration,*
4. cultural systems by latency or *pattern maintenance.*

The communication media are symbolic because they have to fulfil the control task not just once, but continuously. For Parsons, social control media are above all money and power.

Following Parsons, Niklas Luhmann expands this scale and makes communication the basis of systems theory. Through communication

---

[450] McLuhan, Marshall, Fiore, Quentin, The Medium is the Massage, New York 1967, p. 26

[451] Arnold Gehlen, Der Mensch und die Technik, Anthopologische und sozial-psychologische Untersuchungen, S147 ff.

[452] In the style of Zuboff, op. cit.

the system regulates itself and ensures its self-preservation. For Luhmann, public opinion is embedded in the public sphere which he understands as the "form of the themes for political communication, by its suitability as a structure for the communication process", this is shaped by "attention rules", or in common language, the extent to which 'issues' are raised.[453] Due to the scarcity of the resource attention, the 'newsworthiness', the status of who is communicating and the preoccupation with a fixed set of current crises, for example, control the rules of political communication. Luhmann understood systems theory in a consistently descriptive way. According to him, therefore, public communication does not aim to produce a generally accepted reasonable result, but rather merely connects the political system to other specialized systems in society "to adapt the thematic structure of the political communication process to the respective decision-making needs of society and its political system".[454]

In other words: The system theory assumes a constant coupling of the subsystems and thus an exchange of media in a dynamic process, but cannot establish norms that go beyond the mere self-preservation of the system. It rejects this from the outset. In doing so, it reduces politics to social engineering technology.

If, on the other hand, people and societies are regarded as systems that produce norms that they can voluntarily follow, then political theory becomes about determining or criticising these goals themselves. Cybernetics and systems theory are limited to describing the control processes in technical and social systems. However, they cannot give

---

[453] Quoted in Bernstein, J. M. 1994. The Frankfurt School: Critical Assessments. Taylor & Francis.

[454] Author translation of Sarcinelli, Ulrich 2013: Public opinion. In: Andersen, Uwe / Woyke, Wichard (eds.): Hand Dictionary of the Political System of the Federal Republic of Germany. 7th, updated edition. Bonn: Federal Agency for Civic Education. 'http://www.bpb.de/nachschlagen/lexika/handwoerterbuch-politischessystem/202080/oeffentliche-meinung'.

a convincing answer to the question of who should legitimately be the helmsman, how this is determined and controlled, and where the boat should be headed.

2.2. Darwinism and Social Darwinism

The theory of evolution by natural selection first developed by Darwin and Wallace, describes the process by which organisms change over time through a process of mutation and selection on the basis of adaptive advantage. The fundamental principles at work during evolution, variation, inheritance, selection and time, are the fundamental developmental principles of life and a universal explanation for biological development. Darwinian thinking, and its applications for areas outside evolutionary biology, to human issues such as the problems of mind, language and knowledge has been explored by D. C. Dennett, among others[455]: These applications of Darwin's logic try to explain also non-biological developmental processes, in which a certain level of variation exists on which selection can take place.

An early universalization of Charles Darwin's was the development of a theory of Social Darwinism. Late nineteenth and early twentieth century thinkers such as Herbert Spencer began to transfer Darwinian thought from nature to society. Social Darwinists believed that the struggle for existence was equally fundamental in nature and society. Here, the law of the strongest applies, only the fittest are able to multiply, with the continual destruction of the individuals that are the least capable.[456] Philosophically, it was criticized early

---

[455] Dennett, Daniel C., 1996. Darwin's Dangerous Idea: Evolution and the Meaning of Life. Simon and Schuster. Richard Dawkins, 1989. The Selfish Gene. Oxford University Press.

[456] See article on Social Darwinism at https://www.bpb.de/politik/extremismus/rechtsextremismus/214188/was-ist-sozialdarwinismus#footnode14-14.and https://www.britannica.com/topic/social-Darwinism

on that Social Darwinism is based on the naturalistic fallacy that the evolutionary principles that apply in nature, should be chosen by humans, or indeed, sped up.[457] Social Darwinism, and the notion that humans should be selected for breeding on the basis of 'fitness' characteristics, became, among other things, the basis of eugenics, and one of the sources of National Socialist ideology.

Social Darwinism was a widespread popularization of Darwin's ground-breaking discoveries at the time, for those who wanted to transfer Darwin's thinking to the management of society. Which is not surprising if one considers that Darwin adopted terms like "struggle for existence" and "survival of the fittest" from the philosopher Herbert Spencer, who in turn originally wanted to found a theory of society with these.[458]

Spencer was a radical market liberal, and considered the evolution of societal organization towards market utopias as a parallel to evolution of species by natural selection: just as the blind natural process with the struggle for existence at its centre inevitably leads to a higher development of living beings, the blind market mechanism and the free play of forces would necessarily lead to the progress of society. Since Darwin originally drew on models of society when formulating his theory, it was supposedly easy to transfer his revolutionary theory of nature back to society.

Here we encounter a process of mutual ascriptions typical of universal theories. Nature and society are each interpreted in the light of certain values, and at the same time the assertion that everything is value-

---

[457] The naturalistic fallacy was described by the English philosopher George Edward Moore in his work Principia Ethica and, according to Moore, describes the mistake of defining good as a natural characteristic. He applied this to Hume's law, according to which one should not simply infer ethical commandments from the world of being. The short version of the is-and ought problem is that no norms follow from facts.

[458] Spencer, Herbert. 1875. The Principles of Biology. Appleton.

free hides the influence of these basic assumptions on results of the theory. A critical theory is then needed to decipher the genealogy, i.e., the process of creation.

The discovery of the laws of mutation and selection as the basic laws of the natural development of species represents one of the key findings in the history of science. That this was seen by Social Darwinists as a source of inspiration for ideal social systems, should have set alarm bells ringing, for the assertion that the survival of the fittest must also apply in society and among people the basic ideological assumptions in the National Socialists' crude world view.[459]

When today representatives of the Californian ideology, like Kurzweil, define living beings as "cellular automata" and ultimately as algorithms whose goal is the optimization of information processing, they are committing similar misconceptions as Social Darwinism. And they pay the price for the fact that they can neither justify nor reasonably criticise standards. For a in naturalistic world view, in which the sole aim is to optimise information processing, it is impossible to distinguish between power and validity and thus the question of whether and under what circumstances power is legitimately exercised or not cannot be answered.

However, exposing this distinction has far-reaching consequences. Today, in global politics and also in the theory of society, the thesis

---

[459] Basic assumptions that still hold true in the German population today. In the long-term study "Fragile Middle - Hostile Conditions", Zick and Klein found out in 2014 that Social Darwinist ideas are still widespread among the German population. https://www.uni-bremen.de/fileadmin/user_upload/sites/ade/Fragile_Mitte_Feindselige_Zustaende.pdf

The authors of the study also show that, in addition to Social Darwinian assumptions, "market-conforming and market-like extremism" often occurs, which makes the value standards of economic efficiency absolute and which, via the ideology of self-optimization and competition, is increasingly reaching the center of society. For the United States, see also https://inequality.org/research/american-social-darwinism/

of the rule of power is increasingly confronted with the thesis of the rule of law. We Europeans rely on the law, America hardly any more, Russia and China immediately on power.

2.3. Neoliberal Game Theory

Game theory was developed during the Second World War in order to be able to better calculate the behaviour of the enemy.[460] In trying to derive universal laws to deduce likely paths of behaviour, this connects it with cybernetics and Norbert Wiener's attempts to calculate the trajectories of anti-aircraft projectiles. Game theory assumes that each participant will pursue his or her own interests as far as possible. Conventionally, in game theory the human being is modelled as *homo economicus*; a rational agent, or utility maximizer. Instead of relying on morality, in a world like this, according to Schirrmacher, "one must assume the worst behind everything."[461]

After the war, game theory first conquered the social and societal sciences and then the world of finance, using game theoretical methods to predict the behavior of market participants and plan actions for their own benefit. This marked a long development of analysis using utility maximization, developed from utilitarian philosophers, to create a 'science of government'. Schirrmacher sees the principle of the method in the fact that reason is reduced to calculation: "Learning to act rationally meant: Learn to act and think in such a way that you always start from the self-interest of everyone."[462]

---

[460] Schirrmacher, Frank. EGO: The Game of Life. 1st edition. Cambridge, UK ; Malden, MA: Polity, 2015. p. 26.

[461] Schirrmacher, Frank. EGO: The Game of Life. 1st edition. Cambridge, UK ; Malden, MA: Polity, 2015. p. 23.

[462] Schirrmacher, Frank. EGO: The Game of Life. 1st edition. Cambridge, UK ; Malden, MA: Polity, 2015, p. 26.

In the logic of the neoliberal world view this attitude is then the core of all rationality and also morality, which in the end, however, leads to the loss of all morality. "A world view that sees the inescapable logic of self-interest at work behind all human activity produces egoism like on an assembly line."[463] The self-interest of all should then lead through the action of the *invisible hand* to the greatest happiness of the greatest number. In a world of rational egoists, the highest possible prosperity for all and an ideal community would result from itself: Harmony as a whole as a direct result of the actions of selfish actors. Schirrmacher recognized that even today there is a connection between *homo economicus* and computer science, because the scenarios of game theory can easily be transformed into algorithms.

In the Californian ideology, *optimization* means maximizing one's own advantage and benefit: Profitability, productivity, functionality, efficiency, advantage - all maximized for the individual, or, depending on your point of view, for the corporation in question. *Optimizing* further in the sense of cybernetics means that dominating power is exercised through a control software and that everything is done to ensure the smooth flow of information.

For Social Darwinists optimization means that in the struggle of people and social systems the stronger will prevail, viewed from the perspective of information capitalism this will be whoever processes information best, i.e., most directly and quickly. Whoever has the fastest computer is the master. This logic is also followed by the statement of Russia's President Vladimir Putin about AI: "Whoever takes the lead in this field will become ruler of the world"[464] he said to students in 2017.

---

[463] Schirrmacher, Frank. EGO: The Game of Life. 1st edition. Cambridge, UK ; Malden, MA: Polity, 2015. p. 11.

[464] 'Putin: Leader in Artificial Intelligence Will Rule World'. 2017. CNBC. 4 September 2017. https://www.cnbc.com/2017/09/04/putin-leader-in-artificial-intelligence-will-rule-world.html.

What Cybernetics and systems theory have in common as sources of Californian ideology is that they do not contain any criteria that would make it possible to reflect on what the "good life" might be outside their respective system logics. Rather, the systems are purely concerned with self-preservation and optimization, which the control processes serve. "Optimization" then means improving information processing. The system thus controlled is able to handle more complexity and run better. No matter where, no matter in which direction.

Today, this "no matter where", the lack of a direction where the journey should actually go, is glossed over by the propagation of private-sector ethics and self-regulation by corporations. Projects such as *AI for Good* pretend that there is a generally accepted good that corporations can determine either themselves or in cooperation with expert ethicists. In fact, an entire industry of philosophers and scientific ethicists who live from and work for industry has been created in this way. With these efforts, companies are pursuing the goal of replacing legal regulation with self-regulation. This is clearly demonstrated by the background to financing of the Chair of Ethics at the LMU Munich. After Facebook declared in January 2019 that it wanted to fund the Chair of Ethics with 7.5 million euros "completely independently" of its results. Later it emerged that Facebook could terminate the contract at any time if they did not like the "ethics". "In other words: if the research or the published results do not go in the direction of Facebook, the funds can be stopped at any time," wrote the Tagesspiegel.[465]

---

[465] S. Christian Kreiß: A meaningful secret contract with Facebook, Tagesspiegel, 19 December 2019 https://www.tagesspiegel.de/wissen/ethik-institut-an-der-tu-muenchen-ein-vielsagender-geheimer-vertrag-mit-facebook/25351058.html.

No one can ignore the dangers of this technology, not even the GAFAM companies themselves, and so, in the eyes of GAFAM, it is better to have your own ethical principles than to be threatened by laws and sanctions imposed on you by others. But there is utter silence on Popper's criticism of the helmsman in Plato's philosophy. For here the guild of ethicists in GAFAM services succumbs to a *moral hazard*, a rationality trap: if it were to follow Karl Popper, it would no longer have its unique selling point as a helmsman. Because in a democracy, every voice, every vote weighs the same.

C3 Dataism and social dataism - the Californian ideology today

The word dataism was popularized by Noah Yuval Harari. It also refers to the world view in Silicon Valley in conscious satire of Dadaism. This is the term we want to use to describe the current form of Californian ideology, after it has finally moved away from the hope for freedom of its early days and mutated into a kind of substitute religion, with the claim to explaining the world and the will to power. Dataism is the great comprehensive narrative that Californian ideology has given itself in the 21st century.

But why should we deal with the humanity of this ideology? We do so because images of human beings are the essential sources of moral action, from which values are derived that guide people and societies. The cultural image that people have of themselves determines their understanding of the world, their thinking, their actions and their decisions. The human image of *humanism* laid the foundations for modern democracy. It is increasingly questionable whether this self-image can be saved into the 21st century. However, the answer determines whether democracy still has a future under the technological conditions of the 21st century.[466]

What image of man do the protagonists of dataism propagate? And what worldview does it contain?

Jaron Lanier describes dataism as being characterised by the following 6 core beliefs:

[466] For example Hofstetter, Yvonne. Das Ende der Demokratie: Wie die künstliche Intelligenz die Politik übernimmt und uns entmündigt. München: C. Bertelsmann Verlag, 2016. p. 38.

1. Cybernetic information patterns represent the best and ultimate way to understand reality.
2. Humans are nothing more than cybernetic patterns.
3. Subjective experiences do not exist, or are not important because they are a kind of peripheral effect
4. What Darwin describes in biology is *de facto* the only superior description of all creativity and culture.
5. Qualitative and quantitative aspects of information systems will be accelerated inexorably by Moore's Law.
6. Biology and physics will merge with computer science This will create a new biotechnology and a new nanotechnology. To be a human being after this point in time will either be impossible or very different from what we imagine.[467]

Already the term dataism invokes that of Darwinism, merely because they are near homonyms. But there are other parallels. For dataists, technological development is nothing other than the continuation of biological evolution by other means. Both processes appear to be autonomous. Dataism is therefore inherently social dataism, which first wants to explain society by informational rules and then control it. In the final analysis, man and woman appears as an animal controlled by feelings and affect. His and her emotional responses render him/her controllable; he/she can be manipulated with confidence.

However, even the underlying equation of biological and technical evolution is problematic, because the logic of both processes is not really the same.[468] Biological evolution can be seen as a process of progressive differentiation, which takes place through

---

[467] Quoted after Tobias Oertel: Images of human beings in digitalization, 2.4.2019 in 'https://www.tbd.community/de/a/menschenbilder-der-digitalisierung'.

[468] F. Rapp, Analytical Philosophy of Technology, Boston Studies in the Philosophy and History of Science (Springer Netherlands, 1981), https://doi.org/10.1007/978-94-009-8412-7

individualization. The biological genus exists only in the individuals, its diversity and specificity is the guarantee that new forms of development are constantly being sought. Its data set is not stored anywhere by nature, except in the individuals.

On the other hand, technical development has a generally unifying tendency.[469] It is based on general laws of nature, whose modes of action underlie the technical artefacts. At the same time, it amounts to a convergence of technology, because it spreads according to universal and global principles. Contrary to biological evolution, technical development aims at uniform patterns, not at individual living beings. It transforms the individual into a universal pattern through abstraction and formatting.

Biological evolution owes its existence to an ultimately blind natural process. In contrast, technology is consciously planned by humans and owes its values to them, even if this is not always immediately visible. The "adaptation" of humans in technical evolution does not take place in a natural environment, but in a cultural and increasingly technical world shaped by humans. Only that which adapts to this world and then *optimizes* and differentiates itself "survives".

A further basis for the world view of dataism as the current expression of the Californian ideology is quantum theory. From its thesis that all matter ultimately consists of information and energy, dataism derives the radical basic assumption that the world ultimately consists only of information.

From these elements, the greatest possible narrative of anything and everything that has consequences for ethics is created: If the *purpose* or *value* of biological organisms is to process information, dataists

---

[469] Rapp in E. F. Byrne and Joseph C. Pitt, Technological Transformation: Contextual and Conceptual Implications (Springer Science & Business Media, 2012). P.254

conclude that it is valuable to know who processes information best or fastest.

Selfish genes[470], are increasingly being replaced by selfish algorithms in the digital world view. The struggle for existence becomes a struggle for the calculation and control of the world through data analysis.

In this world view, woman and man has no choice but to upgrade the machines to form as autonomous structures as possible and to connect to their superior information processing program through an interface in order not to lose touch[471] - thus following a supposedly compelling logic of evolution. Technological upgrading seems to be without alternative, because technology in dataism is ultimately the continuation of evolution by other, more powerful means.

However, there must be no doubt: Like all claims about an inevitable course of history, this one, too, originates from people, and indeed from people who want to be in charge of the development trajectory. If Larry Page says that Google will change the future of humanity, then he means, as Franklin Foer rightly writes, that Google wants to actively change the course of evolution.[472]

The goal of an optimized transhuman human being is to be achieved by all kinds of implantations of technology into the human organism in order to minimise the distance between humans and intelligent AI robots and cyborgs. This is how Larry Page thinks to direct implants

---

[470] Dawkins, Richard. The Selfish Gene. Oxford University Press. 1989.

[471] According to Sergeij Brin: "If all the information in the world were directly connected to our brain, or to an artificial brain that is more intelligent than our own, we would be better off" Quoted from Franklin Foer, World Without Mind: The Existential Threat of Big Tech (New York: Penguin Press, 2017), p. 52.

[472] Foer, Franklin. World Without Mind: The Existential Threat of Big Tech (New York: Penguin Press, 2017), p. 53

of his service into the human brain: "Perhaps in the future there will be a small version of Google that we can simply plug into our brain."[473]

For the social world, the ethics of optimizing information processing means that problems are first identified as being solvable by algorithms, and these are then interpreted as communication disturbances of the system. Communication disturbances prevent the feedback processes from running smoothly. Since in cybernetics every piece of information serves to control the system, the problems interpreted as communication disturbances must be eliminated in the sense of optimal process control. Algorithms can best solve these problems due to their superior information processing capabilities.

To do this, you need access to all data with as few limits as possible. Private and personal data, which remains private, naturally stands in the way of this process of data optimization. The meaning of the word private is derived from the Latin *privatum*, the own and *privus*, existing for itself. The private in the sense of the *own, existing for itself* resists the robbery of data. But this contradicts the demand of dataism for optimization of the processing procedures.

That's why Google & Co. are simply reversing the meaning of the robbery: Since there is no reason in their world view to withhold information from the big machine they run, they unceremoniously annex privacy and personal data and take whatever data they can get. Those who refuse to do so "rob" them of the opportunity to solve problems.

The word meaning of data is also derived from Latin: date is "the given thing." Data should therefore be given voluntarily, according to the strict sense of the word, and its use requires the consent of the person who gives it.

---

[473] Foer, Franklin. World Without Mind: The Existential Threat of Big Tech (New York: Penguin Press, 2017), p. 53

Basically, dataism is based on a circular procedure and on a bet: namely, that it is possible to completely control the world by expanding computing speed and data processing, because after all the world consists of nothing else but information that can be expressed and captured in data.

If technoevolution, like bioevolution, is ultimately presented as a self-organizing autonomous process. Like dataism, it will not be subject to controls by human values: it is viewed as outside such norms, "beyond good and evil" as Nietzsche put it.

In the meantime, the theory is being propounded in the GAFAM environment that the entire universe is constructed like a quantum computer and that the world as we know it, and women and man, is nothing more than its calculation.[474] While this book was written, Google announced a historical breakthrough on the way to "quantum superiority", as the development of a functioning quantum computer, lacking, somewhat, in modesty.

According to Google, the self-developed quantum chip *Sycamore* performed a calculation task in 200 seconds that would have taken the fastest supercomputer of our time 10,000 years to complete. As always with neologisms in the context of Californian ideology, the term *Quantum Supremacy* should be used with caution. Not only does it evoke the racist term *White Supremacy,* which even the creator of the term noticed[475] it also serves marketing purposes and - like many terms coined in Silicon Valley - is intended to prepare mankind for the coming of a superior power. As with AI, the bets are on:

---

[474] Seth Lloyd, in Tegmark, Max. Our Mathematical Universe: My Quest for the Ultimate Nature of Reality. First edition. London: Allen Lane, 2014. p. 22

[475] The creator of the term, John Prescell, goes into this: "John Preskill Explains 'Quantum Supremacy,'" Quanta Magazine. https://www.quantamagazine.org/john-preskill-explains-quantum-supremacy-20191002/.

Whoever is the first to put a stably functioning quantum computer into operation will not only master the encryption programs but will also be able to evaluate data with a completely different quality.

So, will we soon be decoding the *galactic computer* with high-performance quantum computers or is this the next Californian hype? In the eyes of the quantum evangelists, it should be possible to unlock the last secrets of the universe. Then, so the expectation, we will finally be in the position of God, who looks at his creation and embodies this creation at the same time. A dream that has often been dreamed. Whenever attempts were made to realize it, it led straight to disaster.

When the French Revolution turned into Jacobin terror, Hölderlin warned: "After all, it made the state a hell that man wanted to make it his heaven."[476] What Hölderlin refers to is the terror of the Jacobins. We can easily extend to the redemption fantasies of the ideologies in the 20th century. The warning should also be taken seriously for the technological salvation plans of the digital revolutionaries.

Max Horkheimer refers to his 1944 investigations into the ideology of the National Socialists and Stalinists, which appears in German under the title *Zur Kritik der Instrumentellen Vernunft*, in the original American *Eclipse of Reason*. Today we are again dealing with such a shadowing of reason. It is precisely those thinkers who believe themselves to be closest to progress that are creating a partial eclipse of reason that threatens the future of humanity. For with humanism which the dataists want to overcome, humanity would also disappear. Today we can at least learn so much from history that the proclamation of the superhuman, requires the designation of subhumans and descends into the actions of inhumans. With

[476] Hölderlin, Friedrich. Hyperion or the Hermit in Greece. Chapter 10, op. cit. p. 319.

mankind as we know it, in digital posthumanism humanity would be lost, as would human dignity.

3.1. What does the "end of the outside" mean for our analysis?

If *Whole Earth* is the system in which nature and culture are controlled by the same cybernetic laws and by information, there is no longer any outside to the constraints of system logic and therefore no place for criticism.

In the meantime, the process has even reversed itself: the colonization of the inner world has long since begun. When the outside world has been completely conquered, it is the turn of the inner world. The inner world is to be dealt with according to the same logic of conquest that the settlers used on their way to the West: Stake out, survey, decree and claim. After the seemingly complete subjugation of nature by technology, Californian ideology now takes the path to subjects' innermost lives.

Through comprehensive data of life, which are collected and evaluated, the desires and will potentials of the subjects can first be tapped and then manipulated. Nanotechnology also enables the internal colonization of humans by technology in a very physical sense. And in the cyborg, prosthetic technology is developing from metaphor to reality: the machine-human being as the completion of a technical utopia in which technical implants are supposed to optimize human possibilities.

Without the outside world, in a system that sees itself as comprehensive, there are no more norms that would exceed or question its meaning. Deviations can still be tolerated as non-binding opinions, but this does not change the triumph of "the system". That sounds boring, but boredom it must not be in the entertainment economy. Because deviations are tolerated, as already mentioned, if they do not endanger the big picture. The stage is set for the entertaining productions of political clowns, behind whose unpredictability, which is owed to the dictates of absolute entertainment, the supposed logic of compulsion rules.

3.2.The erosion of language, self-determination and democracy and its consequences

> *"The rhetoric of the technical elite has corrupted language."*

Josef Weizenbaum[477]

The erosion of language has played a decisive role in the triumphal procession of the *Californian ideology*. The function of language(s) to understand and open up the world is being undermined in Dataism. Statements that refer to moral correctness or aesthetic authenticity no longer have any meaning in such a conception of language. The openness and ambiguity of natural language is seen as a disadvantage compared to the apparent unambiguity of cybernetic signals. The potential for creativity, dreams, criticism and the reinvention of oneself and the world fade into the background.

Neither technology nor mathematics or physics can produce the language they need to describe their own discoveries or inventions.[478]

---

[477] Weizenbaum, Joseph. Computer Power and Human Reason: From Judgment to Calculation. Penguin Books, 1993. Original quote in German edition p. 328.

[478] We follow Wolfgang Frühwald in this view. See among others: Sprachen der Wissenschaft oder über die Verwandtlichkeit von Poesie und Wissenschaft, in: ders.Das Talent, deutsch zu schreiben, p. 7 ff Köln 2005.

It is precisely during the development of new technology when we need new images of language, new words, new sentences to describe the changes. What cannot be explained by language is hardly accessible to humans. This must be remembered when companies declare that they cannot explain how AI comes to its results and decides. AI can only make "decisions" on the basis of "old" data. But new ways only result from the openness of language and human imagination. If language no longer has any reference to truth or moral correctness, but is a mere signal that serves to attract attention, this opens up opportunities for its abuse and the weakening of democracy as we experience it today.

3.3. Confucius and the great harmony

2,500 years ago, Confucius taught that "Harmony is present in creation". All human order only consists in clearing the way for harmony to develop in human society.[479] By drawing on quotations from Confucius, the current debates on the relationship between technology and democracy tend to give the impression that technical solutionism is very easily connected to Asian philosophies and religious teachings.

For apparently the core statements of *Californian ideology* fit very well with Asian philosophies such as *Confucianism, Buddhism* and *Shintoism*. They promise the subject relief from the toil of life if he or she fits into the harmony of creation. Can this model not also serve as a model for subordination to a superior AI, which apparently

---

[479] See also Chinese Philosopher Zhao Tingyang, who claims for a Chinese world order analogous to the Tianxia concept in classic Chinese philosophy. Tingyang criticizes the "ego trap" in Western thinking, which would render it unsuitable as the basis of a world order in the 21st century.: Alles unter dem Himmel, Vergangenheit und Zukunft der Weltordnung, Berlin, 2020 or Li, Chenyang. 2006. 'The Confucian Ideal of Harmony'. Philosophy East and West 56 (4): 583–603.

represents something like the world spirit? However, in these comparisons, the cosmic concepts of order of Asian philosophies are all too quickly assumed to have a cybernetic structure. But this is not correct, even if there are differences between the European and Far Eastern concepts of logic and truth.

Confucianism in China, since its elevation to state doctrine by the Han-dynasty, has the character of an "imperial ideology, (which) bears the features of a political religion."[480] Nevertheless, the teachings of Confucius a core which is based on humanization through education and training.[481]

Morality or humanity (*rén*) is contrasted with the concept of li, which encompasses the traditional rites and customs. The required integration of the individual into the superordinate structures of state and society presupposes his humanity and his connection with tradition. "Harmony is said to be necessary for the peaceful life of individuals, the family and the state."[482] But according to Confucius a perfect insertion into the big picture is not possible, merely a perpetual approximation: "To see a god-man is not granted to me, if it were granted to me to see a noble one, it would be all right."[483] In Confucianism, too, it is not possible to blend into the harmony of heaven without preserving the autonomy of the individual and a conception of human dignity.[484] Therefore the individual must be

---

[480] Habermas, Jürgen. Auch eine Geschichte der Philosophie: Band 1: Die okzidentale Konstellation von Glauben und Wissen. Band 2: Vernünftige Freiheit. Spuren des Diskurses über Glauben und Wissen. Berlin: Suhrkamp Verlag AG, 2019 p.385

[481] Habermas, Jürgen. Auch eine Geschichte der Philosophie: Band 1: Die okzidentale Konstellation von Glauben und Wissen. Band 2: Vernünftige Freiheit. Spuren des Diskurses über Glauben und Wissen. Berlin: Suhrkamp Verlag AG, 2019 P.390

[482] Yao, Xinzhong. Yao, Hsin-chung. An Introduction to Confucianism, (Cambridge University Press, 2000). P.172

[483] Confucius Talks, Frankfurt 2008, quoted after Habermas op. cit. p. 390.

subjected to a process of continuous education. This is why Max Weber called Confucianism "a cultural and educational religion".[485] As far as a reference of the Californian ideology to Confucius or other currents of Asian thought is taken, this is thus based on a misinterpretation.

A look at Buddhism also demonstrates the misinterpretation of eastern philosophy in this context. The Berlin-based philosopher Byung-Chul Han explains the fundamental differences between Western and Eastern thinking using the example of Zen Buddhism.[486] This is a "religion without God". The subject is indeed "a nobody", but the cosmos is also a "void" in which there is no room for a central, controlling instance. He argues that the concept within Buddhism of 'nothingness' means that for the "category of power" there is no location or essence in which power can be concentrated.[487] "The category of power is also inappropriate to the Buddhist nothing, because it is an expression of the 'substance' or the 'subject'......It (the nothing) does not represent a 'doing-acting power' (Hegel). It doesn't do anything." Byung-Chul Han's interpretation shows that Buddhism is protected by the absence of domination and power from being appropriated by digital attempts at legitimation: "The absence of the Lord releases Buddhism (...) from any economy of domination (...) No one represents 'power'. "[489]

---

484 According to Habermas: "This corresponds to the egalitarian concept of autonomy and human dignity that we ascribe to everyone, regardless of social status." Habermas, Jürgen. Auch eine Geschichte der Philosophie: Band 1: Die okzidentale Konstellation von Glauben und Wissen. Band 2: Vernünftige Freiheit. Spuren des Diskurses über Glauben und Wissen. Berlin: Suhrkamp Verlag AG, 2019. p. 392

485 Quoted after Habermas, Jürgen. Auch eine Geschichte der Philosophie: Band 1: Die okzidentale Konstellation von Glauben und Wissen. Band 2: Vernünftige Freiheit. Spuren des Diskurses über Glauben und Wissen. Berlin: Suhrkamp Verlag AG, 2019. p. 387.

486 Byung-Chul, Han. Philosophie des Zen-Buddhismus. Stuttgart: Reclam Philipp Jun., 2002.

487 Janning, Finn. 2018. 'THE PHILOSOPHY OF BYUNG-CHUL HAN: Relax. Do Nothing. Become No One.' The Mindful Word. 30 October 2018. https://www.themindfulword.org/2018/byung-chul-han-philosophy.

489 Author translation from Byung-Chul Han, Philosophie des Zen-Buddhismus. Stuttgart: Reclam Philipp Jun., 2002. p.17.

Even though the role of the individual in Far Eastern philosophies may not correspond to the image of the subject in Western philosophy, we must not draw the wrong conclusions from this fact. The "harmony of creation" must not be equated with the machine logic of cybernetic systems, to which the individual must submit. To use the abridged idea of an immediate harmony of heaven, earth and man to legitimize the rule of man-made algorithms over individuals is a straightforward abuse of Far Eastern philosophical traditions.

The argument that Asian culture is alien to Western individualism and thus to the Western concept of human rights and that it gives priority to the community over the individual is not new. There is a longstanding debate about "Asian values".[490] In the debate it was argued that Asian cultures were unaware of the separation of law and ethics and would in principle place the community above the individual.[491]

At that time, the discussion was not yet determined by questions of an ethics for AI systems, but by the fundamental discourse on universal human rights, which were allegedly incompatible with Far Eastern ideas, because the Buddhist and Confucius-influenced cultures of Asia did not "know" individual rights. The arguments in the AI debate today are largely similar, as are the interests of those who put them forward.

Today, as then, the concept of universal basic and human rights is denounced as the result of a Eurocentric concept of reason which has

---

[490] E.g. Said, Edward W. 2003. Orientalism. 25th Anniversary Ed with 1995 Afterword Ed edition. London: Penguin Books.

[491] The Singapore government's announcement on shared values in 1991 triggered a debate of this kind, which was discussed at the Vienna human rights conference and by Habermas in: J. Habermas, The Postnational Constellation: Political Essays, 1st MIT Press Ed edition (Cambridge, Mass: MIT Press, 2001)

aimed at the subjugation of other cultures, and which in reality is not concerned with general values but only with its perfidious claim to power. In this interpretation, human rights are nothing more than an ideology to conceal the *West's* claim to power.

A contemporary foundation of universal human rights can, in any case, do without religious or metaphysical justifications. It is based solely on the concept of autonomy, which demands that citizens make their own laws, and on the assumed symmetry in reciprocal social relations, which also underlie the ideal of a just economic exchange. Like the latter, they are dependent on legal protection.

No priority of heteronomy or oppression can be derived from Asian culture. Nevertheless, attempts will continue to demand the abdication of mankind to the AI, either with utilitarian or Confucian ethics. For the Californian ideology, with its growing influence of its own, increasingly considers it unnecessary to borrow from other cultures or religions in order to increase its effectiveness. It is now self-confident enough to assume the status of a religion for itself.

C4 Internet as religion, AI as god

4.1.By their data you shall recognize them

You have searched me, Lord,
and you know me.
You know when I sit and when I rise;
you perceive my thoughts from afar.
You discern my going out and my lying down;
you are familiar with all my ways.
Before a word is on my tongue
you, Lord, know it completely.
You hem me in behind and before,
and you lay your hand upon me.
Such knowledge is too wonderful for me,
too lofty for me to attain.
Where can I go from your Spirit?
Where can I flee from your presence?

This is not an ode to Google, but Psalm 139. If you leave out the biblical context, it reads like an awesome adoration of the world's greatest search engine, which knows everything about everyone and claims to know what you will think or feel next. The psalm ends with the insight of the individual that there can be no escape from this divine omnipresence.

The religious intuition expressed here is: omnipotence arises where omniscience reigns. So, the psalm could also be used in the *Church of Google* which was founded in 2006 by Matt MacPherson. At least as a website *The Church of Google* and of course, as a satirical project. Or is it? The website states: "We at the Church of Google believe the search engine Google is the closest humankind has ever come to directly experiencing an actual God."[492] The *Church of*

Google has even postulated ten commandments of Googlism. The first commandment reads: " Thou shalt have no other Search Engine before me, neither Yahoo nor Lycos, AltaVista nor Metacrawler. Thou shalt worship only me, and come to Google only for answers." The tech giant tolerates no other gods beside itself, and this is where the satire obviously ends because the striving for supremacy is also cloaked in GAFAM's self-image as a religion. Without any irony, former Google boss Eric Schmidt once said: "There's a particular religion that we all represent, and it goes something like this: 'if you take a large number of people and you empower them with the communication tools and opportunities to be creative, society gets better"."[493] Apart from the fact that Schmitt's formula of society as optimized by the formula of great mass plus tools is certainly unintentionally linked to Lenin's definition of communism, according to which "communism is Soviet power plus the electrification of the whole country"[494] - the communication tools for this religion are of course provided by Google.

Like every religion, Dataism needs in addition to a claim to divinity, a means to ensure the religious following of its believers. On the one hand, these means consist of the control of the technical possibilities with which the communication within the "network community" is organised. On the other hand, they are also the pandering to the wishes of the individual members. In this way, the members of the community are educated to follow the convenience of the platforms, which they are becoming increasingly dependent on.

---

[492] https://churchofgoogle.org/Proof_Google_Is_God.html

[493] Eric Schmidt, speaking at a small technology dinner at the Aspen institute, quoted in Ferenstein, Greg. 2015. 'What Silicon Valley Really Thinks About Politics, An Attempted Measurement'. Medium. 6 November 2015. https://medium.com/the-ferenstein-wire/what-silicon-valley-really-thinks-about-politics-an-attempted-measurement-d37ed96a9251.

[494] Lenin when describing the GOELRO Plan – the first-ever Society plan for national economic recovery and development. Quoted in Carlin, Wendy, Mark Schaffer, and Paul Seabright. 2013. 'Soviet Power plus Electrification: What Is the Long-Run Legacy of Communism?' Explorations in Economic History 50 (1): 116–47. https://doi.org/10.1016/j.eeh.2012.07.003.

The fact that people increasingly perceive digital technologies in a quasi-religious way has to do with networking itself. In a sense, it represents the equivalent of the Holy Spirit for Christianity. The social network can be seen as the counterpart of the Christian community, in which members cultivate their faith and support each other with encouragement in the form of *Likes*. Solutionism is also very religious; this is the part of the Californian dogma that promises to be able to solve all conceivable problems on earth with technical means. If that is not a promise of salvation! It is spread by speakers who seriously call themselves "evangelists" at Ted X conferences and other show congresses. The fact that the *company* is researching how to overcome death completes the religious package. It can include the promise of immortality in the trade of digital indulgences. Personal user data and self-determination end up in the offertory.

But in the long run no religious community can live on dreams of the future alone, and certainly not one whose credo is the fulfilment of wishes in the here and now. Therefore, what is important for the persuasive power of the *Californian ideology* is what technology can achieve in our everyday lives. Most striking is Google's omniscience. The word *ungoogle-able* has long been synonymous with *non-existent*. And since Google also knows the backside of the cloud, which is unknown to the users, i.e., the data from which, for example, behavioral predictions are made, the power is certainly given a universal and thus divine aspect.

The omnipresence of Google is linked to the magical little screen that we all carry around with us at all times. A truly symbiotic relationship has been created, but the power resides with the Google god. And can the person who is both recognized by the almighty platform, and also increasingly manipulated by it, not also rightly consider herself/ himself its creation? The creator is whoever writes the algorithms, which are nothing but instructions for action. Users are the ones who follows these instructions for action, like believers in the Word of God.

The quasi-divine view of the individual human being behind the veil of not-knowing is completed by the apparent care with which Google accommodates and increasingly anticipates the thoughts and wishes of its users. Anyone who knows about the health status of a human more than the people affected, who can predict pregnancies before they are visible, and can tell by the voice or their gait whether someone is suffering from depression or even wants to commit a crime - anyone who has this power and can not only predict the future of people, but can even bring it about - is he not really God? [495]

In view of these overwhelming proofs of divinity, it should not come as a surprise that more and more digital faith communities are being founded, even beyond Ray Kurzweil's development department. Former Google manager Anthony Levandowski founded the sect *Way of the Future,* their most important creed is the belief in unconditional progress. After all, Levandowski once headed Waymo, the department that is researching the self-propelled car for Google. He is considered an expert in robotics and AI. The magazine *WIRED* published 2016 tax office documents of the sect, where it was stated: Way of the Future will focus on "the realization, acceptance, and worship of a Godhead based on Artificial Intelligence (AI) developed through computer hardware and software."[496]

---

[495] Facebook is also trying to achieve this status of omnipotence. Recently, it was revealed that Facebook was tapping information from two well-known menstrual apps, likely to improve its ability to predict pregnancy. This includes sensitive data such as "unprotected sex" or moods such as "anxious" or "in love". Even users who do not have a Facebook account are affected. This makes it possible to create forecasting sites that can be sold to manufacturers of baby products. https://www.privacyinternational.org/long-read/3196/no-bodys-business-mine-how-menstruations-apps-are-sharing-your-data

[496] Harris, Mark, 'Inside the First Church of Artificial Intelligence | Backchannel'. Wired. 15 November 2015. https://www.wired.com/story/anthony-levandowski-artificial-intelligence-religion/.

It is easy to dismiss the founding of the church as an elaborate tax-avoidance scheme or silly jape. But Levandowski is a fervent supporter of Singularity. He is certain that the superior intelligence will come and will ultimately dominate the world. He told a WIRED journalist: "What we want is the peaceful, serene transition of control of the planet from humans to whatever. And to ensure that the 'whatever' knows who helped it get along."[497]

There it is again: the advance adaptation to and submission to the technological being that man has created but which he fears as soon as it is superior to him. We already know the fear of being punished for previous iniquities from Kurzweil's fictitious dialogues with "Molly" about the entry into the singularity. Also, Levandowski wants people to enter his church for fear of the power of the coming AI: "We believe that it is important for machines to know who is well-disposed towards them and who is not. We want to show them, by recording who did what and for how long for a peaceful, respectful transition."[498] The sin register will be kept on the servers to prepare for the final judgment. Eternal salvation is only to be obtained by those who practice and submit in the "transition" to world domination by the AI God.

However serious the founding of the church may be, we are interested in the point where the *Californian ideology* changes from scientific zeal to pure superstition. Levandowski may be a laughingstock. But on the website of his new "church" was the central creed that the majority of GAFAM supporters would sign:

---

[497] Harris, Mark, 'Inside the First Church of Artificial Intelligence | Backchannel'. Wired. 15 November 2015. https://www.wired.com/story/anthony-levandowski-artificial-intelligence-religion/.

[498] Harris, Mark, 'Inside the First Church of Artificial Intelligence | Backchannel'. Wired. 15 November 2015. https://www.wired.com/story/anthony-levandowski-artificial-intelligence-religion/.

> *"We believe the creation of "super intelligence" is inevitable"*[499]

It is this idea of the inevitability that is the central element of Californian ideology.

What is decisive for us is the consequence of this doctrine that human action, for example in the form of legislation or political decisions, can at best slow down the fate of man and the planet, but cannot change it. The message of the Silicon Valley evangelists is to prevent legislation as far as possible and to lobby social and political institutions to their side.

Let us return once more to Psalm 132: The believer can entrust herself/himself to the loving Father, who knows everything and directs everything - even and especially in the difficult hours of life, when her/his own fate appears dark. With Google & Co. we should be much more careful with this trust. The more thoughtlessly we entrust ourselves to their services, the less free and dependent we become. If you want to go back to the Middle Ages, step through this gate.

The promise of the monotheistic religions so far has been renunciation in this world against eternal happiness. The new deal of dataism on the other hand promises absolute fulfilment of needs here and now in exchange for personal data, freedom and self-determination. A bad pact for the people, a good one for the market-dominant corporations.

---

[499] Quoted in Grant, August E., Amanda F. C. Sturgill, Chiung Hwang Chen, and Daniel A. Stout. 2019. Religion Online: How Digital Technology Is Changing the Way We Worship and Pray [2 Volumes]. ABC-CLIO.

4.2. The *God-Man* as the superhero of the GAFAM master narrative

The thesis that science can become the new religion does not originate from Silicon Valley, but from Carl Friedrich von Weizsäcker who drew different conclusions on the consequence of this change. In his lecture *The scope of science in our time* he argued that the belief in science replaced religions.[500] Von Weizsäcker derived the demand for a close connection between science and morality from the enormous application possibilities of modern science. In contrast, representatives of the *Californian ideology* try to replace morality with science.

California ideology wants to be the new religion. Their cathedrals are the algorithms, their scriptures are the program codes. Their doctrine: Creation consists essentially of information; life is information processing in the form of algorithms. Man stages himself as a god-man, as *Homo Deus*[501] who is endowed with three divine attributes:

1. Omniscience
2. Omnipotence
3. Immortality.[502]

In order to achieve this, man must complete three technological steps:

First, is the total digitalization of man and nature. This is done through the *Internet of Things,* connected to 5G networks and Big Data servers. Because for dataism, being nothing more than data,

---

[500] Raiser, Konrad. 2014. Carl Friedrich von Weizsäcker: Major Texts on Religion. Springer.

[501] 'See A New Religion | Christoph Quarch | House of Beautiful Business', 2019, https://www.youtube.com/watch?v=WsdI0NPdjHw. Quarch, Christoph. "A Guidebook for Entrepreneurship in the Third Millennium," Blog - Argo Venture Studio (blog), March 25, 2019, https://blog.argo-venture-studio.com/guidebook-entrepreneurship-third-millennium/.

[502] Harari, Yuval Noah. 2016. Homo Deus: A Brief History of Tomorrow. Random House.

the analog world is transformed into information by the *Internet of Things* and each person who donates their data. It can then be processed and made available by algorithms. Soon the algorithm knows you and your world better than you know yourself.

The optimization of the data is carried out by AI, which takes over the data processing of the organisms. Homo sapiens becomes superfluous and is replaced. Finally, cyborg technology (bio-, gene-, nano-technology) is used to achieve immortality - but probably more for the small elite of optimized superhumans than for the masses of biological humans who have become useless. Welcome to the brave new world of GAFAM.

C5 Critique of the Californian ideology's view of man and world

Why is it important to think about people when thinking about AI? Harari in his proclamation of *Homo Deus* suggests: Man is only a temporary phenomenon of evolution. Homo Sapiens is considered a species that "must be surpassed"[503] as Nietzsche put it, that "will disappear" as Foucault wrote.[504]

Man is to be replaced by AI, a "divine" reason that is already changing the world in practically all areas of life. While we are subject to daily control by algorithms and are increasingly shaped by them, *Californian ideology* can also occupy our thinking. As we adapt to "thinking" computers, we and our thinking become more and more like them. When the prophets of the AI proclaim that "AI will be our equal in 2062,"[505] what they are more accurately saying is that we humans will be adapted to the AI by that time, that we will be its equal. This will have devastating consequences for freedom and self-determination.

Our self-image, as well as our world view, determines how we think and how we act. Therefore, we must ask ourselves today how digital ideology has already changed our thinking and how AI can change our thinking even more radically.

---

503 "I teach you the superhuman. Man is something that hath to be surpassed.", Nietzsche, Thus Spake Zarathustra, Zarathustra's Prologue, https://www.gutenberg.org/files/1998/1998-h/1998-h.htm#link2H_4_0004

504 With what Foucault considers man as a subject of human knowledge: "Man is an invention whose recent date shows the archaeology of our thinking quite openly....If these dispositions were to falter....then one can very well bet that man will disappear like a face in the sand at the seashore" Foucault, Michel. 2002. The Order of Things: An Archaeology of the Human Sciences. Psychology Press. p.462

505 This is from the German substitle of the translation of Toby Walsh's book

The analysis of the dazzling concept of Artificial Intelligence showed that it is based on a mechanistic image of mankind that was passed on almost unchanged from the early Enlightenment to the discourses on technology in Silicon Valley.[506] If the term Artificial Intelligence is supposed to mean that a super-intelligence superior to humans in every respect has been created, this idea owes its existence to a sleight of hand: first human thinking is mechanized, then the superiority of the machine in mechanistic thinking and thus the superiority over humans is established.

The dangers that lie in this ideology have already been explained. In the worst case, the visions of a takeover of power by large-scale technical systems could prove to be a self-fulfilling prophecy. That is, if we are not able to see through them and counter them with an alternative self-image that aims at human sovereignty even under digital conditions.

This becomes clear when we turn to questions of morality. Even if we could perfectly implement ethics and morality in the algorithms of an Artificial Intelligence, this would lead to a completely immoral world. Because the machines would possibly be superior to us in a way that would be very unpleasant for us. Because they could do little with the inner contradictions that characterize humans, they would make short work of these contradictions and imperfections.

It is the same human contradiction and inconsistency that precludes a rigid application of moral laws that makes us creative and opens the door to an ever new, different and human future. There is a risk that we only become fully aware of this after Silicon Valley has closed that door.

---

[506] Nilsson, Nils J. The Quest for Artificial Intelligence. A History of Ideas and Achievements. Cambridge University Press, New York 2009.

5.1. A new natural law - without rights

The Californian ideology drafts a natural law without rights, only with duties.

Human dignity is linked to human's unique position, the idea of the image of God is reflected in inalienable human dignity. "God created man in his own image" - in the imitation of the divine the special dignity of each individual human begins at birth; each individual possesses inalienable human rights. The Californian ideology replaces the image of God with the likeness of God. Thus, the dignity of women/man is no longer seen in his individual uniqueness, which lies precisely in her/his peculiarity, and includes her/his imperfections. If women/man is equal to God, her/his dignity no longer lies in her/his specialness, but in his optimization, in her/ his perfection. The dignity of being and the latitude for uniqueness, has been replaced with the compulsion to conform to an all-encompassing ideal.

Human dignity and human rights are thus undermined by the AI, in which the AI makes more and more factual decisions and restricts personal freedom. When AI takes the place of God, the roles of creation of women/man and creator are reversed, with far-reaching consequences. With AI a creation of man becomes God-like, obtaining powers which are infinitely superior to those of women/man and can thus dominate her/him: *Deus est machina*. From the deification of the machine follows the religious instruction to submit to it. One can call this process the dialectic of digitalization. Developed as a self-empowerment tool, this tool ends up abolishing women/man, or at least, the self-image she/he had hitherto of herself/himself and her/his dignity.

If life is defined as data processing, then living is not more than the sum of data. Comparable to a chip running a certain program, there is no way to go from this reduced image of women/man to a legal entity endowed with inalienable rights. If AI as a whole can become humanoid, then a human-likeness of the technology becomes apparent. Therefore, it seems logical at first glance to discuss rights for robots.[507] The more "autonomous" they become, the more human-like these rights should be. But this is a fallacy, because the precondition of equality is not given.

To extend human rights to an artificial creation of man, which possesses a potentially superior resource of knowledge and power, is tantamount to the end of human rights for humans. For in natural human dignity of all people, the idea of equality is as firmly anchored as that of unmistakable individuality.[508] In the debate on "robot ethics", the demand is made that robots be granted rights because of their "autonomous" decision making capacity; after all, animals are no longer categorised as automatons or objects, but as living beings with rights.[509] Apart from the difference between a living being and a robot, which remains, even if the robot is "more intelligent" in the interest of better computer capacity : If a being enjoys these rights, which is genuinely superior to humans in questions of knowledge, authority and power, the rights of the then inferior humans are waste paper.

By ceding natural rights and dignity to an artificial entity, it threatens an end them for human beings at the same time.

---

[507] Hilldebrandt, Mirelle. Law for computer Scientists, forthcoming. https://lawforcomputerscientists.pubpub.org', and Loh, Janina. Roboterethik: Eine Einführung. Suhrkamp Verlag AG, 2019.

[508] Bloch, Ernst. Natural Law and Human Dignity. MIT Press, 1986.

[509] Loh, Janina. Roboterethik: Eine Einführung. Suhrkamp Verlag AG, 2019.

5.2. Where is the invention of the future?

The ability to imagine a different and better future has always been linked in philosophy to the ability to dream and to critique. Hannah Arendt saw in man's ability to make a new beginning again and again, the actual expression of his freedom. A freedom that is not only freedom from constraint, but includes "the freedom to be free".[510]

But the supposedly "superior" AI systems only use existing, old data, their forecasts are extrapolations of the past, which causes several problems:

- The quality of their conclusions and forecasts depends on the quantity and quality of the underlying data. Incorrect or incomplete data can lead to wrong conclusions. Biased recommendations can be derived from biased data; as the cases of discrimination against African-Americans by the use of algorithms in the US justice system or against women in algorithmic application procedures for jobs have demonstrated.[511]
- However, these mechanisms not only reproduce and reinforce prejudices and stereotypes. Decision-making algorithms also determine how the future is shaped. In more and more areas, they should help people to cope with increasing complexity and to come to better decisions, with which they then intervene deeply in people's lives. Where human inconsistency and empathy have so far led to mixed results, these systems will systematically reinforce prejudice and discrimination through "perfection" on a massive scale.

---

[510] Arendt, Hannah. 2018. The Freedom to Be Free: From Thinking Without a Banister. Knopf Doubleday Publishing Group.

[511] West, S.M., Whittaker, M. and Crawford, K. (2019). Discriminating Systems: Gender, Race and Power in AI. AI Now Institute. Retrieved from "https://ainowinstitute.org/publication/discriminating-systems-gender-race-and-power-in-ai-2

- The praise for AI, which is increasingly used in cases of large amounts of data and high complexity, has an inherent disadvantage: As a future scenario, it can only generate projections of data from the past. Even if these are largely free of distortions and provided with appropriate parameters: it is always the past that is projected and reproduced. Human creativity and, above all, human freedom, which creates something new from criticism and dreams using the power of imagination, is not in the foundational structures of even the most elaborate AI. If decisions in large bureaucracies were transferred from people to AI on a broad front, one would also have to ask how flexible such elaborately programmed systems could be in the event of a change of policy. As a consequence, the widespread use of AI, which is met with blind trust, can lead to a general standstill, which slows down and prevents further development of society if we do not take countermeasures and open up new paths for criticism and imagination.

Part D

How GAFAM Undermine Law and Democracy

From the very beginning, the relationship of the giants of the Internet to the law has been ambivalent, and this ambivalence continues to persist. Until a few years ago, the tech companies tried to evade any legal regulation. Only recently, experience with legislation on data protection, net neutrality, copyright and the enforcement of competition law in the European Union has nuanced the tactical relationship of the corporations to the law.

Today, instead of openly opposing every legal regulation in principle, a toolbox of arguments and procedures is used to either prevent regulations altogether or, if this is not successful, at least to delay them as long as possible in accordance with firms' interests and for content optimization. At the same time, the corporations are trying to opt for self-regulation, arguing that democratically legislated law is not necessary because the rules they themselves have set already (allegedly) act in public interest.

Whether in Washington, Brussels or Berlin - whenever legislative initiatives are announced in the US Congress, the European Parliament or the German Bundestag to regulate certain aspects of the business models made possible by the Internet or digital technology in the democratic process, GAFAM's lobbying machines are set in motion. Traditionally these have aim of either preventing the law altogether or at least making it soft enough avoid harm to business interests.

But GAFAM is not limited to targeted lobbying against specific draft laws. Rather, they deliberately breed a culture of enmity against the institution of democracy, indeed a contempt for democracy and its representatives (D1). In this way they prepare the way for populists. They only want to set the rules themselves and see themselves as guardians of the Holy Grail of global Internet and technology. They still see themselves in the tradition of John Perry Barlow: against governments and parliaments, even if they have recently been saying something different for tactical reasons and presenting themselves as friends of democracy. But we must measure them by their actions, namely their campaigns to prevent and soften laws (D2).

Even once a law is passed that binds them, which in America, unlike in Europe, almost never happens, GAFAM do not abide by it or at least undermines the legitimacy of the law wherever they can. (D3).

D1 How GAFAM promote a culture of contempt for democracy and law

Targeted lobbying against a particular law is preceded by the constant bombardment of democracy, the softening not only of one law but of the entire political system, the perpetual bad-mouthing and ridiculing of democratic processes and other methods of undermining the functioning of democracy in the US, Europe and Germany.[512]

The GAFAM companies want to make people forget that the general law passed by the parliamentary legislator is the noblest form of action in democracy. For democracy is expressed in a law passed after proper debate and by majority vote. We cannot deplore the crisis of democracy and at the same time malign laws in general.

The spread of neo-liberal ideology has led to a widespread instinctive equation of laws with over-regulation. But the requirement that a law be passed by the democratically legitimate parliamentary legislature is precisely a safeguard against the encroachment of governments and the regulatory frenzy of civil servants. In the American constitutional tradition, the sentence *No Taxation without representation* describes exactly this. In Germany, the principle of essentiality laid down by the Federal Constitutional Court has the same function.[513] Everything that is essential, either because it

---

[512] Khan, Nora, "Fred Turner: Silicon Valley Thinks Politics Doesn't Exist - 032c." 30 July 2019. https://032c.com/magazine/fred-turner-silicon-valley-thinks-politics-doesnt-exist Goldberg, Jonah. Suicide of the West: How the Rebirth of Tribalism, Populism, Nationalism, and Identity Politics Is Destroying American Democracy. New York: Crown Forum, 2018. Mounk, Yascha. The People vs. Democracy: Why Our Freedom Is in Danger and How to Save It. Cambridge, Massachusetts ; London, England: Harvard University Press, 2018. Mounk, Yascha, and Roberto Stefan Foa. "The End of the Democratic Century," August 19, 2019. https://www.foreignaffairs.com/articles/2018-04-16/end-democratic-century

[513] Scope of the doctrine of essentiality, elaboration of the Scientific Services of the German Bundestag, 2015, document WD 3 -3000 -043/15, https://www.bundestag.de/resource/blob/413272/8101846ab75a4f07c94c27de20962ed6/WD-3-043-15-pdf-data.pdf

interferes with the fundamental rights of citizens (such as taxation) or because it is of fundamental importance for the functioning of democracy and the community, must be regulated by a parliamentary law. It cannot simply be decided by the government.

This principle of essentiality is spreading worldwide in the current struggles for democracy. For example, the attempt by Prime Minister May and then Prime Minister Johnson in Britain to wrest the Brexit decision from Parliament and make it a matter for the government alone has led to two important judgments by the British Supreme Court, which made it clear that the important decisions about the Brexit cannot be taken by the government alone, but are subject to parliamentary approval.[514] In the EU, the jurisprudence of the European Court of Justice prevents the government of the Member States or the Commission from taking alone decisions that are actually subject to co-decision by the European Parliament, precisely because they are so important that they require the broader democratic legitimacy that direct election of the European Parliament confers.[515]

GAFAM, on the other hand, have a tradition of completely rejecting democratic decisions that would be binding on them. In his 1996 Declaration on the Independence of Cyberspace, the political manifesto of the American tech scene which still guides GAFAM,

---

[514] Verfassungsblog. "The High Court's Brexit Decision: A Lesson in Constitutional Law for the UK Government." https://verfassungsblog.de/the-high-courts-brexit-decision-a-lesson-in-constitutional-law-for-the-uk-government/.; http://www.bailii.org/uk/cases/UKSC/2017/5.html; https://www.supremecourt.uk/cases/docs/uksc-2019-0192-judgment.pdf

[515] Saurer, Johannes. "EU Agencies 2.0: The New Constitution of Supranational Administration beyond the EU Commission." Comparative Administrative Law, August 25, 2017. https://www.elgaronline.com/view/edcoll/9781784718657/9781784718657.00050.xml., European Parliament v. Council, para 64ff, 'https://eur-lex.europa.eu/legal-content/EN/TXT/?qid=1534186617433ri=CELEX:62010CJ0355' , C-293/12 and 594/12/ , ECLI:EU:C:2014:238, Digital Rights Ireland, para 54ff, 'http://curia.europa.eu/juris/documents.jsf?num=C-293/12'. For German language contributions see: Henning Rieckhoff, Der Vorbehalt des Gesetzes im Europarecht, 2007; Sonja Röder, Der Gesetzvorbehalt der Charta der Grundrechte der Union im Lichte einer europäischen Wesentlichkeitstheorie. 2007, doi.org/10.5771/9783845203478

the lyricist of the *Grateful Dead*, John Perry Barlow, poetically denied democratic legislators, parliaments and governments any legitimacy to regulate matters concerning the Internet.[516] Today, the *Electronic Frontier Foundation (EFF)*, an Internet NGO co-founded by Barlow, proudly presents this text. At the same time, it systematically conceals the fact that it was essentially funded by GAFAM and continues to benefit from them today.[517] From this self-empowerment of the masters of the Internet and digital technologies, mixed with neo-liberal and market-fundamentalist theories, the dogma of the unbroken global Internet developed. Only the self-regulation of the actors and the market are effective means to achieve goals that serve the common good, and therefore any intervention in the market and any regulation of the Internet by laws must be rejected, they claim. This concoction of technological absolutism and neoliberal ideas has led to the structural under-regulation of the Internet and has allowed GAFAM to create quasi-monopolies, according to the theory of Peter Thiel, one of the first investors in *Facebook* and the "Palantir" security data powerhouse derived from *Facebook* methods[518], which he co-founded.[519]

---

[516] Davos, Switzerland, 8 February 1996, https://www.eff.org/cyberspace-independence; see also Chapter C 1.2 below.

[517] The non-transparent financial report of the EFF does not show the proportion of total funding comes from contributions from GAFAM, its investors and employees. It results from an accumulation of income from foundation funds since the foundation of the EFF, individual donations from GAFAM employees subsidized by the companies, direct company contributions and contributions from foundations of GAFAM companies, 'https://www.eff.org/files/annual-report/2018/#FinancialsModal'; references to significant contributions, for example from Google, but also from other GAFAM, can be found at Levine Y, "Silicon Valley's Astroturf Privacy Shakedown," The Baffler. 3 July 2018. https://thebaffler.com/salvos/all-effd-up-levine; Vella, M. "Google and Facebook's New Tactic in the Tech Wars." Fortune. https://fortune.com/2012/07/30/google-and-facebooks-new-tactic-in-the-tech-wars/. https://en.wikipedia.org/wiki/Electronic_Frontier_Foundation#Financial; https://www.documentcloud.org/documents/4488896-CfA-the-Lobbyist-in-the-Garage-5-30-18; https://www.nytimes.com/2017/08/30/opinion/google-influence-think-tanks.html and https://musictechpolicy.files.wordpress.com/2010/09/google-shill-list-2.pdf and https://www.ipwatchdog.com/2018/06/04/report-engine-eff-shills-google-patent-reform/id=98007/.

[518] https://www.palantir.com/.

[519] Masters, Blake, and Peter Thiel. Zero to One: Notes on Start Ups, or How to Build the Future. 01 edition. London: Virgin Books, 2015, Greene, Tristan. "Study: Trump's Paid Peter Thiel's Palantir $1.5B so Far to Build ICE's Mass-Surveillance Network." The Next Web,

John Perry Barlow's declaration of the independence of cyberspace was completely blind to the concentration of power in the economy, which was already an issue in America when he wrote the manifesto. His sole concern was to keep state and democracy out of the Internet. This is even though lessons should have already been learned in the United States from the experience of monopoly in railway and telephone networks. The exploitative concentrations of power in network industries without regulation and the monopolist *AT & T* was smashed by a settlement with the US Department of Justice in 1984.[520]

The fundamental doubt about the suitability of the law to regulate technology continues to be eagerly cultivated in public, even outside Silicon Valley in the good tradition of neoliberalism and more recently for AI, despite the bad experience with the legal vacuum of the Internet.[521]

GAFAM's explicit and implicit assertion that parliamentarians and governments do not understand the Internet and new technologies such as AI, and therefore have no legitimacy to create rules for them, is not linked to a self-reflection on how little technologists understand democracy and the function of the rule of law, and the need to protect

---

August 12, 2019. https://thenextweb.com/artificial-intelligence/2019/08/12/study-trumps-paid-peter-thiels-palantir-1-5b-so-far-to-build-ices-mass-surveillance-network/. Peter Thiel on monopolies: "We always believe that capitalism and competition are almost the same thing. But the great capitalists of history have built unique companies. That was the case in the late 19th century with the Rockefellers, Carnegies and Mellons, and it's the same now in the Internet revolution. Google today is the best example of such unique monopoly companies. It is highly profitable and will continue to grow as far as you can see. The extreme counter-example is a restaurant that is in extreme competition with other restaurants, but is not particularly capitalistic because the operators make very little money from it". In Thiel, Peter. “Competition Is for Losers.” Wall Street Journal, September 12, 2014, sec. Life and Style. https://www.wsj.com/articles/peter-thiel-competition-is-for-losers-1410535536.

520 Goldman, David. “The U.S. Government’s Long History of Suing AT&T,” 24 November 2017. https://money.cnn.com/2017/11/24/news/companies/doj-vs-att/index.html.

521 Delcker, Janosch. “Europe’s AI ethics chief: No rules yet, please”, Politico, 30 November 2018. https://www.politico.eu/article/pekka-ala-pietila-artificial-intelligence-europe-shouldnt-rush-to-regulate-ai-says-top-ethics-adviser/

fundamental rights in a world where technology increasingly tends to undermine all three pillars of constitutional democracy.

The argumentation figures of technology corporations and activists against new law[522] show again and again that they still place technology before and above democracy. They do not recognize or do not want to recognize that their crusade against the law puts democracy in a pincer movement between technology on the one hand and populists and dictators on the other. They are delegitimising democracy and democratic law, putting authoritarian populists in the saddle all over the world, in the US, Britain, Hungary and Poland, and help them to grow in Germany, France and elsewhere.

It is no coincidence that authoritarian populism is sweeping the world parallel to the Tech-discourse of the California Ideology , the Internet with its communication model and digital technologies. The new automated *electronic public sphere*[523] is playing into the hands of populists, as is the new communication environment of the Internet. Populists can best communicate their ideology in short messages adapted to the new agora of political discourse, the mobile phone screen. Trump's rule by tweet is the best example of this.

Constructive political ideas and concepts that deal with the complexity of modern societies and strive for a democratic, inclusive debate

---

[522] See Big-Tech-Lobby campaign against new laws – for example in the article of Meyer, David. "Inside the ePrivacy Regulation's furious lobbying war, 31 October 2017, https://iapp.org/news/a/inside-the-eprivacy-regulations-furious-lobbying-war/; and Fang, Lee. "Google and Facebook are quietly fighting California Privacy Rights Initiative, TheIntercept.com, 26 June 2018, https://theintercept.com/2018/06/26/google-and-facebook-are-quietly-fighting-californias-privacy-rights-initiative-emails-reveal/; Brody, Ben. "Big Tech worried as California Law Signals US Privacy Push". Bloomberg, 29 June 2018, https://www.bloomberg.com/news/articles/2018-06-28/big-tech-worried-as-california-law-brings-privacy-push-to-u-s.

[523] Pasquale, Frank A. "The Automated Public Sphere." SSRN Scholarly Paper. Rochester, NY: Social Science Research Network, November 8, 2017. https://papers.ssrn.com/abstract=3067552.

cannot be explained in a tweet or only on a mobile phone screen. They need more text to read, and most importantly: they need direct debate between people, not single-source authoritative channels.

There are therefore three forces of the delegitimization of democracy that GAFAM are responsible for: First, the constant assertion that democracy is incapable of solving problems. Second, the establishment of a communication environment which, even from its basic structure, favours the undifferentiated messages of the populists. Third, the business model of the agitation economy, which prefers extremist messages and sends them viral, because agitation leads to more clicks and more clicks lead to more money.[524]

The accumulation of economic and technical power is joined by a new narrative, the ideology of total technical feasibility. The new thing about this ideology is the assertion that all political problems of this world, i.e., the great questions of freedom and dictatorship, war and peace, social justice, environmental destruction and climate change, and even the question of the conditions for the continued existence of mankind on earth, can only be answered by technology and the new Artificial Intelligence. This conviction is expressed in many ways, from the withdrawal of the US from the Paris Accord on Climate Change, the books of *Eric Schmidt*[525], the former Google boss, and the creation of the global problem-solving branch of *Google Jigsaw*[526] in New York, to the foundation activities of Internet capitalists like Bill Gates, who always aim to show that technology is the solution, or at least that they can solve the problems better than democratic states or the international community. Dictatorial and

---

[524] see below, Chapter E 5 ff. and Margetts, Helen. Political Turbulence: How Social Media Shapes Collective Action, 2015.

[525] Schmidt, J. Eric, Eric Schmidt, and Jared Cohen. The New Digital Age: Reshaping the Future of People, Nations and Business. Alfred A. Knopf, 2013.

[526] https://jigsaw.google.com/.

autocratic ambitions of this world, also within Europe and in the US, are strengthened by the new technology.[527]

However, Silicon Valley did not consider that the weakening of democracy leads to a weakening of the law and thus to a central functional condition of capitalism. That capitalism is ultimately dependent on the law, even if it only serves the purpose of striving for a level playing field for all market participants, now also seems to transpire to Mark Zuckerberg who has recently been calling for more legislation to regulate the Internet. Just before his latest Project, the Facebook currency *Libra,* he superficially seemed to be ready to subject to the democratic process, and thus to law and justice.

After the Libra project failed and the American election campaign found supporters for the demands of Democratic Senator Elizabeth Warren, who wanted to go head-on to break up *Facebook,* Mark Zuckerberg openly took up the fight against democracy and law again. He not only proclaimed this struggle to his staff.[528] *Facebook* also deleted ads by Warren explaining the need to break up *Facebook.*[529] The law is only supposed to be used as a tool for preserving wealth and the current structure of GAFAM, according to Zuckerberg's statement to staff.

---

[527] See also Fred Turner https://032c.com/magazine/fred-turner-silicon-valley-thinks-politics-doesnt-exist and https://harpers.org/archive/2019/01/machine-politics-facebook-political-polarization/

[528] Newton, Casey. "In Leaked Audio, Mark Zuckerberg Rallies Facebook against Critics and the US Government." The Verge, October 1, 2019. https://www.theverge.com/2019/10/1/20756701/mark-zuckerberg-facebook-leak-audio-ftc-antitrust-elizabeth-warren-tiktok-comments.

[529] Lima, Cristiano. "Facebook Backtracks after Removing Warren Ads Calling for Facebook Breakup." Politico. https://politi.co/2TKyxX1.

1.1. Voluntary rather than binding law, private rather than state control

The GAFAM want to determine for themselves the rules by which they work. This applies to the Internet, but also to all digital technologies based on the Internet, including AI. There are only two ways to do this: either you have governments and parliaments under control and thus ensure that they only issue rules that fit GAFAM, if at all. Or, even better, you can ensure that governments and parliaments accept that GAFAM sets the rules for themselves or, in any case, that they are only subjected to rules with which they voluntarily agree to. That would be the abdication of democracy because it would grant these companies a privilege that no citizen is granted in this way, namely that they only have to comply with rules that they have personally agreed to. Better still, these rules of self-government are often incomplete, because they do not contain any provision for sanctions in the event of non-compliance.

Moreover, there are hardly any practical examples of self-regulation that have really helped users and other interested parties and are working effectively to the benefit of others than those they are supposed to regulate in the first place.

1.2. How GAFAM keep the German state out through its digital business association "Bitkom"

The activists of technological neo-liberalism have repeatedly succeeded in incorporating self-regulation in the form of opening clauses in private certifications or codes of conduct in parliamentary legislation. Such clauses are often fought over for a long time in the European legislative process, especially by German ministries. And in the end, self-regulation usually does not come to anything because the companies do not reach agreement or because the planned self-regulation does not meet the legal requirements after all. Incidentally, the small and medium-sized enterprises that characterise the German economy are usually not involved in such self-regulation systems or their development. The advocates of self-regulation therefore play into the hands of the large companies in cases of doubt. Bitkom e. V., a German GAFAM lobby association, is an example of how a few large companies want to keep the state out by means of complicated structures of self-regulation.

Articles 40 to 42 GDPR provide for systems of voluntary codes of conduct and certification schemes for industry. To ensure that these systems do not become a loophole to circumvent the law, data protection authorities must carefully consider which codes and certification systems they recognise. However, they have not been able to do this at all due to a lack of qualified personnel. The hope that codes and certification systems would relieve the burden on data protection authorities will, however, only be fulfilled, without calling into question compliance with GDPR, if the codes on which these systems are based, and also their implementation on a regular basis, are strictly monitored for unquestionable compliance with GDPR. This requires qualified personnel.

All GAFAM are members of Bitkom, together with a large number of other international groups. A former head of *Microsoft* in Germany is Bitkom president.[530] Interestingly, the largest German Media Company Bertelsmann AG has left Bitkom. It obviously felt no longer represented by this association. It is not astonishing that Bitkom systematically not only promotes private rule-making but also

---

[530] Berg, Achim. " Ehemaliger Microsoft-Deutschlandchef wird neuer Bitkom-Präsident." 20 June 2017. https://www.horizont.net/tech/nachrichten/Achim-Berg-Ehemaliger-Microsoft-Deutschlandchef-wird-neuer-Bitkom-Praesident-158983.

private supervision. Together with other associations and interested companies, Bitkom has also already advocated self-regulation and privately financed supervision in the legislative process for the GDPR. Together with a few large companies in the industry, Bitkom has set up its own organization to promote self-regulation, which in turn includes Bitkom itself, *Microsoft* and *Google* as founding members: the registered non-profit association *Selbstregulierung Informationswirtschaft e. V.*[531] An e. V., a non-profit association, needs an idealistic purpose under German law for recognition that is not of an economic nature. It is quite astonishing that an association founded by a few large commercial enterprises, which is committed to ensuring that these companies can evade democratic law and are only subject to self-regulation, is recognised as a non-profit association. Is it an idealistic purpose worthy of recognition in a democratic society to exempt commercial enterprises from democratic laws and state supervision?

This registered association, in turn, has founded a subsidiary company, *Scope Europe bvba,* which is to act as an "independent" monitoring body for codes of conduct in accordance with Article 41 GDPR. In this way, the technology industry is attempting to take the rules and the application of the rules into its own hands, not only evading binding legislation but also public enforcement. For the fact that a structure set up by a few interested companies is supposed to function as an independent monitoring body shows how companies and their associations imagine independence: as the absence of statehood as in *failed states*. Let us hope that the data protection authorities will not let this pass. This example also shows how GAFAM imagines society: Not democratically set rules, but

---

[531] "Mitglieder: SRIW e.V." https://sriw.de/der-sriw/struktur/mitglieder/. and "Selbstregulierung Informationswirtschaft e.V. (SRIW) | LobbyFacts Database." https://www.lobbyfacts.eu/datacard/selbstregulierung-informationswirtschaft-ev?rid=618569919406-26

self-regulation by industry. And no direct state control of whether the rules are observed by the corporations, but rather, if at all, only a self-policing by institutions that are financed and controlled by the businesses themselves. The state organs are kept weak and with scarce resources, they in the view of the corporates should receive as few funds and personnel as possible, so that even they themselves wish to be relieved of the burden of control through handing over tasks to privately financed supervision bodies. This is a revolution in the tech industry from above, which undermines the principle of binding everyone to democratically established law. Without this enormous amount of concealment, everything could be much simpler and also more efficient: If only GAFAM would pay their share of taxes properly and accept that they are subject to democratic law like everyone else. With the proper taxes paid by GAFAM, the data protection authorities could also be better equipped, and GAFAM could then save itself concealment games and the expense of financing its own private supervision.

1.3. How the ethical discourse initiated by GAFAM is delaying the legal regulation of AI

For Artificial Intelligence, the American GAFAM started a discourse on ethics early on, in line with *Californian ideology*. They were so successful that both the European Commission and the German government initially engaged in this discourse, which was planned by the corporations as an alternative or at least a delay for new law. Both the EU and Germany set up ethics groups with representatives from the scientific community and interest groups, in Europe also from the GAFAM companies themselves.

Now that both advisory bodies have presented their recommendations, it is clear that Artificial Intelligence as a general-purpose technology, which is supposed to be as widespread as electricity one day, needs legal regulation. This was foreseeable from the outset, but the industry has won again more than a year without legal regulation. Even before the work of both bodies began, there were 15, some say 70, different sets of ethics for autonomous systems and AI developed by industry, science and civil society already.[532]

While the committees were still working, Google dissolved its own ethics committee[533].

1.4 The moral hazard for academia in dealing with GAFAM

Nevertheless, philosophers and ethicists are currently booming among tech companies. And they face a moral dilemma, a moral hazard: when they argue for legal regulation of AI, they give up their sovereignty over the issue, as governments and parliaments then take up their work, certainly supported by ethicists and philosophers, but also by many other experts from other disciplines. And when they plead for legislation, they lose friends in the tech industry who make life and work more pleasant, from invitations to parties and company visits to research funding and exclusive insights into the latest state of the art technology.

---

532 Jobin, Anna, Marcello Ienca, and Effy Vayena. "The Global Landscape of AI Ethics Guidelines." Nature Machine Intelligence 1, no. 9 (September 2019): 389–99. https://doi.org/10.1038/s42256-019-0088-2; Hagendorff, Thilo. "The Ethics of AI Ethics: An Evaluation of Guidelines." Minds and Machines 30, no. 1 (March 1, 2020): 99–120. https://doi.org/10.1007/s11023-020-09517-8; "Overview of ethics codes and principles for AI | Rathenau Instituut." https://www.rathenau.nl/en/digital-society/overview-ethics-codes-and-principles-ai.

533 Levin, Sam. "Google Scraps AI Ethics Council after Backlash: 'Back to the Drawing Board.'" The Guardian, April 5, 2019, sec. Technology. https://www.theguardian.com/technology/2019/apr/04/google-ai-ethics-council-backlash.

The Technical University of Munich recently received a $7.5 million grant from *Facebook* for the AI Ethics Chair,[534] and in Berlin *Google* is funding the Humboldt Institute for Internet (HIIG) and Society.[535] One week after the first meeting of the German government's *Data Ethics Commission,* which was charged with proposing rules for AI, the *Humboldt Institute* in Berlin had nothing better to do than invite to an event on AI with *Google's* chief lobbyist, Nicklas Lundblad, without making its guests aware in the invitation that not only the guest is coming from Google, but also money for the hosts.[536] Google critic Shoshana Zuboff also spoke at the Institute, but only after her criticism was taken up by Springer publishing house and she was invited to the award of the Axel Springer Prize with a laudatory speech by Ursula Von der Leyen, today President of the EU Commission.[537]

Also Ms. Zuboff was not informed in the invitation that *Google* was co-financing her stage at the *Humboldt Institute* for Internet and Society. Even during a speech at the German Jurists Association in Freiburg in the presence of Members of the Federal Constitutional Court and other highest Courts , in which the thesis was put forward that the law could no longer regulate the Internet, the lecturing director of the Institute did not consider it appropriate to point out to the judges present that the Institute was financed by *Google,* although she put forward theses that could not have been better formulated

---

[534] Kreye, Andrian. "Facebook an der Uni - Der falsche Gönner." Süddeutsche.de. https://www.sueddeutsche.de/bildung/facebook-tu-muenchen-ethik-1.4630044; Köver, Chris. "Warum Facebook ein Institut für Ethik in München finanziert." netzpolitik.org (blog), January 21, 2019. https://netzpolitik.org/2019/warum-facebook-ein-institut-fuer-ethik-in-muenchen-finanziert/.

[535] "Google: Gute Beziehungen - Brand Eins Online." https://www.brandeins.de/magazine/brand-eins-wirtschaftsmagazin/2018/geduld/google-gute-beziehungen

[536] https://www.hiig.de/en/events/lunch-talk-a-fast-moving-technology-like-ai/.

[537] Axel Springer Award, "'With Courage and a Clear Attitude': Shoshana Zuboff Receives the 2019 Axel Springer Award – Axel Springer Award." http://axel-springer-award.com/2019/11/07/with-courage-and-a-clear-attitude-shoshana-zuboff-receives-the-2019-axel-springer-award/.

by *Google's* lobbyists. It is not right that this institute should still be allowed to bear the name of the famous explorer Alexander von Humboldt. It should rename itself to *Google* Institute.

In order to advance the ethics discourse and keep it as far away from politics and democratic rule making, the companies themselves have also founded various research institutes and think tanks or platforms. Here, they maintain a dialogue with civil society and science, apparently in the public interest. It is interesting to note, however, that practically nothing positive is ever said from these institutions about democracy or legislation. The topic is systematically kept small, instead they cultivate technical solutions or the *ethics community*, seemingly harmless, but in fact to forge coalitions and alternatives in the system competition with democratic and legal solutions. By the way, GAFAM makes those who could also be involved in the democratic process familiar with the advantages of alternatives to legislation, i.e., self-regulation and technical solutions, at an early stage and in a pleasant atmosphere. And anyone who has ever seen the better nicer offices and equipment of the Google-funded *Humboldt Institute* in Berlin, for example in comparison to the state-funded *Weizenbaum Institute*, can imagine: The scarce equipment of state research can also drive good people into the arms of GAFAM.

One of the think tanks funded by GAFAM and other industry is the *Partnership on AI*. It was co-founded by Mustafa Suleyman of *DeepMind*, the British AI developer bought by *Google*.[538] The *Partnership on AI* systematically avoided including political and governmental actors in the committee structure. GAFAM, fully represented on the Board of Directors,[539] wanted to keep the scientific

[538] https://www.partnershiponai.org/about/; https://deepmind.com/blog/announcements/announcing-partnership-ai-benefit-people-society.

[539] https://www.partnershiponai.org/team/.

community and civil society among themselves so that democratic regulatory options would not even have to be included in the discourse. This is the practical implementation of the principle that also underlies another Alphabet/Google subsidiary, namely *Jigsaw*[540] in New York: We solve the world's problems through technology, not through democracy. Legal rules adopted through democracy, which are binding and enforceable against Big Tech and subject to sanctions are rejected systematically. The dogma in these circles of idealism is far from democratic - if we can even assume that all actors are idealistic. The idealism of a Mustafa Suleyman, still present at *DeepMind* in his early days, was probably already too much for the new owners and supervisors of *Google*, who wanted to buy not only his company, but probably also his idealism.[540a]

1.5 GAFAM and the NSA form a Private Public Partnership for surveillance

Shoshana Zuboff impressively describes the emergence of a culture of lies at *Google*. It continues to exist to this day. The founders of *Google* realized that they could collect much more personal data through their search engine than was necessary to operate and develop the search engine. They decided to systematically expand this business of mass collection of personal data and earn more money than they had ever dreamed of. It was clear to them from the outset that they were ripping off users. They were penetrating deep into people's lives and soon created profiles of each individual user, fed not only by the content of search queries but also by other behaviour on the Internet that they were now observing. They knew that if people

---

[540] https://jigsaw.google.com/

[540a] https://www.cnbc.com/2022/01/28/mustafa-suleyman-deepmind-co-founder-quits-google-ai-role-to-be-vc.html

knew how *Google* was observing them and exploiting their trust in the company, their business would suffer and perhaps be regulated by law. Because people in America also value their privacy. That's why *Google* kept its business model secret for so long. This created a culture of lying and secrecy that is still practiced today and is now continuing in the development of AI. *Facebook* followed a similar path, and as Shoshana Zuboff describes, the ideas of collecting and commercially exploiting personal data were brought from *Google* to *Facebook* by a few exchanges in personnel. Facebook adapted exploiting personal data for the social network.

The mixture of huge commercial success, tech idealism to the outside, and an internal culture of lies and secrecy made the tech companies vulnerable to patriotic demands by American intelligence agencies such as the NSA after the attacks of 9 September 2001. The lie of respect for users and their data did not have to be reinvented. It already existed before 9/11 and it was easy to integrate the fact that data now also flowed to the intelligence services, in particular the NSA, into this culture of lying.

As the world learned only late through Edward Snowden,[541] companies of the technological-economic complex collaborated in different ways with the NSA and other US Intelligence Agencies by secretly sharing user data and content with US services. In some cases, the Intelligence Services, such as the CIA, were directly involved in the companies through the venture capital firm In-Q-Tel[542] and were even among the initial investors.

---

[541] Szoldra, Paul. "This Is Everything Edward Snowden Revealed in One Year of Unprecedented Top-Secret Leaks." Business Insider. https://www.businessinsider.com/snowden-leaks-timeline-2016-9..

[542] Paletta, Damian. "The CIA's Venture-Capital Firm, Like Its Sponsor, Operates in the Shadows." Wall Street Journal, August 30, 2016, sec. Tech. https://www.wsj.com/articles/the-cias-venture-capital-firm-like-its-sponsor-operates-in-the-shadows-1472587352. ; www.IQT.ORG; https://news.crunchbase.com/news/heres-20-q-tel-investments-said-taking-cias-money/; https://www.crunchbase.com/organization/in-q-tel/investments/investments_list#section-investments

Especially for the NSA, the high market share of *Google, Facebook* and US Cloud providers globally were and are a fantastic starting point for the broadest collection of data about individuals, groups, companies, their innovations and businesses models, the world has ever seen. Google has a search engine market share in Europe above 90%, and so have US Cloud services providers. If financial flows are now added through new GAFAM payment systems, these will be additional valuable sources for US Intelligence.

Shoshana Zuboff describes GAFAM's enormous, insatiable hunger for more and more personal data, as ever finer personality profiles enable ever greater commercial value to be created, particularly in the market for behavioural prediction. The US Intelligence Services as well as any Intelligence Service in the world are also starving for personal data and behavioral predictions. It is therefore not surprising that this parallel interest has led to comprehensive cooperation between GAFAM and the US Intelligence Services, from the development of collection to the exploitation of data.[543]

The job market for data and analytics also caused many programmers and data scientists to oscillate between private and public employers. This is how the public-private partnership of the Tech surveillance complex in America, which comprises about 107,000 employees in the 16 different Intelligence Services in the US, developed: partly directly in the public sector, partly with private service providers working for the state, such as Edward Snowden, who was also employed by such a private service provider to the NSA. The total

[543] Greenwald, Glenn, and Ewen MacAskill. "NSA Prism Program Taps in to User Data of Apple, Google and Others." The Guardian, June 7, 2013, sec. US news. https://www.theguardian.com/world/2013/jun/06/us-tech-giants-nsa-data; Nesbit, Jeff. "Google's True Origin Partly Lies in CIA and NSA Research Grants for Mass Surveillance." Quartz. https://qz.com/1145669/googles-true-origin-partly-lies-in-cia-and-nsa-research-grants-for-mass-surveillance/. Bernau, Patrick. "Spähprogramm „Muscular": So hat die NSA die Google- und Yahoo-Nutzer ausspioniert." FAZ.NET. https://www.faz.net/1.2641705; HuffPost UK. "Emails Show Google Cooperated With The NSA For Years," May 6, 2014. https://www.huffpost.com/entry/nsa-google_n_5273437.

budget of this complex is estimated at between 500 billion and one trillion dollars annually.[544]

Bypassing or deliberately breaking the law, telling legislators half the truth or mocking them, as we have recently seen in the hearings of Mark Zuckerberg on *Facebook* and the *Cambridge Analytica* scandal, became a sport on both sides of the Atlantic. Digital companies, digital activists and digital engineers and programmers patting each other on the back, feeling like the kings of the world. We can see from the haughty attitude of people like Mark Zuckerberg towards the US Congress and the parliaments of the whole world: Snowden's *wake-up call* on crossing the boundaries of individual's natural rights had no effect in this respect. What Snowden uncovered, and this is the real scandal, was legal under US law. No one was held responsible for the mass collection of personal data that he revealed. In fact, nothing has changed in reality, only cosmetic changes have been made to the law in the US. In fact, we can never be sure what really happens to our data in the United States: because of the instruments that American law gives to the State, in particular the secret jurisdiction and the criminalised obligation of confidentiality that the State can impose on companies when tapping data, the so-called Gag orders contained in National Security letters. And because Europe has no technical means of control and inspection of what happens to Data of Europeans transferred to the US by GAFAM.

The *EU-US Privacy Shield* was a step towards similar data protection rules in the EU and the US when processing data from Europe in the

---

[544] Shane, Scott. "New Leaked Document Outlines U.S. Spending on Intelligence Agencies." The New York Times, August 29, 2013, sec. U.S. https://www.nytimes.com/2013/08/30/us/politics/leaked-document-outlines-us-spending-on-intelligence.html; Defense One. "The US Intelligence Community Is Bigger Than Ever, But Is It Worth the Cost?" https://www.defenseone.com/ideas/2015/02/us-intelligence-community-bigger-ever-it-worth-it/104799/.

US.[545] A book of its own could be written about the lobbying efforts to soften this set of rules. There are lively discussions about European data sovereignty,[546] there is a desire for more EU cloud capacity and there is the German project of a European cloud called Gaia-X[547]. They all show that, in spite of the EU-US Privacy Shield, no lasting trust in data protection in the US has been established. In the US, the lack of protection of personal data and privacy remains a major issue, to which the New York Times has now devoted a permanent reporting series.[548] As early as the year 2000,[549] US citizens had little confidence in data protection and respect for privacy in the US. Since then, confidence in data protection in the US has fallen even further.[550]

1.6. No more election campaigns without Facebook and Google

While Twitter does without political ads, *Facebook* and *Google* are itching for political money: the total campaign spending for digital advertising in the US presidential election campaign 2020 is expected

---

[545] ‹https://ec.europa.eu/info/law/law-topic/data-protection/international-dimension-data-protection/eu-us-data-transfers_en.

[546] 'https://techcrunch.com/tag/privacy-shield/.

[547] Bedingfield, Will. "Europe Has a Plan to Break Google and Amazon's Cloud Dominance." Wired UK, January 27, 2020. https://www.wired.co.uk/article/europe-gaia-x-cloud-amazon-google.

[548] "N.S.A. Bill's Reforms Could Create More Privacy Concerns." Savage, Charlie. "N.S.A. Triples Collection of Data From U.S. Phone Companies." The New York Times, May 4, 2018, sec. U.S. https://www.nytimes.com/2018/05/04/us/politics/nsa-surveillance-2017-annual-report.html; Warzel, Charlie. "Opinion | 4 Things That Keep the N.S.A. Up at Night." The New York Times, September 10, 2019, sec. Opinion. https://www.nytimes.com/2019/09/10/opinion/nsa-privacy-gerstell.html.

[549] Lohr, Steve. "Compressed Data; Survey Shows Few Trust Promises on Online Privacy." The New York Times, April 17, 2000, sec. Business. https://www.nytimes.com/2000/04/17/business/compressed-data-survey-shows-few-trust-promises-on-online-privacy.html.

[550] NW, 1615 L. St, Suite 800, Washington, and DC 20036US202-419-4300 | Main202-857-8562 Fax202-419-4372 | Media Inquiries.

to reach 2.9 billion dollars.[551] In August 2019, *Facebook* raised its prices enormously, from $9 to $279 per email address generated, but only for the Democrats. Because on the Democratic side many candidates competed with each other in the primary. That drove up prices. For Republicans they remained cheap, with only one serious candidate, the incumbent. So, *Facebook's* ad auction system punished the party that gives its voters a real election choice and favours Trump.[552] For money-making, Democracy is good enough for Mark Zuckerberg, yes, he put in place the price system of *Facebook* ad auctions that is downright perverse for the democratic process. At the same time, he refuses to reject political ads that contain lies. This is what happens when unregulated capitalism meets democracy.

Not only in Germany, but also in Belgium and other EU countries fascists and populists are the biggest buyers of campaign ads on *Facebook* and *Google*.[553] *Facebook* and its employees earned money from the ads because they earned commissions for selling ads. These were abolished by Facebook in the US only in May 2019. Whether commissions for ads were also abolished in Europe remains unclear.[554]

The recently introduced dynamic archives of *Google* and *Facebook* for campaign ads are extremely confusing. They neither transparently consolidate the ad expenditures of the same party according to

---

551 Seetharaman, Deepa. "How a Facebook Employee Helped Trump Win—But Switched Sides for 2020." Wall Street Journal, November 24, 2019, sec. Tech. https://www.wsj.com/articles/how-facebooks-embed-in-the-trump-campaign-helped-the-president-win-11574521712.

552 Haggin, Emily Glazer and Patience. "Facebook Ad Prices Surge Due to Barrage by Democratic Hopefuls." Wall Street Journal, August 29, 2019, sec. Business. https://www.wsj.com/articles/facebook-ad-prices-surge-due-to-barrage-by-democratic-hopefuls-11566984601.

553 Luxen, Martina. "Wahlkampf: Massive Ausgaben für Online-Kampagnen." vrtnws.be, 17 May 2019. https://www.vrt.be/vrtnws/de/2019/05/17/wahlkampf-massive-ausgaben-fuer-online-kampagnen/.

554 Horwitz, Emily Glazer and Jeff. "Facebook Curbs Incentives to Sell Political Ads Ahead of 2020 Election." Wall Street Journal, May 23, 2019, sec. Tech. https://www.wsj.com/articles/facebook-ends-commissions-for-political-ad-sales-11558603803.

uniform rules, nor do they allow parties and candidates who are actually facing each other in an election to be easily compared in terms of ad expenditures, content and targeting of their ads.

Facebook and *Google* do not understand the democratic processes and create a false transparency that does little to help the public get orientation in the actual election campaign. The archive formats as they now exist can only be a start, even if one or the other company would do better. Elections are so essential for the functioning of democracy that there is no way around the legal regulation of the election campaign and the transparency and comparability of online advertisements for elections.

Even though GAFAM now generally claim that they are in favour of legal regulations, they always contradict them when governments launch initiatives in this direction. In the US Congress, they have still prevented any regulation that would create more transparency in election advertising. Google is even boycotting campaign advertising in some of the US states that have passed legislation, arguing that because of transparency, there are no profits to be made from campaign ads. Nevertheless, it is already becoming apparent that more and more US states are passing laws on transparency of campaign ads.[555]

---

[555] Noti, Adav. "Where the Patchwork of Digital Regulation Could Lead." 10 June 2019. https://www.campaignsandelections.com/campaign-insider/where-the-patchwork-of-digital-regulation-could-lead. and Born, Kelly. "How States Are Experimenting with Digital Political Advertising Regulation: Interview with Campaign Legal Center's Erin Chlopak." Hewlett Foundation (blog), May 28, 2019. https://hewlett.org/how-states-are-experimenting-with-digital-political-advertising-regulation-interview-with-campaign-legal-centers-erin-chlopak/.

See also tools (in German) for journalists: 'https://www.journalisten-tools.de/recherchieren/anzeigen-archive-von-facebook-und-twitter-als-daten-quelle/'; 'https://netzpolitik.org/wp-upload/136204668-Facebook-Fuer-Politiker-Download.pdf'.

How to use Facebook in the election campaign - Politics and communication 'https://www.politik-kommunikation.de/ressorts/artikel/so-nutzt-man-facebook-fuer-politische-kommunikation-503365169';

https://www.politico.com/story/2017/10/26/facebook-google-twitter-trump-244191;

https://campaignforaccountability.org/work/partisan-programming-how-facebook-and-googles-campaign-embeds-benefit-their-bottom-lines/;

https://www.bloomberg.com/news/articles/2018-09-20/facebook-to-give-less-direct-support-to-trump-in-2020-campaign

1.7. GAFAM want to set the rules themselves

The technical-economic complex has not only adopted the neo-liberal view of law as a cost factor and a restriction hostile to innovation. It has also developed its own theory of the undesirability and impossibility of regulating the Internet and digital technologies by law, which follows the *Californian ideology*. Recently, the attack on institutions took the form of the Facebook Libra dream, a currency without central banks and without monetary law.

Blockchain is the technical incarnation of the neoliberal. It claims to establish a system of *trustless technology*, which one should trust without people being involved. But the tech companies and their activists do not manage to establish a direct path between individual and information without any rules, even through Blockchain. They create a new path with new rules that they set. Because *Facebook*, *Google* and others are new institutions, with their own community rules, controlled by their own interests. They want to put an *International Law of Google*,[556] as a seminar at New York University was called, in place of the rules of democracy. *The Internet and Jurisdiction* think tank in Paris,[557] which is run with the financial support of the Internet giants and a few states, also serves this goal, as do the *Global Network Initiative (GNI)*[558] and the *Internet Governance Forum (IGF)*, which was set up by the United Nations in 2006, but which has never actually produced any regulations that set any limits to the Internet giants or economic interests on the Internet. Politics itself is reinforcing this trend: in 2019, the *IGF, the Internet Governance Forum* of the United Nations in Berlin, was co-financed

---

[556] NYU Law, "Institute for International Law and Justice Colloquium: International Law of Google - Writing Credit | NYU School of Law." https://its.law.nyu.edu/courses/description.cfm?id=21774

[557] https://www.internetjurisdiction.net

[558] https://globalnetworkinitiative.org/

by the German Federal Ministry of Economics and Technology. It has evolved from a civil society platform into a mixed civil society - business forum and serves primarily to avoid binding rules.[559]

But with what legitimacy does GAFAM deny itself law and democracy? They have no democratic legitimacy. Do they achieve better results because of this? Some believe so. For example, by saying that *Google Search* was not invented by democracy and would never have been invented by democracy, so *Google* is better than democracy. But that is a misconception, because there is a division of labour between business, science, the press, the arts and society on the one hand and politics, government and parliament and the judiciary on the other. The separation of these areas, with rules on responsibility, but also certain independence from each other, with rules on *checks and balances* and separation of powers: they all have a purpose and are almost constitutive of a free society. Only dictatorships attempt to bring all these areas into line and to centrally control the entire system in its function. The simultaneous presence and influence of the GAFAM companies with their money in the economy, politics, science, journalism and culture is a problem, even without the system control ambitions they have with regard to the new technology of the global Internet, AI and quantum computers. But with this technology and their control ambitions, they become a real dictatorial danger.

In democracy, nothing happens quickly; every political dispute requires perseverance and lasting solidarity. The classic of the modern company speech from Silicon Valley is the Mark Zuckerberg-Motto *Let's move fast and break things*, i.e., the request to technicians and business developers to move forward ruthlessly fast in order to reduce the *time to market*. On the other hand, they try to slow down the political system by saying:

---

[559] https://www.intgovforum.org/en/content/igf-2019

Yes, let's have a debate! And then they instigate debate forever without any consequence, the ethics of AI debate being the most recent example. At the same time, GAFAM itself create facts by its own technical actions. And then, at the next opportunity, they criticise again the slowness of the democratic process and the democratic state. This is how democracy is destroyed by GAFAM.

Recently, some people from Silicon Valley suddenly demand laws, including Mark Zuckerberg and representatives of *Microsoft*. They can do so with the certainty that the US Congress is clearly no longer capable of passing laws in the public interest. Either nothing will happen, or the corporations will have such a comprehensive influence on the legislation that their exploitative and liberty-robbing business models will not have to be changed in any way, or their very existence will be threatened.

The power the corporations have gained in American politics has not only been described many times in the nascent American literature on the downfall of democracy,[560] but is also shown very concretely in the example of the privacy legislation in California and Washington State, which has been beaten to a pulp by *Google* and *Facebook* in California and *Microsoft* in Washington.[561]

---

[560] Levitsky, Steven, and Daniel Ziblatt. How Democracies Die: What History Reveals About Our Future. UK US Canada Ireland Australia India New Zealand South Africa: Viking, 2018. Goldberg, Jonah. Suicide of the West: How the Rebirth of Tribalism, Populism, Nationalism, and Identity Politics Is Destroying American Democracy. New York: Crown Forum, 2018. Mounk, Yascha. The People vs. Democracy: Why Our Freedom Is in Danger and How to Save It. Cambridge, Massachusetts ; London, England: Harvard University Press, 2018. Mounk, Yascha, and Roberto Stefan Foa. "The End of the Democratic Century," August 19, 2019. https://www.foreignaffairs.com/articles/2018-04-16/end-democratic-century

[561] Fang, Lee. "Google and Facebook Are Quietly Fighting California's Privacy Rights Initiative, Emails Reveal." The Intercept (blog), June 26, 2018. https://theintercept.com/2018/06/26/google-and-facebook-are-quietly-fighting-californias-privacy-rights-initiative-emails-reveal/. Brody, Ben. "Big Tech Worried as California Law Signals U.S. Privacy Push." Bloomberg.Com, June 28, 2018. https://www.bloomberg.com/news/articles/2018-06-28/big-tech-worried-as-california-law-brings-privacy-push-to-u-s.

1.8. GAFAM want their own courts

The privatization of the judiciary, which large companies have long been operating through private dispute resolution bodies, reaches an absurd climax with the establishment of a *Facebook Supreme Court*. Here, a company first of all presumes to be able to do business in many countries around the world without wanting to comply with the respective national law. It took a long time until *Facebook* accepted that Europe would not consider its self-invented *community rules* adequate, but that it must abide by the law of the land.

No sooner had *Facebook* swallowed that than it makes another attack on the rule of law and democracy by pretending to be able to make decisions about what to take off the net in a centralized, global body. *Facebook* is serious about its internal justice project: the *Supreme Court* will be given USD 130 million annually, or approximately EUR 116 million, for the first six years.[562] By comparison, the annual budget of the German Federal Constitutional Court is just under 35 million euros a year.[563]

Facebook acknowledges that the decision-making body of five judges should include one from the region affected by the case in question. But *Facebook* apparently does not understand that the mere establishment of such a body reveals its completely misguided ambitions as a world state with its own jurisdiction. With this body, *Facebook is going* against the basic principle that in free, democratic states the independent judiciary decides according to democratically set rules, also when it comes to questions of content control on

---

[562] Horwitz, Jeff. "Facebook Pledges $130 Million to Fund 'Supreme Court' for Content." Wall Street Journal, December 12, 2019, sec. Tech. https://www.wsj.com/articles/facebook-pledges-130-million-to-fund-supreme-court-for-content-11576166992.

[563] https://www.bundesverfassungsgericht.de/DE/Das-Gericht/Gericht-und-Verfassungsorgan/gericht-und-verfassungsorgan_node.html#:~:text=Das%20Bundesverfassungsgericht%20untersteht%20als%20Verfassungsorgan,Jahr%202022%20und%2040%20Mio.

the Internet. That is a good thing and must remain so. Because recognising a worldwide *Facebook court* as legitimate is a step towards relinquishing state sovereignty in a core area, namely jurisdiction. Facebook and all GAFAM must recognize that, given their size and importance in the marketplace as well as for the formation of opinion in a free society, they have a special responsibility and must submit to special rules. These rules, namely for shaping democratic public opinion, are set in a democracy by parliaments and applied by independent judges. There is no place for a GAFAM World Court on *Facebook* or elsewhere, quite apart from the fact that the *Facebook Supreme Court* faces difficult times for many other reasons.[564]

1.9. Two schools of self-regulation in America

We have already described the connection between the self-regulation and neoliberalism. This is the cynical school of self-regulation, the cynical school of ethical discourse, which collects arguments with which one can fight against democratic and binding legal regulations: Look, we give ourselves good rules, we do not need a law.

But there is also a school of self-regulation and ethics that call for codes of conduct and technological solutions out of despair over American democracy. If the legislature in Washington is not able to adopt the necessary laws, we must give them to ourselves, they argue. This desperate school of self-regulation usually also makes it clear that ethics and self-regulation can and should never take the place of democratic law.

---

564 Verfassungsblog. "Difficult Times Ahead for the Facebook, Supreme Court". https://verfassungsblog.de/difficult-times-ahead-for-the-facebook-supreme-court/. See also with correct criticism, but what we consider the wrong proposals for the expansion of the competence of the Facebook Oversight Board: Ghosh, Dipayan. "Facebook's Oversight Board Is Not Enough." Harvard Business Review, October 16, 2019. https://hbr.org/2019/10/facebooks-oversight-board-is-not-enough.

Unfortunately, however, a melange from both schools of thought spill over the Atlantic to us in Europe and Germany and is enthusiastically taken up here by neo-liberals and unfortunately also by some useful idiots who do not realize that they are being harnessed to advance the interests of GAFAM.

But democracy still works here, as we can see from the fact that laws are still being passed that affect the Internet. Just think of the "Act to Improve the Enforcement of the Law in Social Networks", short Network Enforcement Act (Netzwerkdurchsetzungsgesetz, short NetzDG)[565] in Germany, the GDPR, the Copyright Directive and the Directive on Network Security in the EU. Of course, these laws are not perfect; they must and may be criticised. Perfect laws do not exist in democracy, if only because of the compromise nature of laws, which results from the need to win a majority. There were almost 4000 amendments to the GDPR in the European Parliament, which is a good thing, as it signals democratic debate that is actually taking place.

On the other hand, in the US Congress, practically no law regulating the Internet has been passed since Obama. It may therefore be right and proper in the US to call for self-regulation at home. For Europe and Germany, however, we have to be careful that the frivolous discourse on self-regulation and ethics, combined with the sublime or explicit assertion that democracy is too slow or incompetent, does not damage democracy in the same way as happened in the US long before Trump's election to the presidency and continues to the present day.

---

[565] The full official English title is "Act to Improve Enforcement of the Law in Social Networks", short Network Enforcement Act, see official translation at https://www.bmjv.de/SharedDocs/Gesetzgebungsverfahren/Dokumente/NetzDG_engl.pdf?__blob=publicationFile&v=2; Official Text in German of the„ Gesetz zur Verbesserung der Rechtsdurchsetzung in sozialen Netzwerken, short Netzwerkdurchsetzungsgesetz or NetzDG" at https://www.gesetze-im-internet.de/netzdg/BJNR335210017.html;

Let us remember one thing at the end: the absence of law and self-made rules, thus self-regulation, only help the powerful. Anyone who strives for a free society with solidarity, equal opportunities and democracy must not bet on it.

D2 Preventing laws or diluting their impact wherever possible

In Washington, Brussels, Berlin, Paris, not a week goes by without GAFAM parties, events by universities or think tanks financed by GAFAM. Critics are also invited, especially the most extreme, who are not politically effective. GAFAM pretend that they are tolerant and always open for a conversation and a good discussion. Let's have a debate is the battle cry of the clever lobbyist, because the longer and more complex the debate becomes, the less likely it is that an oh-so-annoying law will be passed by the Bundestag or the European Parliament. And the work of GAFAM to avoid, dilute or make ineffective parliamentary laws is moving on, systematically, and with the help of their party crowd.

2.1. The fight against the Network Enforcement Act (NetzDG) in Germany

The German Network Enforcement Act (NetzDG)[566] was preceded by clear announcements from the German Federal Government. In October 2015, Federal Minister of Justice Heiko Maas and GAFAM set up a Task Force *Together Against Hatred in the Net* to address the challenge in partnership with GAFAM. At EU level, GAFAM had signed a code of conduct on 31 May 2016 to combat hatred and incitement to violence on the Internet.[567] But progress

[566] The full official english title is "Act to Improve Enforcement of the Law in Social Networks", short Network Enforcement Act, see official translation at https://www.bmjv.de/SharedDocs/Gesetzgebungsverfahren/Dokumente/NetzDG_engl.pdf?__blob=publicationFile&v=2; Official Text in German of the „Gesetz zur Verbesserung der Rechtsdurchsetzung in sozialen Netzwerken, short Netzwerkdurchsetzungsgesetz or NetzDG" at https://www.gesetze-im-internet.de/netzdg/BJNR335210017.html;

[567] European Commission - European Commission. "European Commission and IT Companies Announce Code of Conduct on Illegal Online Hate Speech." Text. https://ec.europa.eu/commission/presscorner/detail/en/IP_16_1937.

was insufficient. At the end of 2016, Gerd Billen, State Secretary in the Federal Ministry of Justice and Deputy to the Minister, said as regards *Facebook* only additional political pressure would help.[568]

On 17 March 2017, Maas published the draft NetzDG law in a press conference. It was probably no coincidence that already on 27 March 2017, Dr. Thomas Helck, a former public prosecutor, just promoted to head of the German criminal law department at the international Law Firm *Freshfields*[569], the highest-turnover law firm in Germany[570], published an article against the NetzDG under the heading *Großer Löscheinsatz geplant* (Large-scale deletion operation planned on the Internet), where he compiled criticism of the law.[571] Today, the same law firm is also appearing for Facebook in proceedings against the Federal Office of Justice, which is tasked to ensure compliance with the NetzDG.[572]

On 27 June 2017, a few days before the German Bundestag (Parliament) vote on the NetzDG Law, *Facebook* published a paper by former Liberal Member of the British Parliament, Richard Allan, then Facebook's chief European lobbyist.[573] In it, Richard Allan explained how constructively and transparently Facebook was already dealing with the issue of incitement to hatred and violence

---

[568] Pimpl, Roland."Justiz-Staatssekretär Gerd Billen: 'Gegenüber Facebook hilft nur Druck.'" 17 November 2016. https://www.horizont.net/medien/nachrichten/Justiz-Staatssekretaer-Gerd-Billen-Gegenueber-Facebook-hilft-nur-Druck-144199.

[569] White & Case. "Dr. Thomas Helck Joins White & Case as a Partner in Frankfurt | White & Case LLP," https://www.whitecase.com/news/press-release/dr-thomas-helck-joins-white-case-partner-frankfurt.

[570] https://de.statista.com/statistik/daten/studie/172574/umfrage/anwaltskanzleien-in-deutschland-nach-umsatz/.

[571]https://www.politik-kommunikation.de/gesetz-des-monats/grosser-loescheinsatz-im-internet-geplant-388585723.

[572]https://www.juve.de/nachrichten/verfahren/2019/07/erstes-bussgeld-zum-netzdg-freshfields-vertritt-facebook-gegenueber-den-justizbehoerden.

[573] His former party leader, Nick Clegg, is the global head of public affairs

on the Internet.[574] The message was clear: we do not need a law, we do it well ourselves. But all this was untrue, which became apparent soon after the law came into force.

Once the NetzDG Law was in force, the Federal Office of Justice fined Facebook €2 million for not being transparent in its reports on the handling of deletion requests and for hiding the forms used to submit requests to *Facebook*.[575] The Brussels EU Commission was also not satisfied with Facebook because the actions against disinformation to which *Facebook* and all GAFAM committed themselves were not sufficient.[576] However, Brussels cannot sanction violations without a legal basis, as it exists in Germany. This is the disadvantage of voluntary codes of conduct.

When the NetzDG law was adopted in the last session of the Bundestag before its dissolution for the 2017 federal elections, Members of the Bundestag were almost exclusively concerned with one question: Who actually makes the rules here, us or *Facebook*? The law, a proposal by Federal Minister of Justice Maas, was adopted in this last session with unusual speed and a large majority.

There is a growing realization that *Facebook* is not following the law properly and that the law is not enough. Now the German conservative party CDU no longer wants to be fooled by the GAFAM

---

[574] Scott, Mark, and Melissa Eddy. "Facebook Runs Up Against German Hate Speech Laws." The New York Times, November 28, 2016, sec. Technology. https://www.nytimes.com/2016/11/28/technology/facebook-germany-hate-speech-fake-news.html.'; Allan, Richard, Hard Questions: Who Should Decide What Is Hate Speech in an Online Global Community?, June 27, 2017, 'https://newsroom.fb.com/news/2017/06/hard-questions-hate-speech/'

[575] Mühlbauer, Peter. "Verstößt Facebook fortgesetzt gegen das NetzDG? Hohes Bußgeld beantragt." Heise online. https://www.heise.de/tp/features/Verstoesst-Facebook-fortgesetzt-gegen-das-NetzDG-Hohes-Bussgeld-beantragt-4132803.html.

[576] Boffey, Daniel. "EU Disputes Facebook's Claims of Progress against Fake Accounts." The Guardian, October 29, 2019, sec. Media. https://www.theguardian.com/world/2019/oct/29/europe-accuses-facebook-of-being-slow-to-remove-fake-accounts

companies. The new Federal Minister of Justice Lambrecht from the Social Democratic Party SPD is also tightening the screw. Both demand an obligation for companies to report criminal acts in their networks, such as incitement of people to hatred and violence, to a newly established office in the Federal Criminal Police. On 30 October 2019, the Federal Government adopted a whole package of measures against right-wing extremism and hate crime. This includes the obligation to transmit the IP addresses of the senders of such illegal texts with the report.[578]

From a European EU perspective, the voluntary regulations continue to be viewed positively. It remains unclear whether Facebook and the other GAFAM, despite or precisely because of the German NetzDG, observe the voluntary Code of Conduct signed in Europe and how they would have behaved towards Europe without the NetzDG in Germany.[579] The NetzDG's lesson for the GAFAM, however, was that total opposition to laws no longer works in Germany today, at least not when the absence of laws benefits GAFAM. Because ultimately all democratic parties have to admit that they do not want the country to be governed by these corporations but prefer to pass laws in the Bundestag, the german Parliament. The majority in favour of the NetzDG in the Bundestag showed one thing above all: the vast majority of parliamentarians felt the need to demonstrate that democracy in Germany still operates through the Bundestag.

[578] Deutschlandfunk. “Es geht darum, dass wir Grenzen aufzeigen”, 3 November 2019, https://www.deutschlandfunk.de/bundesjustizministerin-lambrecht-spd-zu-hass-im-netz-es.868.de.html?dram:article_id=462210' and 'https://www.bundesregierung.de/breg-de/suche/gegen-extremismus-und-hass-1686442.

[579] Commissioner Vera Jourova on Twitter: (...) all IT companies fully meet the target (...), 4 February 2019, 'https://twitter.com/VeraJourova/status/1092382162735775750', with reference to https://commission.europa.eu/strategy-and-policy/policies/justice-and-fundamental-rights/combatting-discrimination/racism-and-xenophobia/eu-code-conduct-countering-illegal-hate-speech-online_en; see also https://www.spiegel.de/netzwelt/netzpolitik/vera-jourova-und-das-netzdg-eu-justizkommissarin-zweifelt-am-deutschen-gesetz-a-1188703.html

Shortly after the adoption of the NetzDG by the Bundestag, Richard Allan, a British liberal and former Member of Parliament in Westminster, left *Facebook*. *Facebook* was accustomed to prevent the adoption of laws because this worked regularly in Washington.

2.2. The fight against data protection laws in the US and Europe

By joining forces, GAFAM succeeded in the US in preventing the American counterpart to the GDPR, the Privacy Base Line Act proposed by President Obama in 2012.[580] Obama pushed for the law twice, without success.[581]

What succeeded in Washington, namely, to prevent a horizontal law on the protection of privacy and personal data, was also attempted by GAFAM in Brussels. Unfortunately, they were supported in their work against GDPR by President Obama and those in the US who are still trying to get his privacy initiative through Congress.[582] For even though the US government itself had envisaged a law on privacy, it supported GAFAM in Europe in their fight against the binding rules of the GDPR. Politics can be so contradictory.

Rarely has Brussels seen such a massive, coordinated, and well-funded deployment of GAFAM lobbyists against a law.[583] Only

---

[580] Whitehouse.gov. "We Can't Wait: Obama Administration Unveils Blueprint for a 'Privacy Bill of Rights' to Protect Consumers Online," February 23, 2012. https://obamawhitehouse.archives.gov/the-press-office/2012/02/23/we-can-t-wait-obama-administration-unveils-blueprint-privacy-bill-rights.

[581] Bachman, Katy. "Does Obama Privacy Push Have Oomph?" Politico. https://www.politico.com/story/2015/01/obama-cybersecurity-privacy-initiatives-114184.html.; Singer, Natasha. "Why a Push for Online Privacy Is Bogged Down in Washington." The New York Times, February 28, 2016, sec. Technology. https://www.nytimes.com/2016/02/29/technology/obamas-effort-on-consumer-privacy-falls-short-critics-say.html.

[582] Liebelson, Dana. "Ex-White House Official Joins Group Fighting 'Excessive' Online Privacy Laws." Mother Jones (blog). https://www.motherjones.com/politics/2013/03/daniel-weitzner-internet-privacy-coalition/.

[583] Liebelson, Dana. "First SOPA, Now Your Privacy: Facebook, Google Flex Lobbying Muscle in Europe." Mother Jones (blog). https://www.motherjones.com/politics/2013/03/google-facebook-sopa-privacy/.

recently did it become known what methods and connections *Facebook* was using in its lobbying campaign against GDPR.[584] Lies have been told, and not just once. A glaring example was a study commissioned by the US Chamber of Commerce, which claimed that the GDPR would lead to a welfare loss of up to 3.9 percent of the EU's gross national product, or $4,566 per household in Europe.[585] The history of the dirty fight against the GDPR and the self-sacrificing efforts of civil society,[586] including American civil society,[587] to ensure proper data protection through EU legislation is yet to be written.

Brussels learned a lot during this time, including the meaning of the word "astroturfing", a method of lobbying popular in Washington. "Looks like grass but is not grass" was the explanation of the experts. You create an organization, preferably a non-profit organization, with a nice name, for example, one that sounds like it has the common good at heart. But the work of this organization is exactly the opposite of what it looks like. A good example of this, according to the Financial Times, is the *European Privacy Association.* Comprised of former Members of Parliament and a university lecturer, its name suggests a commitment to privacy. The website says that it takes care of the concerns of small and medium-sized enterprises in Europe, and that it is a pan-European think tank on data protection and privacy.[588] In reality, Microsoft and Google funded this organization.[589] *Follow the*

---

584 Cadwalladr, Carole, and Duncan Campbell. "Revealed: Facebook's Global Lobbying against Data Privacy Laws." The Observer, March 2, 2019, sec. Technology. https://www.theguardian.com/technology/2019/mar/02/facebook-global-lobbying-campaign-against-data-privacy-laws-investment.

585 Data, T. The Economic Importance of Getting Data Protection Right: Protecting Privacy, Transmitting Data, Moving Commerce, ECIPE, 2013, https://www.uschamber.com/sites/default/files/documents/files/020508_EconomicImportance_Final_Revised_lr.pdf.

586 See 'https://edri.org/' and the member associations of this association of digital civil society in Europe and their staff.

587 Whittaker, Zack. "Privacy Groups Call on US Government to Stop Lobbying against EU Data Law Changes." ZDNet. https://www.zdnet.com/article/privacy-groups-call-on-us-government-to-stop-lobbying-against-eu-data-law-changes/.

588 http://www.europeanprivacyassociation.eu/.

589 Fontanella-Khan, James, "Brussels: Astroturfing Takes Root," June 26, 2013. https://www.ft.com/content/74271926-dd9f-11e2-a756-00144feab7de; also see https://en.wikipedia.org/wiki/European_Privacy_Association

money, always ask who is actually funding the voice, the lawyer, the professor, the think tank. This is something you had to get used to in Brussels and Berlin unfortunately, given the deep pockets of the GAFAM lobbyists. Even with organizations that at first glance look like they come from civil society.

Viviane Reding was in the lead on GDPR as Vice-President of the European Commission. She comes from the small State of Luxembourg and is a member of the Conservative EPP group, which also includes the German CDU. She told the British conservative newspaper *The Telegraph* that the lobby against her proposals on data protection, the GDPR, was the most aggressive she had ever seen.[590] Despite all the lobbying campaigns: she remained firm on the substance of her proposals and protected her officials under pressure. In the European Parliament, she could rely on the Green MEP Jan Albrecht, now Minister in the German State of Schleswig-Holstein, and he could rely on her. Edward Snowden's courageous revelations about the NSAs brought the turnaround in the legislative negotiations. Viviane Reding called it the *wake-up call.*[591] The activist Max Schrems from Vienna and the journalist Richard Gutjahr helped to create transparency with their website *Lobbyplag.eu*,[592] with graphic comparisons of lobby letters and amendments to the draft GDPR tabled in the European Parliament. These comparisons between lobby proposals and amendments tabled provided clarity not only about

---

590 Warman, Matt. "EU Privacy Regulations Subject to 'Unprecedented Lobbying,'" February 8, 2012, sec. Technology. https://www.telegraph.co.uk/technology/news/9070019/EU-Privacy-regulations-subject-to-unprecedented-lobbying.html. also see US tech groups criticised for EU lobbying, Dembosky, April. Fontanella-Khan, James"US Tech Groups Criticised for EU Lobbying," February 3, 2013. https://www.ft.com/content/e29a717e-6df0-11e2-983d-00144feab49a.

591 Reding, Viviane. "Opinion | Protecting Europe's Privacy." The New York Times, June 17, 2013, sec. Opinion. https://www.nytimes.com/2013/06/18/opinion/global/viviane-reding-protecting-europes-privacy.html.

592 Lobbyplag. "Data Sale Fest in Brussels." 10 March 2015. https://www.gutjahr.biz/en/2015/03/lobbyplag-dataleaks/

the activities of the lobby towards the European Commission, but also about those in the Parliament and the Council of Ministers. Some Members of Parliament and ministers tabled lobbying amendments in Brussels which contradicted the positions announced at home. Double-speak was popular at the time, especially in Germany.[593] A website like *Lobbyplag* is necessary in every legislative procedure.

[593] "New Platform Tracks Lobbyists Copy-Pasting Their Way into EU Legislation | Euronews." https://www.laquadrature.net/2013/02/14/euronews-new-platform-tracks-lobbyists-copy-pasting-their-way-into-eu-legislation/. For details see Ehmann / Selmayr, Datenschutzgrundverordnung, 2nd edition, 2018, introduction Rz 45ff, p. 116ff; Uwe Ebbinghaus and Thomas Thiel, Europäische Datenschutzreform, Machtprobe mit Silicon Valley, FAZ, 11 March 2014; Stefan Schulz, Die Informationsfreiheit und das Prinzip Big Data, FAZ, 25 May 2014.

D3 The toolkit of the lobby against the law

The clever lobbyist does not deny the concerns of public interests. She or he only shows how the public interest is supposedly better served without law. There is a whole toolbox of arguments that are usually unwound at various stages of the legislative process. Always the same, often wrong and unfortunately often clouded with lies or with false impact assessments and decorated with apocalyptic predictions.

3.1. Better no law than a bad law

The saying "Better no law than a bad law" and the plea for *smart regulation*[594] usually comes first. However, if you listen more closely, you will understand that *Smart Regulation* usually *does not* refer to a *regulation* or a *law*, but rather to a technical problem solution, self-regulation or voluntary agreements. But without forgetting to elegantly praise the parliaments for the fact that their criticism helped to improve products and services, as the chief lawyer of Google kindly said.[595] Parliamentary democracy is thus shrunk to the status of a consumer protection organization.

Apple is also a master advocate of self-regulation, provided it fits into the business model and helps to avoid binding legal regulation. In 2009, on the impulse of the European Commission, *Apple* signed a

[594] Google. "How We're Supporting Smart Regulation and Policy Innovation in 2019," January 8, 2019. https://blog.google/perspectives/kent-walker/principles-evolving-technology-policy-2019/

[595] Google. "How We're Supporting Smart Regulation and Policy Innovation in 2019," January 8, 2019. https://blog.google/perspectives/kent-walker/principles-evolving-technology-policy-2019/

voluntary agreement with Samsung and Nokia to bring joint chargers with uniform connections to market within two years. The diversity of charger connections is not only a nuisance for consumers. It is leads to massive amounts of pollution, as many more chargers are sold than would be necessary with standardized connections. But they still do not exist. Nor is there a binding regulation on this.[596] The EU Commission has always been run by a liberal-conservative majority in the area of industrial policy, which in principle thought it was good not to legislate. Even though experience shows that self-regulation does usually not work.

"Better no law than a bad law" - words from the same hymnbook as the advocates of the Machiavellian *Let's move fast and break things* reveals an anti-democratic attitude, in which the entrepreneur is allowed to do everything, even if it causes damage. Perfection is expected from democracy, which it cannot achieve, because every law is seen as a bad law by some people, due to the compromise which every law must be, due to the procedures of lawmaking.

3.2. The test of Erdogan

A popular GAFAM argument against laws in Berlin and Brussels is the claim that a law here provides dictators somewhere in the world with a pretext to pass a similar, but actually abusive law. This would destroy the globally unbroken Internet and with it the freedom that it brings. We call this argument the *Erdogan test*. If you think this argument through to the end, it means that no law at all can be enacted in a democracy. The fact is that the potentates in Moscow,

[596] Fanta, Alexander "[Investigation] How Apple Lobbied EU to Delay Common Smartphone Charger." 29 April 2019, https://euobserver.com/science/144538; Fanta, Alexander "MEPs slam Commission over common charger delay", 8. November 2019, EU Observer, https://euobserver.com/science/146557. But see now https://www.europarl.europa.eu/news/en/headlines/society/20220413STO27211/usb-type-c-to-become-eu-s-common-charger-by-end-of-2024

Beijing and Ankara can, of course, use any law in a democracy as a pretext for enacting a similar one in their own countries, but one that is so vague that it encourages dictatorial arbitrariness and would not stand up in a constitutional state.

The important difference is that here, whether in Germany or in Europe, the legality of a law is measured against a catalogue of fundamental rights and the principle of proportionality. The case law of the European Court of Human Rights in Strasbourg applies which is that fundamental rights must never be restricted more than is necessary and proportionate in a liberal democracy in order to protect important interests of the common good. At the same time, the principle applies that limitations to fundamental rights require a sufficiently specific law as their legal basis. And the examination of whether these requirements are met is carried out by independent judges, first within a state like Germany, with the Federal Constitutional Court as the highest court, and if necessary, also by courts outside the country, such as the European Court of Human Rights in Strasbourg and the European Court of Justice in Luxembourg. But these guarantees do not exist in the dictatorships of this world.

We must not accept the *Erdogan test,* put forward by GAFAM lobbyists and their traveling salespeople, for this argument would ultimately lead to the elimination of democracy and the rule of law everywhere and would throw us back into the state of lawlessness desired by John Perry Barlow and GAFAM.

3.3. No horizontal legislation - only specific legislation

If the complete rejection of any legislation is not capable of winning a majority, GAFAM generally advocate narrow sectoral special rules with little regulatory and binding effect. The experienced lobbyist switches to argumentation level two, which is convincing at first glance, particularly for the technical elite of engineers: a general law on the problem cannot work, only special laws that precisely address the problems identified are conceivable. We have just heard "One size fits all doesn't make sense" again from *Google's* chief executive on the regulation of Artificial Intelligence.[597]

This argument was also used against the GDPR, but without success. For the American method of limiting laws to very narrowly applicable sectoral rules, for example to protect children or to protect health data or credit data, conceals the fact that this creates wide regulatory gaps. But that is precisely what GAFAM like about the narrow special laws. Because a large part of their business activities remains unregulated in this way. Normal citizens can no longer find their way around the rules of the various special laws in the US, for example on data protection. Even the GDPR is complicated. Imagine if there were now a different data protection regulation for every area of life. Then citizens would be even less able to exercise their rights and that is exactly what the GAFAM companies want.

A popular technique of the Silicon Valley lobby, which it has taken over from Wall Street and the oil lobby, is to make the debate increasingly complex and intricate with the aim of fragmenting and reducing the scope and binding effect of laws. The lobbyist's handbook says that no one is as interested in legislation as her or his own client, who after all pays to prevent it. Money and knowledge of the subject matter determine the ability and stamina to participate in discussions. If their complexity and number is increased, many

---

[597] Bradshaw, Tim, "Google Chief Sundar Pichai warns against rushing into AI regulation", FT, 20 September 2019. https://www.ft.com/content/b16e6ee8-dbb2-11e9-8f9b-77216ebe1f17.

people who are not so interested leave the discussion. Civil society, in particular, is often not able to follow the multitude of topics.

Microsoft's demand to regulate facial recognition is an example of this. Facial recognition is one of many methods of collecting biometric data. Biometric data is one of the many forms of personal data that identifies people or makes it identification possible. Democracy is overwhelmed when we make different laws for different types of personal data. That is why it was right to adopt GDPR as a General Data Protection Regulation in Europe that covers all personal data, including biometric data, and data obtained by facial recognition.[599] We in Europe need no special regulation of facial recognition, we just need to enforce GDPR properly.

Fragmentation of legislation undermines the legitimacy of democracy. But new tricks are emerging: there are many friends of the fragmentation and thus multiplication of special laws in politics and administration. After all, politicians and civil servants who have played a leading role in a successfully completed legislative process can make a career out of it. The friends of neo-liberalism are always in favour of making laws as narrow as possible if they cannot be prevented. They also have no problem with the fact that the legislator itself often cannot keep up with this. After all, Members of Parliament only have 24 hours in a day, and even the most conscientious and well-resourced ones are overloaded at some point. This is the new tactic of GAFAM and its neo-liberal supporters: to overburden the legislator in the technical world to such an extent that it can no longer deliver.

---

599 https://edpb.europa.eu/sites/edpb/files/consultation/edpb_guidelines_201903_videosurveillance.pdf

3.4 Limitation of laws and long transition periods until they enter into force

Many companies demand long periods of time until the regulation comes into force, on the grounds that they need time to prepare for it. It is quite astonishing that it is precisely those companies that want to be masters of the short *time to market* and *move fast and break things that* demand years for implementing laws. This is even more astonishing when the new law is not a revolution, as in the case of GDPR for data protection, but a further development of rules that were previously binding. For the EU already had the 1995 directive on data protection on the books, which has practically the same rules on the subject as the GDPR.[600] Anyone who properly complied with the 1995 directive at the time did not have to make substantial investments in adapting to GDPR. Anyone who ignored the 1995 directive, which unfortunately had virtually no sanctions for non-compliance, and thus behaved illegally for years, broke into a sweat. But anyone who does not comply with existing law should, at any rate, be unable to claim a long transitional period when sanctions are tightened. After all, that would mean rewarding those who did not respect the law in the first place.

[600] Directive 95/46/EC of the European Parliament and of the Council of 24 October 1995 on the protection of individuals with regard to the processing of personal data and on the free movement of such data, OJ L 281, 23.11.1995, p. 31-50, 'http://data.europa.eu/eli/dir/1995/46/oj.

D4 Laws do not apply to us

In a *Google Books* story, Scott Rosenberg describes[601] Google's early self-image as an engineering company without respect for the law. The attitude "better to ask forgiveness than permission" brought *Google Books* into conflict with copyright law just as Uber came into conflict with labour law and public transport regulation. *Google* later learned a lesson, involving lobbyists and lawyers earlier in product development to play the sometimes necessary "game" of politics and law.

But it was not only the Californian ideology that led to the disregard of laws. The doctrine of *disruptive innovation* that is prevalent in business schools across the world eventually even legitimized the violation of the law.[602] The heroes of the *destructive Internet* not only fought against every new regulation by governments and parliament. They also took the liberty of simply breaking the law: Not only intellectual property rights like *Google* Books and local traffic rules like Uber, but also tax law and competition law. *Apple* did not pay its taxes properly in Ireland and was required by the European Commission to pay EUR 13 billion in back taxes.[603] It fought this decision and even got away with it before the first

[601] Rosenberg, Scott. "How Google Book Search Got Lost | Backchannel." Wired, April 11, 2017. https://www.wired.com/2017/04/how-google-book-search-got-lost/.

[602] See: Biber, Eric, Sarah Light, J.B. Ruhl, and James Salzman. "Regulating Business Innovation as Policy Disruption: From the Model T to Airbnb." Vanderbilt Law Review 70 (October 1, 2017): 1561. ; and Toolan, Narayan. "3 Ways in which the Fourth Industrial Revolution breaks the law", World Economic Forum, 13 April 2018 https://www.weforum.org/agenda/2018/04/three-ways-the-fourth-industrial-revolution-is-disrupting-law/.

[603] European Commission, State aid: Ireland granted Apple illegal tax benefits of up to 13 billion euros, European Commission press release, 30 August 2016, 'http://europa.eu/rapid/press-release_IP-16-2923_en.htm', full text of the decision: https://ec.europa.eu/competition/state_aid/cases/253200/253200_1851004_674_2.pdf

Instance General Court of the European Union which annulled the Commission decision.[604] The European Commission decided whether to appeal this judgement and took the case to the European Court of Justice where the case is still pending.[605]

Facebook simply took the liberty of not telling the European Commission the full truth about the merger of personal data when *Facebook* and *WhatsApp* merged, which led the European Commission to impose a fine of 110 million euros on *Facebook*.[606] *Google* was fined a total of €8.3 billion for breaches of European competition law.[607]

So, when a legal act is adopted, GAFAM make its application more difficult, by boycotting, or simply failing to follow the law. If they are caught not complying with the law, they fight to the last instance in court, using the unlimited financial resources they command. This makes life difficult for the enforcement authorities. And this practices contains a warning to them by GAFAM: If you sanction us, you will be dealing with us for years afterwards, so better to just leave us alone. In fact, the GAFAM groups often achieve their goals in Germany and Europe in this way, in flagrant disregard for the law. Because public authorities become shy about enforcement against GAFAM.

---

604 https://curia.europa.eu/jcms/upload/docs/application/pdf/2020-07/cp200090en.pdf; The appeal of the European Commission is still pending.

605 https://news.bloombergtax.com/daily-tax-report-international/eu-appeal-of-15-billion-apple-tax-case-faces-uphill-battle;

606 European Commission, Mergers: Commission imposes fines of EUR 110 million on Facebook for providing misleading information about the acquisition of WhatsApp, European Commission press release, 18 May 2017, available at 'http://europa.eu/rapid/press-release_IP-17-1369_en.htm'; full text of the decision: 'http://ec.europa.eu/competition/mergers/cases/decisions/m8228_493_3.pdf

607 European Commission, Commission imposes fine of EUR 2.42 billion on Google for abuse of its dominant position as a search engine by giving undue preferential treatment to its own price comparison service, 27 June 2017, 'https://europa.eu/rapid/press-release_IP-17-1784_de.htm'; Commission imposes fine of 4.34 billion euros against Google for illegal practices in Android mobile devices to strengthen Google's dominant position as a search engine, 18 July 2018, 'https://europa.eu/rapid/press-release_IP-18-4581_de.htm'; Commission imposes a fine of 1.49 billion euros on Google for abuse of a dominant position in the market for online advertising, 20 March 2019, 'https://europa.eu/rapid/press-release_IP-19-1770_de.htm'.

D5 Permanent fight against laws in court: Google's fight against the right to be forgotten

The struggle of the tech giants against law and order is not limited to the political arena. They also fight against the binding effect of laws in the courts themselves. A prominent example is the *Google Spain* case before the European Court of Justice. The case is interesting for the question of the responsibility of GAFAM and for the relevance of Artificial Intelligence. A Spanish citizen had contacted the data protection authority and demanded that Google delete an entry about his personal financial difficulties from 1998. While the Spaniard did not question the report, he considered it incompatible with data protection laws that *Google* refused to delete the report after more than 15 years. The man felt that this violated his right to be forgotten. Google initially questioned the applicability of EU data protection law.[608] *Google* argued that only Californian law was applicable and that only a judge in California had jurisdiction, as the answer to search queries made in Europe came from Californian servers. Secondly, *Google* denied that its search engine operations could be considered "data processing" under data protection law,[609] which would have relieved Google of any responsibility for search answers. *Google* simply claimed that the selection process of the search engine's answers was fully automated in the form of an algorithm, and therefore *Google* had no control over the search answers, nor could it be held responsible for them under European data protection law. This argument didn't convince the ECJ who

---

[608] Case C-131/12 Google Spain SL v Mario Costeja González [2014] ECLI:EU:C:2014:317, para 47, 'https://eur-lex.europa.eu/legal-content/EN/TXT/?uri=CELEX%3A62012CJ0131

[609] Case C-131/12 Google Spain SL v Mario Costeja González [2014] ECLI:EU:C:2014:317, para 22.

made it clear that "An Internet search engine operator is responsible for the processing that it carries out of personal data which appear on web pages published by third parties".[610]

Google has revealed its gloomy views on the rule of law here: First, automation through algorithms that provide a service to individuals is intended to protect the intermediary who has developed and offers this technology from any responsibility. And what no longer applies to automated systems, of course, no longer applies to AI. Second, if there should be any form of legal responsibility at all, then please let there be only one global system, which is of course dominated by US law and may only be interpreted and applied by US judges. That would be ideal for *Google*, because then it would no longer have to worry about the regulations in other countries around the world, which would save a lot of money and effort for *Google*. The citizens of this world would then also find it almost impossible to pursue claims against *Google* with the help of the judiciary, because non-Americans can hardly afford to sue *Google* in America, quite aside from the fact that Americans also very rarely do so because they cannot afford it either.

It is therefore all the more gratifying to know how the ECJ reacted to these arguments, its response signalled a "return of law"[611]. It acknowledged the right of citizens to legal protection in an increasingly automated and globalized world. The Court granted the Spanish plaintiff a right to non-indexing of his data by *Google*, - based on primary law and a long European legal tradition of protecting privacy and personal data.

[610] Case C-131/12 Google Spain SL v Mario Costeja González [2014] ECLI:EU:C:2014:317, para 41, 'https://eur-lex.europa.eu/legal-content/EN/TXT/?uri=CELEX%3A62012CJ0131.

[611] Kühling, Jürgen. Rückkehr des Rechts: Verpflichtung von »Google & Co.« zu Datenschutz, Europäische Zeitschrift für Wirtschaftsrecht, 2014, S. 527 .

The *Google* Spain judgement has set important precedents against separating the responsibility for the automation and autonomous development of Artificial Intelligence from the original creation and commissioning of such programs. One can speculate on whether *Google's* representatives, when they denied responsibility for the search algorithm, were already thinking of setting a precedent to shield the company from responsibility for autonomous algorithms and Artificial Intelligence in the future. In any case, the ECJ's answer in this case was clear: this cannot happen.

Google turned the tables and held hearings across Europe on how to implement the right to forget. It purported to do this because it wanted to respect the judgment. But the real issue was to mobilise the Internet freedom activist community against the ruling to prevent the ruling from being codified in the new data protection regulation GDPR.[612] *Google* did not succeed.

But that's not all. To this day *Google is fighting* against the right to be forgotten before the European Court of Justice (ECJ). In another lawsuit, this time against a decision by the French data protection authority, *Google has* defended itself against the order to delete data worldwide.[613] *Google* wants to ensure that data that it stores outside Europe does not have to be deleted. However, if *Google* were allowed to delete only data stored in Europe in response to deletion requests from citizens in Europe, this would mean the end of data protection and any informational self-determination because Google and the other GAFAM secure almost all data outside Europe.

---

612 Kauffmann, Sylvie. "Opinion | Goodfortheworld.Com." The New York Times, December 19, 2014, sec. Opinion. https://www.nytimes.com/2014/12/20/opinion/sylvie-kauffmann-google-europes-favorite-villain.html; the videos of the hearings and transcripts can be found at 'https://archive.google.com/advisorycouncil/'

613 Kelion, Leo. "Google Wins Landmark Right to Be Forgotten Case." BBC News, September 24, 2019, sec. Technology. https://www.bbc.co.uk/news/technology-49808208

However, the application of the right to deletion cannot depend on where in the world a GAFAM company collects and stores our personal data. The Court did not question the right to worldwide deletion, but made it clear that there is a special right vis-à-vis search engines, namely the right to have information about the named person that is available somewhere on the Internet not shown by *Google* when a name search is made (a right to 'de-link' on search engines, which is different from the normal right to deletion of data under GDPR). Only this special right to not be shown on a search engine was limited to Europe by the ECJ, and only unless there are specific other reasons or national law requiring worldwide non-disclosure of the information in question. The ECJ further stated that such a worldwide order of non-disclosure is not contrary to EU law but is not included in GDPR for every case. This second ruling is not a victory for *Google*, and *Google* will not let up in its efforts to soften data protection in Europe, and in particular the right to be forgotten. If it were to succeed, *Google* would have taken an important step on the road to total surveillance capitalism: data would never be deleted again, the company could make money with this data indefinitely and exercise endless power over people and states with this data. It must never come to that. The general right of every human to request and obtain deletion of his or her personal data wherever it is stored in this world must not only be maintained for Europeans, but strengthened and adopted in more jurisdictions.

Part E
Public Sphere and Democracy

Public Sphere and Democracy

> *"The public sphere [...] stood or fell with the principle of universal access "*[614]

German Philosopher Jürgen Habermas' account of "The Structural Transformation of the Public Sphere", written in 1962, leaves nothing to be desired in terms of clarity. It reads today as extremely topical if one considers the networks of the Internet as new media, which they undoubtedly are, just like Theodor Adorno's 1967 lecture on "Aspects of the new right-wing Extremism".[615]

The *public sphere* began as a place for the meeting of ideas, and where debates about public matters could take place; a function, that according to Habermas provided an important counterweight to executive authority. Habermas in 1962 described the decline of the media since the end of the 19th century, and how the press created only the appearance of a public sphere. According to his analysis, from the 19th century on critical journalism was increasingly replaced by manipulative advertising. This results in a degradation from a true public sphere to a pure enactment of a public sphere, and the dissolution of journalism into *public relations*. The latter attempts to sell private interests as general public interests conceal true intentions. Worse still, public "reasoning tends to turn into consumption".[616]

---

[614] Habermas, Jürgen. 2015. The Structural Transformation of the Public Sphere: An Inquiry Into a Category of Bourgeois Society. John Wiley & Sons. p.89

[615] Adorno, Theodor, Hoben, Wieland, 2020, Aspects of new right wing Extremism, Wiley & Sons, 2020; see also https://www.dw.com/en/what-thinker-theodor-adorno-understood-about-the-far-right-50-years-ago/a-49908686

[616] Habermas, Jürgen. 2015. The Structural Transformation of the Public Sphere: An Inquiry Into a Category of Bourgeois Society. John Wiley & Sons. p. 177.

How can the decline of the public sphere be accounted for? By subjecting the press to a political economy of supply and demand, it becomes economically dependent on advertising revenues and is thus forced to enter into dependencies that increasingly corrupt it. The perfidious thing about this is that PR does not only target the consumer in order to influence the purchase decision, but public opinion as a whole. In doing so, it deceives about its true intentions: "The sender of the message hides his business intentions in the role of someone interested in the public welfare".[617]

Further distortions and deformations take place in the media commercialized in this way: Humanity is reduced to *human interest*; emotions are addressed more strongly than the intellect. This has consequences: "The strict separation of facts and figures is increasingly being abandoned".[618] When the boundaries between fact and fiction begin to blur, under the "pressure of commercial interests...a power-dominated (vermachtet) public sphere is rising"[619] in which "the press becomes corruptible under the pressure of commercialization."[620]

If the public was already so distorted and "power-dominated" in the 19th and 20th centuries, one might believe the influence of commercial interests on news and opinions seems to be bearable for democracy. With regard to the excesses of *Fake News* and *Hate Speech* today in the digitalised public sphere, one could say: we have seen this before, it's not such a problem. We could also suggest that

617 Habermas, Jürgen. 2015. The Structural Transformation of the Public Sphere: An Inquiry Into a Category of Bourgeois Society. John Wiley & Sons. p. 193.

618 Habermas, Jürgen. 2015. The Structural Transformation of the Public Sphere: An Inquiry Into a Category of Bourgeois Society. John Wiley & Sons. p. 204.

619 Habermas, Jürgen. 2015. The Structural Transformation of the Public Sphere: An Inquiry Into a Category of Bourgeois Society. John Wiley & Sons. p. 247.

620 Habermas, Jürgen. 2015. The Structural Transformation of the Public Sphere: An Inquiry Into a Category of Bourgeois Society. John Wiley & Sons. p. 221.

Habermas' analysis is outdated anyway, as it comes from a time when television was just beginning its triumphal march and the Internet is not even mentioned.

Habermas is focused on analyzing the classical mass media. He argues with Brecht that the electronic media in particular, such as first radio and then television, have also distorted public communication because of their monological character, because they deprive the public as passive addressees of their own voice: "*They don't talk back*".[621] "Under the compulsion of '*don't talk back*', the behaviour of the public takes on a different form. The programmes that the new media broadcast, in comparison with printed messages, "curiously restrict the reactions of the recipient...and at the same time take away the distance necessary for mature judgment."[622]

Precisely this one-way street character of the mass media now seems to be overcome in the multi-way medium of the Internet. Instead of "*one to many*" and "*top down*" as in mass communication, now "*many to many*" dominates and everyone communicates without hierarchy and from the "bottom up".

With the *World Wide Web*, the apparently free space of *cyberspace*, in which supposedly anyone can exchange information with anyone without rules, it seemed that the appropriate medium for Habermas' normative concept of the public sphere had finally been found. The Internet seemed to be the medium that would finally remedy his and Brecht's complaints. This is exactly how the first Internet enthusiasts saw it, who recognized the possibilities of worldwide networking

---

[621] Enthusiastically, Bertolt Brecht had called for the technical development of radio towards a two-way medium in order to unfold its full emancipatory and enlightening potential: Brecht, Bertolt. 2015. Brecht on Film & Radio. Bloomsbury Publishing.

[622] Habermas, Jürgen. 2015. The Structural Transformation of the Public Sphere: An Inquiry Into a Category of Bourgeois Society. John Wiley & Sons. p. 205.

for the counterculture of the late 1960s. The term "counter-public", which at the time was critical of the commercialized and deformed public sphere, inspired the enthusiasts of the *World Wide Web*. Welcome to *cyberspace*. Welcome to the pure space of "domination-free communication", as Habermas later called his ideal model.

But the dream of cyber-enthusiasts has not yet come true. Indeed, it has rather turned into its opposite. In contrast to the euphoric expectations of the 1980s and 1990s, the *World Wide Web* seems to have mutated into one of the greatest threats to freedom and democracy. The Californian path led from a grassroots democratic original impulse via the Barlow Declaration to the "Community Rules" and "self-obligations" of *Google, Facebook* and Co. Whether a trend reversal in the form of a renaissance of the law for the Internet will now begin remains to be seen.

The core of Habermas' *Theory of Communicative Action*, the title of his later major work, is that it set out the difference between instrumental purpose-rational and communicative action. While the former refers to the effective use of means to achieve certain goals, communicative action aims at achieving agreement among actors by recognizing reasonable grounds for consensus.

The instrumental purpose rationality of reason provides the basis for the domination of nature and human beings by power that is not democratically legitimized. In contrast, the basic structure of a successful intersubjectivity is found in communicative action, a central building block of democracy. In the instrumental context of domination, means are used to achieve given goals. In communicative action, an understanding of common purposes and goals takes place. Through understanding, the exercise of power can be criticised if it exercises rule that is not legitimised. Conversely, power can be legitimised if it is committed to the interests of the majority and the public good.

E1 Structural Transformation of the Public Sphere

Habermas examines in his habilitation thesis the origins and functioning of the public sphere in modern times. The book with the proverbial title is much more than a historical analysis. It is impressive in its combination of historical-empirical methodology and philosophical reflection even six decades after its creation.

The basic question of the study is: How do citizens arrive at their political convictions; how do they justify their political actions? How and under what circumstances does the formation of opinion and will take place in a society? The central point of the considerations is the modern concept of freedom, from which individual autonomy and political democracy are derived.

Habermas argues that the public sphere is the space in which individualization and socialization of the subjects take place simultaneously and in which, on the other hand, political rule is legitimized in democracy. In modernity, politics is thus dependent on a protected public sphere in a twofold sense: It forms the space in which autonomous citizens blossom and find complete expression and in which they jointly exercise sovereign rule in democratic institutions. The legitimizing power of the public sphere thus forms the basis of "deliberative" democracy, in which by weighing reasonable grounds decisions are formed, which are binding for all, because they are cast in law.

Public sphere is the central normative concept in this theory. It takes the place of the divine grace which is no longer accepted by the self-confident citizens as a source of legitimacy for rule.

The normative idea of the public sphere is certainly an idealization, but an unavoidable one. Ultimately, it is about establishing the legitimacy of sovereign rule. The fact that an idea is not fully realizable does not in itself speak against this idea. On the contrary: From Kant we learn that "regulative ideas" have an irreplaceable orientation function, also and especially because they cannot be completely realized.

The alternative to a normative concept of the public sphere is a Luhmann-style systems-theory understanding in which the public sphere is completely reduced to its functional character.[623] The public sphere is then the space in which the system registers and eliminates "disturbances" of the rather "improbable" communication. Normative resources, i.e., the human capacity to develop ethical rules for the future, which could provide orientation for the control of the system beyond its pure self-preservation, have no place in this theory.

In contrast, Habermas shows that there are indeed types of action rationality that go beyond the purely functional instrumental control of rationality.[624] Especially in the "subsystems" of culture, society, personality, where values, norms and goals are at stake, a cybernetic control system that is exclusively oriented on the difference between given actual and target values cannot help. A comprehensive theory of rationality, however, can.

Habermas develops his model against the backdrop of the modern rise of the bourgeoisie, which, with the development of the capitalist economic system, attains increasing economic power, but initially

---

[623] see also Habermas, Jürgen / Luhmann, Niklas. 2012. Theory of Society, Volume 1. Stanford University Press.

[624] see also: Habermas, Jürgen. 1986. Theory of Communicative Action, Volume 1: Reason and the Rationalization of Society: Reason and the Rationalization of Society Vol 1. New Ed edition. Boston: Polity. p.11

hardly any social or political power to shape the future. In this historical upheaval, a rule based on arbitrariness (*voluntas*) is to be opposed by legislation based on reason (*ratio*). Whereas Machiavelli had based his princely rule on the principle of secrecy, and developed a catalogue of secret practices in his book "The Prince" that were intended to ensure the prince's rule over the unemancipated people,[625] Habermas shows that the counter-principle to this is the idea of publicity.[626] By "arguing" in public space, the citizens, quite practically and independently of the concrete contents of their discourses, are directed against the arcane secret practices of the ruling class. With this term, which Habermas first introduced in a non-religious context, and which originally denotes the obligation of initiates to strictly preserve religious secrecy, Habermas draws a line from religious cults to political ruling practices. In his juxtaposition of secrecy and publicity in the Age of Enlightenment, we also recognize the starting position at the beginning of democracy in Athens. Apparently, in times of upheaval, questions of legitimacy are increasingly being raised. If it is a democratic upheaval, orientation is thought in reasonable reflection and action.

Habermas understands the public sphere equally as principle and institution in the relationship between state and society. As a principle, because only in public discourse can the sphere of private life freely communicate with that of political rule. As an institution, because only in a protected space where reliable information circulates unhindered and is openly debated can opinions be freely formed. The protection of the public space refers to the guarantee of freedom and liberty of citizens and especially the right to freedom of expression. In analogy to the autonomy of the individual, this

---

[625] Machiavelli, Niccolo. 1988. Machiavelli: The Prince. Cambridge University Press.

[626] Habermas, Jürgen. 2015. The Structural Transformation of the Public Sphere: An Inquiry Into a Category of Bourgeois Society. John Wiley & Sons,. p.102.

enables a process of common self-determination to unfold. Just as individual autonomy leads to a sovereignty that is ultimately based on reason, practical reason develops in communities that legitimise the sovereignty of the people through democratic rule.

In order to safeguard this connection between the individual and the common, based on reason, citizens developed a catalogue of inalienable fundamental rights. They are essential to guarantee that the encounter in public space can take place "free of domination". The Declaration of Human Rights codifies the rights of the individual and protects it against coercion by the state. This is for Habermas at the core of the idea of a society which enlightens itself and thus functions according to general laws and rules of some permanence and based on reason. Habermas sees "public opinion as the only legitimate source of this law."[627]

With the transition to a bourgeois society the concept of representation profoundly changes. From the world order of God "represented" in the body of the ruler, which derives its legitimacy from the divine grace, representation becomes a function of the will of the people in parliament. The role and significance of democratic institutions can be derived from the idea of representation. The public mediate the boundaries of the private and social on the one hand and the political on the other, rather than remove these entirely. Thus, the private sphere of the individual and the political structures and institutions of democracy retain their respective status. This tension between the private and public spheres feeds the freedom of the citizen.

The central means for the formation of a democratic order is public discourse, which develops in tandem with the free press. It

---

[627] Habermas, Jürgen. 2015. The Structural Transformation of the Public Sphere: An Inquiry Into a Category of Bourgeois Society. John Wiley & Sons,. p.54

is only with the free press that the change from the courtly order to a bourgeois public sphere can take place. "The rule of law was guaranteed by publicity, namely, by a public sphere whose ability to function is posited by implication together with the posited natural basis of the juridical condition,"[628] as Habermas stated, referencing Kant's political philosophy. He goes on: "Compulsion could then no longer occur in the form of personal rule or violent self-assertion but only in such fashion that 'reason alone has force' [...] [this is] the most extreme counterposition to the principle *auctoritas non veritas facit legem.*"[629] It was with this dictum that Hobbes expressed that authority, not truth, creates the law, which is the principle of decisionism.

Habermas emphasizes the close interaction between the economic and technological in the development of the press: On the one hand "in proportion to the increasing prevalence of the capitalist mode of production, social relations assumed the form of exchange relationships."[630] This has consequences for many areas of social life, including consequences for the legal system, which ensures the free movement of private individuals through the legal institutions of property and fundamental contractual freedoms. On the other hand, the development of the press is only possible through the invention of printing, which is the prerequisite for a wide range of publishing activities.

Deepening specialization is another facet of the development of media. While in the early days of bourgeois media the journalist and

---

[628] Habermas, Jürgen. 2015. The Structural Transformation of the Public Sphere: An Inquiry Into a Category of Bourgeois Society. John Wiley & Sons. p. 115

[629] Habermas, Jürgen. 2015. The Structural Transformation of the Public Sphere: An Inquiry Into a Category of Bourgeois Society. John Wiley & Sons. p. 103

[630] Habermas, Jürgen. 2015. The Structural Transformation of the Public Sphere: An Inquiry Into a Category of Bourgeois Society. John Wiley & Sons. p. 74

the entrepreneur were identical, the progressive commercialization and professionalization of the media led to the division of these roles into that of the journalist and the publisher.

Today, these roles are being merged again, reflecting the opening of a new space: *influencers* and *Youtubers* are journalist, creative director and business entrepreneur all rolled into one. They operate a business, generating revenues from advertising, product placement and merchandising, while simultaneously working as the producers and hosts of their channels, filling them with life and content. This is basically in line with the spirit of the Enlightenment. Kant writes that everyone is "born to be a publicist" who "speaks through writing to the actual audience, namely the world"[631]. In his eyes, however, these expressions of opinion should also be made in a qualified manner, i.e., with reasonable arguments.

The Internet thus offers the infrastructure that enables the full potential of the public idea to be realized: Here, anyone can be both creator and viewer. The basically decentralized structure can enable a free exchange of participants, and everyone has potentially unlimited access to all information.

1.1. Structural Transformation of the public 2.0: Power domination and personalization destroy the public

For the first time, network communication allows for the personalization of the media. But in the pitfalls of "personalization" there is a danger of the suffocation of the shared forum for debate

[631] Kant, Immanuel. Kant: Political Writings, 2nd edition (Cambridge England ; New York: Cambridge University Press, 1991), p.56

through commercialization and the rules of the click economy. If information is individually tailored, individuals cannot rely on it being seen and checked by everyone else. Truth, however, needs intersubjectively shared content and standards. Truths cannot exist as only personally "felt truths", which is more and more talked about today. Shared and verified information is a prerequisite for a functioning public sphere. Part of freedom of information is the certainty that the information we receive is shared with others and is not personalised to the individual on the basis of non-transparent criteria.

The consistent personalization of news destroys the basic function of the public sphere, that information which is relevant to public life can be discussed by all on a common basis. "Common sense" only emerges when there is common ground in a shared public sphere to be debated and contested. But the personalization of the Internet has fragmented that common ground into countless small plots by personalising access to information. This consistent individualization breaks up the public sphere with far-reaching consequences. It threatens to give rise to a state in which people live in completely separate worlds. Without common ground also the basis for collective action is removed. Instead of common ground there is a risk of creating a tragedy of the commons.

The total personalization of digital communication is accepted because it appears that media entities are speaking directly to listeners in a private language.[632] But in reality, it is aimed at capturing the most personal data it can attain. Simultaneously, it creates a form of standardization to which people then adapt. Like

---

[632] In the Philosophical Investigations Ludwig Wittgenstein had showed that there can be no private language in which each speaker knows the meaning of the words only for himself: "The words of this language should refer to what only the speaker can know about; to his immediate, private feelings." (Philosophical Investigations. p.243) But this idea fails to recognize that meaning always arises only in the intersubjective use of language, in the so-called language games. There sensations can be communicated, but only in a language that

Horkheimer and Adorno warned, the "managed world" replaces the enlightened idea of freedom with standardization.[633] In today's data-driven world, freedom is replaced by personalization derived from the previous standardization. Completely individualized and personalized consumption only creates the illusion of uniqueness. Enlightenment today therefor means to enlighten the basic assumptions of digitalization and dataism about oneself.

The recommendation algorithms which produce forecasts are based on data from the past. They form not only an informational filter bubble, but also a time capsule, because they exclude the surprises of the new and the foreign, thus preventing the human from changing and adapting over time. In addition to personalization there are the rules of the click economy: emotionally salient content triggers more clicks and retweets and generates more *followers* than facts and reasonable arguments. Because money is earned by the number of clicks and *retweets,* algorithms prefer emotions to reason. This forcibly moves the public sphere from a space in which reasoned arguments are debated for the sake of democratic decision making, towards one dominated by extreme views, hatred and emotion. Attention is the primary aim, followed by emotional arousal. This occludes the space for rational democratic decision-making. The ultimately economic pull of the attention spiral drives the descent from debate to hate campaigns.

---

starts from common meanings. Attempts to convey individual meaning in a private language are in vain for Wittgenstein, as he shows in his beetle parable: "Suppose each person had a box with something in it we call a bug. No one can ever look into each other's boxes, and each says he knows what a bug is just by looking at his own. [...] The thing in the box is not part of the language game at all, not even as something, because the box could be empty" (Philosophical Investigations p.293). Wittgenstein's late philosophy of language, in which he turned away from the logical atomism of the Tractatus logico-philosophicus, shows that linguistic meaning arises on the basis of a commonly shared intersubjective life practice. Today we see that, conversely, the supposed individualization of meanings through consistently personalized public communication also destroys intersubjective life practice and thus the social community.

[633] See here: Horkheimer, Max, and Theodor W. Adorno. 2002. Dialectic of Enlightenment. Stanford University Press. p. 141.

The message of supposedly total personalization through *targeting* is so ironic that little remains of the person, the unique individual: "you are absolutely special - just like everybody else."[634] The personalization of algorithms, bypasses each persons' unique individuality, because each subject of an algorithm doesn't appear as their full "self", but only as precise "type". The dignity of the human being is missed by this search for ideal types. AI can never recognize the human being, because it is not in the logic of their program codes, it can merely track statistically significant behavior patterns.

The personalization of information is based on the mass collection of personal data. But if the right to informational self-determination is not respected the individual should expect that his behaviour and the data of his life will be constantly recorded and evaluated. In turn their behaviour will adapt to this possibility according to the laws of the Panopticon. Constant observation has a chilling effect that leads to conformism and kills criticism. An accumulation of knowledge about people, whether in the state or in corporations, as it occurs today, is simply not compatible with a democratic social order that relies on the free activity and the self-determination of its members. Nor is it compatible with the idea of an innovative society. Conformism is the enemy of creativity.

1.2. Journalism as the fourth estate - a future model?

In modern societies, the press plays a prominent role in creating a functioning public sphere. Habermas describes how the media promotes a public spirit that develops criteria of criticism in the public sphere. But this does not apply to any press, only an independent

---

[634] Such a line from the song "Something Special" by the band DePhazz.

journalism, which "constantly comments on and criticizes"[635] the government's actions. A journalism free of vested interests is necessary for the press to play its role as the "fourth estate". This position has developed since the middle of the 18th century, providing checks and balances to the state,[636] complementing the separation of powers between the legislature, executive and judiciary.

This special status subsequently creates a peculiar mixture of specific rights and obligations for the press. The freedom of information of journalists who exercise special rights, such as the right to information vis-à-vis companies and authorities or the protection of informants, is, conversely, linked to a series of voluntary commitments which are intended to ensure that these special rights are handled responsibly. The German Federal Constitutional Court (BVerfG) has described "professional journalism", which complies with its duties of care as well as the statutes, in which the press imposes voluntary obligations on itself, for example in the form of the "Press Code of Conduct", as an indispensable component of a democratic order. In the US there are similar Codes of Ethics, for example, by the American Society of Newspaper Editors (ASNE)[637] and others.

The special role of the "fourth estate" has been repeatedly emphasized and strengthened by the BVerfG over the years due to its basic function for democracy.

In a judgment of 25 April 1972, the BVerfG concluded that "free intellectual debate is a vital element of the free democratic order in

---

[635] Habermas, Jürgen. 2015. The Structural Transformation of the Public Sphere: An Inquiry Into a Category of Bourgeois Society. John Wiley & Sons, p. 60.

[636] Habermas, Jürgen. 2015. The Structural Transformation of the Public Sphere: An Inquiry Into a Category of Bourgeois Society. John Wiley & Sons, p. 60.

[637] ASNE, Statement of Principles 20 August 1996, accessible at: https://www.pbs.org/newshour/classroom/app/uploads/2014/03/mediaethics_handout6.pdf

the Federal Republic of Germany and is the very foundation of this order. It is decisively based on freedom of opinion, freedom of the press and freedom of information, which stand independently next to each other as equal guarantors."[638] The essential task of the press is described as criticism and control of government action.

Habermas maintains that the public sphere must be secured time and again as a space of free encounter, where arguments can be measured against each other, and the better ones prevail. The "unconstrained compulsion of the better argument" in the fair exchange among equals makes the most compelling argument win, guiding the citizen and the state. The free exchange of opinions is thus required, so to speak, in analogy to the free exchange of goods - in both cases the consequences should be a positive development, of the economy and of democracy.

At the end of the 19th century, according to Habermas' description, the bourgeois public sphere begins to dissolve in the grip of commercialization. This may, however, have been somewhat too critical an assessment. Habermas himself admitted in the preface to the 1992 new edition of his book that his analysis was originally rooted in a concept that “remained captive of a notion that became questionable in the meantime, i.e., that society and self-organization are to be considered a totality”.[639] Today, his modified theoretical framework focuses more strongly on a sustained interplay between the institutions of the rule of law and the actors of a civil society, each of which remains independent and in a tense relationship.

In a more recent text Jürgen Habermas takes up these questions once again and at the same time locates the theory of the public

[638] Source: Para 118 in the Judgment of the German Constitutional Court of 25.04.1972 - 1 BvL 13/67, with many further references, https://openjur.de/u/175999.html.

[639] Habermas, Jürgen. Further reflections on the public sphere, in Calhoun, Craig J. 1992. Habermas and the Public Sphere. MIT Press.

sphere within a deliberative model of democracy.[640] In it, he turns to Internet communication, which "seems to counterbalance the weaknesses associated with the anonymous and asymmetrical character of mass communication", because at first glance it has "revived the historically submerged phenomenon of an egalitarian public of writing and reading between conversational partners and correspondents."[641]

But Habermas immediately turns from the Internet to the new fragmentary character of the public sphere created by the Internet:

"Internet use has led to an expansion and fragmentation of communications networks. Thus, although the Internet has a subversive effect on public spheres in authoritarian regimes, at the same time the horizontal and informal networking of communications diminishes the achievements of traditional public spheres. For the latter pools the attention of an anonymous and dispersed public within political communities for selected messages, so that the citizens can address the same critically filtered issues and contributions at the same time. The price for the welcome increase in egalitarianism due to the Internet is a decentring of the modes of access to unedited inputs."[642]

In other words, for now, there are no functional equivalents on the Internet for the traditional public sphere structures that collect decentralized messages, select them and synthesize them in edited form for public discussion based on common input. Alternatives that are equivalent to high-quality press on the net have only been sporadically

---

[640] Habermas, Jürgen. 2014. Europe: The Faltering Project. John Wiley & Sons. First published in German as Ach, Europa, Suhrkamp Verlag, Frankfurt am Main, 2008. When the German edition of this book was published, the latest update from Habermas had not yet been published:Ein neuer Strukturwandel der Öffentlichkeit und die deliberative Politik (A new structural change in the public sphere and deliberative politics), Berlin 2022

[641] Habermas, Jürgen. 2014. Europe: The Faltering Project. John Wiley & Sons. p. 161

[642] Habermas, Jürgen. 2014. Europe: The Faltering Project. John Wiley & Sons. p. 53

developed, as financing deep-dives and investigative journalism is hardly possible in the fragmented, small-scale nature of the World Wide Web. In any case, the new publishers are not independent of the platforms, their valuation logic and their algorithms. The mass copying of news without paying for it, but also the mechanism of *targeted advertising*, has undermined the business model of the free press. Today, more than 80 percent of advertising revenues on the Internet accrue to two companies, *Facebook* and *Google*.[643] These companies either use the content generated by the press without paying, or they dictate, through their market power, which articles are made visible at all and what share of the revenue the publisher receive. This is how they earn the brunt of revenues from content and the press, democracy's fourth estate, goes down the drain. For this reason alone, *Google* and *Facebook* bear a heavy responsibility for the crisis of democracy, which has now become a global phenomenon.

However, even the traditional media companies have increasingly abandoned their core task, under pressure from the new digital players. Thus, a large-scale study by the renowned media researcher Siegfried Weischenberg painted a sobering picture as early as 2006: "The public task that (journalism) is supposed to perform according to the highest jurisdiction [in Germany] is now to be looked for with a magnifying glass. In journalism as a whole, the role of critic is increasingly being called into question. The crisis of journalism is increasingly proving to be a crisis of its critical function. It becomes obsolete when the distance is lacking, and the relevance is missing."[644]

The author notes the increase of corrupt structures in travel, motoring and business journalism, as well as the dependence on audience

---

[643] Source: https://www.welt.de/wirtschaft/webwelt/article202021472/Google-und-Facebook-Werbe-Duopol-in-Deutschland-waechst-weiter.html

[644] Weischenberg, S et al., Die Souffleure der Mediengesellschaft: Report on Journalists in Germany, 2006, p. 18

ratings and advertising revenues that prevent a focus on the less lucrative strands of investigative and critical journalism. Thus, the orientation of media companies, lost under economic pressure, can be identified and named as a further factor in the crisis of democracy. Media companies have to decide whether they want to continue to fulfil their essential mandate of creating a critical public and thus their distinguished role in the democratic public sphere, or whether they want to be content with their role as showmen at the Vanity Fair. Whining, and then aligning oneself more and more with the digital competitors, is not enough.

Structural change has long since taken hold of the journalist profession: In recent years, professional journalists have been replaced more and more by amateur bloggers and influencers who very often have a better command of the mechanisms of distribution and marketing than the professional media companies for which most journalists work. They have found means of generating political communication, especially in social networks, which in most cases are far removed from the ideal of rational political debate. Instead, they have found an adaptation to PR in their style and craft. Politicians, too, are increasingly choosing the path of direct communication with citizens, thus following the path of PR instead of exposing themselves to the critical questions of the press.

In 2017, Angela Merkel gave 78 interviews to journalists, in 2018 it was only 22.[645] Instead she has adopted podcasts as a medium that allows her to address the electorate directly. The lawyer Hermann von Engelbrechten-Ilow has filed a lawsuit against the Federal Press Office because he believes that this practice of communicating politics

[645] Der Tagesspiegel, "Merkel zieht sich zurück – von den Medien," 31 October 2019. https://www.tagesspiegel.de/politik/merkel-zieht-sich-zuruck--von-den-medien-4668841.html

without journalistic critical questions is against the constitution.[646] Good reasons can be found for his position, including the BVerfG case law on the Spiegel ruling,[647] in which the court strengthened the role of the press as a necessary mediator between politics and the public. From the freedom of the press pursuant to Article 5 of the German Constitution, the German Constitutional Court requires institutional guarantees for a free press, because the press is an essential factor for the democratic formation of will. Let us quote a passage that summarizes this:

"A free press, not controlled by public authority, not subject to censorship, is an essential element of the free state; in particular, a free and regularly published political press is indispensable for modern democracy. If the citizen is to make political decisions, he must be fully informed, but he must also be able to know and weigh up the opinions which others have formed. The press keeps this constant debate going; it obtains the information, gives its own opinion and thus acts as a guiding force in public debate. Public opinion is articulated in the press; the arguments are clarified in speech and counter-arguments, gain clear contours and thus make it easier for citizens to judge and decide. In representative democracy, the press is at the same time a permanent liaison and control organ between the people and their elected representatives in parliament and government...."

The critical "public task" which is assigned to the press can hardly be fulfilled by organised state power. Press enterprises must be able to form freely in the social space. They work according to private-sector principles and in private-law forms of organization. They are

---

[646] Müller-Neuhof, Jost, "Social-Media-Angebot der Bundesregierung kommt vor Gericht" Tagesspiegel, 17.12.2019 https://www.tagesspiegel.de/politik/unzulaessige-beeinflussung-social-media-angebot-der-bundesregierung-kommt-vor-gericht/25340486.html.

[647] The text of the judgment: BVerfGE 20, 162 - Spiegel - is at http://www.servat.unibe.ch/dfr/bv020162.html

in intellectual and economic competition with each other, in which the public authorities may not intervene on principle."[648]

Politics, which elsewhere verbally deplores the decline of the press, at the same time encourages it by making intensive use of *social media channels* and ignoring critical journalists. The former German Chancellor has long since entered the "unknown territory", the "*Neuland*", as Angela Merkel herself named the Internet once, and unfortunately, she may also have violated the precept of the press being far removed from the state. In any case, her *Instagram* and *YouTube channels* reach millions of followers, without any critical questions from journalists. In this way, the government itself is helping to undermine the public, which has the task of controlling it.

Public 2.0 is now a different public from the democratic public demanded by the German Federal Constitutional Court in two respects: it gets by largely without the critical questioning and weighting of professional journalists, and distribution is carried out by the ultimately non-transparent amplification rules of the large platforms, provided that the content meets their criteria of the attention economy.

Of course, from the perspective of those in power, it is more pleasant to be able to communicate directly to the citizens without being grilled by journalists. But even if politicians, present themselves on *Instagram* while drinking coffee or in everyday situations – Merkel is not one doing that - this banalization of politics only serves to stage fake authenticity and fake transparency. Ultimately, media that focus exclusively on emotionalization serve the populists best.

According to a study by the Friedrich-Ebert-Stiftung on the 2019 European elections, it is above all the new German right-wing

[648] http://www.servat.unibe.ch/dfr/bv020162.html

populist Party AfD that dominates virtual space.[649]. The AfD has more *followers* than the parties in power, SPD and CDU, combined. Only the so-called until recently"fun party" "Die Partei" can keep up with the right-wing populists, not least because of the ongoing promotion it receives on satirical programmes like the public TV Channel ZDF "Heute Show" (Today Show).

So why has virtual space become the rallying ground for those who want to attack democracy or ridicule it? An exhaustive answer would go beyond the scope of this book. Here it must suffice to point out that the strategies of the counter-public sphere from the early days of *cyberspace* are now being used by *anti-establishment* forces on the other political fringes, thus the right, to build parallel public spheres online. One the one hand the difference between the two forms of protest against the dominant order is striking. On the other side both forms are deeply connected to the core of the Californian ideology and the functional mechanisms of the algorithms that determine visibility and invisibility on *social media*. In particular, the "idiosyncratic concept of truth" of right-wing and radical right-wing groups is favored in *social media*, as another study has shown.[650] Activities by fringe groups on networks such as *Facebook* are predominantly fuelled by scare tactics, distortions and conspiracy theories. The platforms offer the ideal amplifiers for this; they can be used very quickly to build up reach and generate approval through "metric manipulation" and *gaslighting* techniques, i.e., the creation of a completely fictitious world impression as perceived reality, often combined with the sowing of mistrust of the institutions of state and society.[651]

---

[649] Fuchs, Martin, Holnburger, Josef:#ep2019 - The digital party strategies for the 2019 European elections https://library.fes.de/pdf-files/id/ipa/15655.pdf

[650] Fielitz, Maik Marcks, Holger: Digital Fascism: Challenges for the Open Society in Times of Social Media https://escholarship.org/uc/item/87w5c5gp

[651] See interview by Bernd Graff with the authors of the study in Graff, Bernd, “Im digitalen Faschismus-Strudel”, Sueddeuteche, 8 September 2-19 p. 17 https://www.sueddeutsche.de/kultur/soziale-netzwerke-faschismus-studie-maik-fielitz-holger-marcks-1.4590356.

A public sphere structured in this way offers an ideal field of activity for liars and agitators, but it is increasingly unable to meet the demands made on a political public sphere in the deliberative model of democracy. Many of the frenzied online debates favoured by the algorithms are far from the factually oriented, balanced and rational ideal. The resulting cacophony of the Internet does not replace the structuring and criticizing function of journalism as the fourth estate checking on power.

Through the business models that shape electronic communication, economic power imperatives intervene directly in the generation and distribution of content. A business model that does splendid business with hate, lies and conspiracy theories because such content attracts the attention of users has corresponding consequences for society. As former Google manager Tristan Harris formulates it: "Free is the most expensive business model we've ever created because if people can't agree of shared facts and trust, that's it."[652] The costs of the supposedly free culture are borne by individuals and society.

If the political public, which struggles for rational decision-making in democracy, is hardly detectable any more, democracy itself threatens to disappear with it. In other words, the Habermasian "communication pathologies" can reach a scale that can destroy a functioning democracy.

Digital change has literally been taking place over the last two decades in the middle of the public sphere, a sphere it partly created and increasingly structures. This is reason enough to take a closer look at the underlying mechanisms of action, the communicative and economic development logics and the power structures.

---

[652] Johnson, Eric. 2019. 'Tristan Harris Says Tech Is "Downgrading" Humanity — but We Can Fix It'. Vox. 6 May 2019. https://www.vox.com/recode/2019/5/6/18530860/tristan-harris-human-downgrading-time-well-spent-kara-swisher-recode-decode-podcast-interview.

1.3. Public 2.0: Democracy in the pincer grip of populism and technology

German writer Eva Menasse said in her 2019 Börne Prize speech that "the public at large, which in Börne's time had only just formed as a civil society counter-public to the state and its organs, is currently completely disintegrating."[653] This fear does not seem exaggerated. The digital public sphere is a public space whose size, architecture and acoustics are subject to constant change, but the parameters are in no way determined by the users, nor can its boundaries really be viewed. The principle of personalization creates a structural difference in Public Sphere 2.0, when compared with Habermas' ideal or the analyzed public of the mass media. The problems associated with mass media were still transparently visible and thus easy to critique. But personalization removes the visible public sphere, and thus opportunities for engagement and criticism.

The consequences are fragmentation and affirmation as well as a lack of power control over the platforms. In the film *The Social Network*, Aaron Sorkin has Sean Parker, Mark Zuckerberg's initial co-creator, say: "First people lived in the village, then in the cities and now on the Internet." People move around the net as if it were a village square or an agora, yet know comparatively little about its structures. The architects of this new living space of the digital public work in secret, according to rules that their companies give themselves, but which are not clear to the inhabitants. Enthusiasm for the new potential for freedom offered by the Internet and digitalization lasted for a long time, followed only much later by disillusionment and serious concern that these technologies would destroy freedom and democracy.

---

[653] Manesse, Eva, "Für Pessimismus ist es zu spät (Dankesrede von Eva Menasse anlässlich der Verleihung des Ludwig-Börne-Preises)", https://www.kiwi-verlag.de/magazin/gastspiel/fuer-pessimismus-ist-es-zu-spaet-dankesrede-von-eva-menasse-anlaesslich-der

Like with radio, the original enthusiasm for the new medium, were subsequently dampened when its impacts were better understood. In the early days of the radio, the ones who used the magic of the new medium most efficiently, were the Nazis. The presence of Hitler's voice in almost every German living room through the "Volksempfänger" (people's Radio) was a decisive factor in the success of the National Socialists.

Radio's *one-to-many* communication structure may be in keeping with the strict hierarchy of the leader state. However, it would be premature to conclude from the apparently decentralised network structure of the Internet that non-hierarchical democratic systems are preferable for democracy. Today, right-wing parties and populists make the most effective use of the possibilities of communication via *social media networks*. New media thus seem to favour new political "movements" in general. Or is there a deeper reason why social media and populism seem to be fuelling each other and democracy is thus caught in the pincer grip of technology and populism?

In the *many-to-many* communication of the Internet, those who organise the flow of information amass significant power. Their interventions increasingly replace the open structure of communication in favor of structures that first of all enable the monitoring and then the marketing of participants, their communication and their entire *lebenswelt* (life-world) behavior. These interventions are bringing about a fundamental change in the public sphere: Not only is public communication increasingly determined by economic imperatives of a few platform operators; the participants themselves have to pay the price for the participation in the form of personal data to an extent that increasingly undermines their freedom and self-determination. The asymmetry of knowledge, authority and power then also leads to a deformation of the public sphere that goes far beyond that which was conceivable in the age of mass media through classical distortions due to commercial interests.

The deformations of the public in the age of mass communication consisted of advertising influence and propaganda or PR. In the digital network age, the deformation sets in deeper in the individual lives of the subjects, and attempts to capture and control them through dense surveillance.

The reports in the mass and tabloid media may have always been partly unbalanced, partisan and manipulative. Nevertheless, identical reports reached a broad audience. The audience could then discuss these reports together and, supplementing their understanding of the situation with other independent sources, make them the basis for their own opinion formation. Only through this process does the published opinion become public opinion. The basis is a critical awareness among the audience that the basis of their criticism is found in criteria that they can form by consuming diverse but jointly shared media content.

In the digital public, however, this reflection driven by shared media disappears as the bubbles become better formed, and separated. This reduces the willingness and ability for critical reflection. This effect also arises due to personalization if as some researchers have recently shown,[654] the bubble effect of social media communication is weighted less in favor of greater diversity. This is because the algorithms generally give higher weighting to content which is more emotional and attracts more attention. The published opinion of certain opinion-forming groups within their respective communities therefore threatens to become the "public opinion" that is loudest and most radical, with the chance of gaining ever larger numbers of *followers* based on the mechanisms of polarization and radicalization.

---

[654] Scharkow, Michael, Frank Mangold, Sebastian Stier, and Johannes Breuer. 'How Social Network Sites and Other Online Intermediaries Increase Exposure to News'. Proceedings of the National Academy of Sciences 117, no. 6 (11 February 2020): 2761–63. https://www.pnas.org/doi/10.1073/pnas.1918279117

In 2007 Habermas took up the theme of his first work once again under the title "On the Reason of the Public Sphere"[655] and placed it in the deliberative theory of democracy. This time he emphasizes even more concisely the importance of a political public sphere for the preservation of democracy. And much more detailed than in his habilitation thesis, he analyses the factors of domination that are at work in the differentiated communication structures of modern democracies in order to enforce particular interests in the community. The increasing flow of political communication has according to him meanwhile led to a "communicative liquefaction of politics". There are still actors at work who are trying to dominate public debate,[656] either by bringing their commercial or personal interests into play. This battle is being fought, with the aim of gaining the scarce resource of human attention. But: if the orientation towards the normative ideal of a rational discourse guided by arguments is still recognizable in mass communication, democracy can live with these distortions. For the published opinion co-determined by unilateral influences is first reflected in the sounding board of an independently arguing public and then formed into public opinion.

The generally known criticism of Habermas' writing is that the concept of ideal discourse free of domination it to idealistic and therefore not realistic. But the concept can also be correct if it is counterfactual. It is part of the core of norms that do not merge into being and that a gap to the should is determining. Therefore, Habermas rightly insists on analysing the normative resources of modern systems of rule. For the fact that discourses are often deformed does not mean that there can be no ideal by which to measure them. Habermas provides helpful prerequisites for an ideal public sphere, which are that (i) no one capable of making a relevant

---

655 Habermas, Jürgen. 2014. Europe: The Faltering Project. John Wiley & Sons p.131 ff.

656 Habermas, Jürgen. 2014. Europe: The Faltering Project. John Wiley & Sons p.156.

contribution has been excluded, (ii) participants have equal voice, (iii) they are mentally free to speak their honest opinion without deception or self-deception, and (iv) there are no sources of coercion built into the process and procedures of discourse.

Because of the interlocking of economic and political interests, the public sphere has always been subject to commercial or particular interests that try to assert themselves, building sources of coercion into the processes of discourse and preventing equality among the participants. The arena of arguments has also always been a battlefield of manipulative techniques, at the latest since the emergence of the Sophists in ancient Athens. The participants in democratic discourses know this and usually reflect on the power structures before forming a public opinion from the widely circulated mainstream opinion. However, a truly disruptive element has been added to this complex dynamic with the emergence of the large platforms: Digital manipulation penetrates deeper under the skin than seemed conceivable in the age of mass media, because it is based on precise personal profiles.

The combination of personalization and universalization, which results from the AI-based analysis of *big personal data,* allows predictions to be made about the behaviour of citizens, which massively restrict their autonomous scope for decision-making, even in the field of politics. The algorithms that are calculated to control the purchasing behaviour of consumers also affect citizen voting behaviours. This takes place as algorithms evaluate the "preferences" of the users, and filter the information to which they will be exposed, thus manipulating citizens, as in the case of *Cambridge Analytica* - and in the recent US election campaigns. The political voice thus becomes a commercial product like any other.

Through these mechanisms, one can see the "intrusion of the functional imperatives of the market economy in the 'internal logic' of the production and presentation of messages.".[657] The result is that what people believe to be true has become the product of the manipulative control to which they are subjected. The data for this manipulation was previously provided by their own use of platforms.

Even in the age of the classical mass media, economic imperatives affected what was published: after all, advertising revenues have been an important economic pillar for the media for a long time. Nor was opinion making, i.e., the manipulation of the public, alien to the mass media. But during this period, the interests embedded in the big newspapers and television channels were, more or less, known, and there were clear legal regulations to ward off predominant power of opinion, which were intended to ensure the diversity of journalistic voices and provide the prerequisites for free opinion formation.

But the greatest change is that even the spectacle of tabloid media could not be targeted at pressing a specific reader's buttons. It was instead, directed at a commonly shared public sphere, available in all newsagents, and visible to any citizen. It was possible for anyone who happened to see the paper to reflect and respond to misinformation. There were obviously journalists and publishers producing questionable content in the published opinion of the analog media. But it was more likely that they were caught and corrected by the sounding board of a shared public sphere before they could become a contribution to public opinion.

This necessary sounding board of shared information and values threatens to get lost in the personalized bubbles of the 2.0 public. The 'general' public is replaced by an ever proliferating, cordoned

[657] Habermas, Jürgen. "Political Communication in Media Society: Does Democracy Still Enjoy an Epistemic Dimension? The Impact of Normative Theory on Empirical Research1," Communication Theory 16, no. 4 (2006): 411–26, https://onlinelibrary.wiley.com/doi/abs/10.1111/j.1468-2885.2006.00280.x

off, number of sub-publics. This is leading to the loss of common convictions, and thus the basis for organizing society together in a democratic community.[658]

The consequence of digitization for individuals is that their private lives are made public, and the major platforms, by collecting data and creating personal profiles, undermine their autonomy and manipulate their decisions. The digitization of public communications has played a significant role in the atomization of societies, which can be observed worldwide.

For the public, it means that the commonly shared space is replaced by a confusing multitude of personalised bubbles in which it is increasingly difficult to communicate about common concerns. At best, social groups with similar profiles can still carry out a kind of partial decision-making, but hardly one that is oriented towards universalistic values of freedom and self-determination.

[658] There are studies, partly financed by Facebook, (see for example: "The Truth behind Filter Bubbles: Bursting Some Myths | Reuters Institute for the Study of Journalism," https://reutersinstitute.politics.ox.ac.uk/risj-review/truth-behind-filter-bubbles-bursting-some-myths. https://www.zeit.de/2017/34/algorithmen-filterblase-meinungen-selbstbetrug), which aim to disprove the formation of the filter bubbles. Even if one admits that through Internet communication also a broadening of perspectives and thus an increase in diversity can be achieved: Our thesis is that in the long run, the filter effects of personalization of information will predominate and have the mentioned disadvantages for public communication.

E2 Power domination and automation destroy the democratic public

2.1. All-powerful algorithms?

> *"The scientific abstraction from concreteness, the quantification of qualities which yield exactness as well as universal validity involve a specific concrete experiences of the Lebenswelt – a specific mode of "seeing" the world. And this "seeing", in spite of its "pure", disinterested character, is seeing within a purposive practical context. It is anticipating (Voraussehen) and projecting (Vorhaben). [...] the science of methodological, systematic anticipation and project. But – and this is decisive – of a specific anticipation and project – namely, that which experiences, comprehends and shapes the world."*

Herbert Marcuse.[659]

Algorithms are electronic instructions for action, arithmetic steps and procedural rules that serve to achieve specific goals - no less, but also no more. Due to the growing importance of algorithms in the control of political processes, we are facing the new danger of an 'algocracy'.[660] So what can algorithms actually do?

They can

- *select*
- but also *weigh*.
- *and above all, decide.*

---

[659] Marcuse, Herbert. 2013. One-Dimensional Man: Studies in the Ideology of Advanced Industrial Society. Routledge. P.168

[660] The Stanford sociologist A. Aneesh coined the term "algocracy", a form of rule in which program codes implement political control, a concept developed in Aneesh, A. 2006. Virtual Migration: The Programming of Globalization. Duke University Press.

The transition from a descriptive to a normative approach in this combination is fluid. And the norms that are generated here are false semblance.

If an algorithm pursues what Larry Page once defined with his famous *Page Rank* as 'relevance', namely to arrange the selected Internet pages according to how often they have been referred to from other sites, and weighting them, with the aim of building a *ranking*, he makes a judgement. To deduce from the mere accumulation of links that something is important and relevant may possibly lead to relevant results in science, but even there it is not undisputed. Larry Page adopted this method from science. His father was a professor of computer science at the University of Michigan. In the scientific community, the citation rate is regarded as an indicator of the importance of a publication. This so-called *impact factor*, listed in the *Science Citation Index*, since the 1960s has been established as a measure of the quality of the science itself.[661] This is despite the fact that it has been extensively criticised in the scientific community. For one, vanity plays a not insignificant role in the counting of the citations. However, it can be assumed that the members of a scientific community not only have their own interests in increasing their citation-rate and raising third-party funds, but that they also pursue research interests that are more 'purely' motivated and oriented towards scientific standards and thus towards truth.

The *Google* search algorithm is quite different, it is even further away from 'truth'. It sorts results using the number of links pointing to the site and time spent by users alongside the producers' social media

[661] The German Research Foundation criticises the impact factor, which is not a sufficient criterion for scientific relevance: "The citation frequency (depends) obviously not only on the reputation of a journal or a working group, but above all on the size of the group of scientists interested in the topic. Specialised journals have lower impact factors than those with a broad readership; a small subject has different quantitative measures than a large one". PDF document Securing Good Scientific Practice (German Research Foundation, memorandum adopted on 3 July 2013).

profile. At the most, it considers the 'truth' of a particular page to the extent that it can be approximated by probability calculations or calculations on a quantitative basis. Instead of an objective reference to truth, the page-Rank puts relations first.[662] The *Page Rank* thus contains the disadvantages of the scientific impact factor, on an exponential scale: since the quality of scientists after publications in scientific journals is calculated down to the decimal point, and careers and research projects are decided by this, many promising fields of research are neglected in favour of those fields and topics that promise the most citations. A kind of SEO-optimization of research is taking place, which can corrupt science by adding third-party funding. Through *Google's* search algorithm, the web determines from within itself what is important and what is not. Only what is measurable and quantifiable is taken into account. Qualities that do not fall into this category cannot even be considered. At this point, the search algorithm replaces the limitation of measurability with a consideration of *rankings*. The fact that something is linked frequently is initially a purely descriptive fact. In the mechanism of *page rank,* however, this becomes a normative *ranking*. This creates the impression that *Google* can identify "relevance" in a comprehensive normative sense.

Now one can argue that the algorithm is nevertheless "objective" and thus relevant, because the scientific community is replaced by the "*wisdom of the crowd*". The algorithm may come close to this in some search fields. But in general, net communication rewards what is quickly grasped and strongly emotionalized rather than in-depth and instructive. Since *Google,* according to its own statements, aims to order the world's knowledge and make it accessible, it is surprising

---

[662] This is also a thesis of structuralism in the 20th century. Claude Lévi-Strauss, for example, believed the error of the humanities and social sciences was that they did not pay enough attention to the relationships between the elements of the respective systems: "The error of traditional sociology, as well as of traditional linguistics, is to have looked at the members and not at the relationships between the members". Lévi-Strauss, Claude. The Savage Mind, Frankfurt 1968, quoted in the Whole Earth Catalogue p. 310.

how little the public learns and discusses the criteria according to which this new order, which controls access to information, takes place.

Through personalization, the information shown is perceived as "appropriate", the selection of messages thus seems to follow a higher power, and thereby gains credibility and authority. A mechanical *Circulus Vitiosus* is created: the more the results are perceived as "suitable", the more credible the selection of the algorithm appears and the more authority it gains.

Studies show that by 2005 *Google* had already overtaken traditional media as the most credible news provider.[663] *Google* does not produce a single piece of news itself - except, of course, corporate news. Nor does the algorithm have criteria relating to "truth" with which it could check the selected information. It makes "recommendations" without knowing whether these are true or credible.

Incidentally, the term "recommendation algorithm" demonstrates the perfidious language policy of the major platforms. Similar to the terms like *Friends or Likes,* bonds are taken in the life world, which we associate with human characteristics in order to transfer them to automatic machine procedures. Friends we like. Whether *Friends* are our friends or *chatbots* and whether they really like their *Likes,* we do not know. We also don't know with which logic the messages are displayed in the *timeline* or in the *news feed*. And the algorithm doesn't "know" all that either. But it is programmed for targets that are not ours: Firstly, to increase the time spent on the site by providing "appropriate" information and entertainment and

---

[663] McDulling, John. Google is now a more trusty newssite, than the websides it aggregates, Quartz.com, 20 January 2015 https://qz.com/329211/google-is-now-a-more-trusted-source-of-news-than-the-websites-it-aggregates

secondly, but even more importantly: to do everything possible to satisfy the sensationalism of the users and to keep offering fresh stimulation. Then it's about getting us to make as strong a "commitment" as possible - and collecting as much data about us as possible, which is then auctioned off as a behavioural prognosis about us on the commodity futures markets of *Targeted Advertising*.

2.2. From the people's radio to the people's amplifier

With the *Volksempfänger* (people's radio) the Nazis created a propaganda- and indoctrination instrument without equal. With it, the voice of the "Führer" could penetrate every living room. Goebbels recognized the radio as a new propaganda instrument and ensured that the Volksempfänger was soon to be found in most households.

Today, with the *social media channels*, we have people's amplifiers which, with their logic of noise, prefer polarizing and spectacular contributions to moderate, reasonable contributions that are designed for understanding and balance. In contrast to Volksempfänger Radio, the Volksverstärker (people's amplifier) of *Social Media* also broadcasts the voices of the people according to the *prosumer principle* of the Internet where everybody potentially is a producer as well as consumer. But for the reasons outlined above, unfortunately preferably the loudest voices as the most audible ones. While the Nazi propagandists had to stage themselves as the voice of the people, today real people have their say. This shift was first cause for celebration, but for the unfortunate reality that preference in terms of transmission is given to those who take the most strident positions. These can be the most radical political perspectives or the most entertaining jokers. As the algorithms, search engines and social networks then amplify these voices via network effects, they create feedback loops in ever larger echo chambers. As a result, the radicals gain traction that goes far beyond their actual importance and contribution to democratic debate and decision-making.

Just as *Alt-Right* and Steve Bannon in the US took possession of these mechanisms of the Internet, so too do the AfD and "Die Partei" in Germany. The latter, represented by Martin Sonneborn in the European Parliament, had the most social media supporters in the European elections. No wonder, since their top candidate is a trained editor-in-chief of a satirical newspaper and filled the party's list for the 2019 European elections with names like "Hitler" and "Bombshell", because it is so funny to make fun out of Hitler and terrorism to attract attention an election. The Internet success of these candidates is an example of how algorithms support dissemination: purely according to the criteria of the attention economy. But if you try to spread reasonable arguments, the algorithms will censor you to the backmost hit pages. *YouTube's* algorithms make sure that users tend to be encouraged and radicalized in their views, because it is in the logic of the attention economy, whose first law is: never bore, always entertain!

This logic was further intensified by the modification of the *Facebook algorithms* at the beginning of 2018. Marc Zuckerberg stated that the contributions of friends and family should be weighted higher than corporate news and that the algorithms should no longer simply extend the time a user spends on *Facebook*, but should aim to increase their engagement. As a result, those who share posts with a *love* symbol instead of a *like* will receive greater visibility.[664] It is above all the viral spread of emotions that Facebook understands and rewards as *meaningful social interactions*. For these algorithms, "high-quality"

---

[664] Wong, Julia Carrie. 2018. 'Facebook Overhauls News Feed in Favor of "Meaningful Social Interactions"'. The Guardian, 12 January 2018, sec. Technology. https://www.theguardian.com/technology/2018/jan/11/facebook-news-feed-algorithm-overhaul-mark-zuckerberg

content is defined as content that triggers the most intense emotions in the *followers*.

Since almost half of all citizens now obtain political information from social networks,[665] the risks for democratic societies are foreseeable. Populists, extremists and entertainers, who are already playing the piano of social media with the greatest virtuosity, could receive even more attention in the future.

Democracy could end up being a victim of the demagogy catalyzed by technology. This too is a dialectic of the digital. After all, it is technology that promises the greatest possible participation and acceleration, and which may ultimately lead to a total political regression. Those who thought that populism and fascism had been overcome are now seeing new populism and authoritarian regimes flourishing all over the world. Whoever thought that the open society would be realized and strengthened by the Internet, now see the tribal society on the rise. Digital technology, the Internet and AI have all played their role in this situation.

2.3. The puppet master, not the puppet![666]

If half of all Europeans do not know what an algorithm is and how it filters and sorts what they see[667]: how many will know that behind each algorithm there are people who use the program to pursue their own interests? At this point, there is no significant difference to the

[665] According to Reuters Intstitute, Digital News Report 2019 p. 42/43, 42% of the population in the US and Europe use the Internet and social media as a source of political information, see: https://reutersinstitute.politics.ox.ac.uk/sites/default/files/inline-files/DNR_2019_FINAL.pdf.

[666] See above Chapter C and Zuboff, op. cit. 30

[667] Bertelsmann Study: Half of the Europeans do not know what an algorithm is: https://www.bertelsmann-stiftung.de/de/themen/aktuelle-meldungen/2019/februar/europaeer-wissen-wenig-ueber-algorithmen/.

age of mass media. Just like publishers and *tycoons* who influence public opinion to a greater or lesser extent with their publications, so do the owners of the platforms.

In the age of AI, on the other hand, the Californian ideology has spread the myth of the freedom of value and the "divine superiority" of these systems. Even the former world chess champion Kasparov, in his historic defeat by *Deep Blue*, expressed his surprise that the machine stopped playing like a computer, ignoring the allure of the queen attack, and adopted a more advanced strategy. Having previously been sanguine about his prospects of winning Kasparov was unsettled by a shift towards a subtler form of play by the computer, and believed that *Deep Blue* had become "superior" to him, "suddenly played like a god for a moment."[668] Finally he gave up because he felt inferior to the program: "Deep Blue's playing had impressed me very much. I thought so much about what the computer was capable of that I didn't realize that my problems were more because I played badly than because it played well."[669]

Users and society today seem to feel the same way about AI and the power of digital corporations. According to one study, 51 percent of users said they didn't know that the messages they receive on social networks are pre-sorted at all, while other surveys show that social media users, for the most part, cannot distinguish between news and comment pieces.[670] They consider opinion-based pieces to be an objective selection of world events, like the news items in a

---

[668] Quoted from Fry, Hannah. Hello World: How to Be Human in the Age of the Machine. Black Swan, 2019. p. 18.

[669] Quoted from Fry, Hannah. Hello World: How to Be Human in the Age of the Machine. Black Swan, 2019. p. 19.

[670] Fischer, Lara and Thomas Petersen, Bertelsmann Foundation: https://www.bertelsmann-stiftung.de/de/publikationen/publikation/did/was-deutschland-ueber-algorithmen-weiss-und-denkt/

Also see : https://www.americanpressinstitute.org/publications/confusion-about-whats-news-and-whats-opinion-is-a-big-problem-but-journalists-can-help-solve-it/.

news broadcast. This is not to deny that news broadcasts are also subject to selection processes. There is no complete objectivity. But at a minimum they are selected by people who, in their profession, have to learn how to check news and facts and present it according to professional journalistic criteria.

It is therefore important that the search and filter mechanisms become transparent and that no more power is attributed to the algorithms than they already have, especially not that of "superior" intelligence. Already in the above shown process of apparent norms, which are generated by algorithms through the selection decisions and the ranking, it can be seen that the norms of a machine generated "correct" weighting only exist if we recognize this norm.

2.4. Power and violence in digital space

> *"In a lot of ways, Facebook is more like a government than a traditional company. We have a large community of people, and more than other technology companies, we're really setting policies"*

Mark Zuckerberg[671]

Scott Galloway, professor of economics at NYU, calls Marc Zuckerberg the "most powerful, dangerous man in the world"[672] because of his influence on now 2.8 billion users in 2020 on *Facebook, WhatsApp* and *Instagram*.[673] The danger lies in the power that Facebook wields and

---

[671] Quoted in Kirkpatrick, David. 2011. The Facebook Effect: The Inside Story of the Company That Is Connecting the World. Simon and Schuster.

[672] Aiello, Chloe. 2018. 'Mark Zuckerberg Is the Most "Powerful, Dangerous Person in the World": NYU's Scott Galloway'. CNBC. 11 May 2018. https://www.cnbc.com/2018/05/11/nyus-scott-galloway-zuckerberg-is-the-most-dangerous-person-in-the-world.html.

[673] Source: https://allfacebook.de/toll/state-of-facebook.

in the company's claim to take the place of politics. Yet the influence of *social media sites* and search algorithms on the public supply of information continues to grow.

The digital public sphere seems to be entirely in line with the attention economy conceived in systems theory in pushing the permanent generation and amplification of public excitation curves. If the "reasoning" audience is only pushed into changing states of excitement, the critical contribution to the formation of opinion is lost.

The manipulation of public opinion becomes a form of structural violence through the personalization of information flows and the replacement of journalistic *gatekeepers* with commercially controlled algorithms. On the one hand, it potentially prevents individuals from identifying and pursuing their interests. On the other hand, with the intersubjectively divided public sphere, which goes further than its own statistical *peer group*, it deprives people of the very stage on which self-understanding can take place. Moreover, digital paternalism with AI has sophisticated monitoring and manipulation possibilities at its disposal, which can intervene deeply in the personality and the ability for individual self-determination.

On the basis of an enormous amount of personal data collected, the combination of attention economy and surveillance capitalism produce new totalitarian forms of rule. The banality of stupidity,[674] which arises from the predominance of apparently personalized communication that is in fact radicalized, turns the contextual

---

[674] Hannah Arendt spoke of the banality of evil in the case of the bureaucrat Adolf Eichmann, who justified the extermination of the Jews in his statements in court by saying that he had acted as a will-less instrument of the Führer's will, Arendt, Hannah. Eichmann in Jerusalem, 1st edition. New York, N.Y: Penguin Classics, 2006.

[675] Horkheimer, Max, Theodor W. Adorno, and Gunzelin Noeri. Dialectic of Enlightenment. Stanford University Press, 2002.

delusion (*Verblendungszusammenhang*)[675] which Adorno pointed to into a contextual stupefaction (*Verblödungszusammenhang*).

The expression of stupefaction refers to the original meaning of the word idiot. *Idiotes* comes from the Greek for a private citizen who kept completely out of public life and only cared about his own affairs. This phenomenon is the reason for the rejection of the *idiot* in Athens' democracy, which punished those who evaded popular assemblies: "In Attic democracy, which was based on informed and active citizens (*politai*), the *idiotai* were little appreciated. One was born and remained an *idiot* if not exposed to the education and training that was considered to create the politically conscious citizen."[676]

Digital filter bubbles and the fragmented society threaten the return of the phenomenon of *idiotai*. The radical tone of online communication is also spurring this development. According to a survey conducted by the Allensbach Institute for Public Opinion Research on behalf of the newspaper Frankfurter Allgemeine Zeitung, only one in five Germans feels free to express their opinion in public, while 80 percent feel restricted in their freedom of expression. The respondents do not cite state paternalism or fear of state sanctions as reasons, but rather the hostility of public debate on the Internet. The way of communication on the Internet thus generates the perception of censorship itself which then spreads "virally" within it.

The counterculture in the 1960s proclaimed that the private was political. In the meantime, the political has become private by falling into the hands of private companies. *Facebook* and co. declare the private to be a commodity and thus not only destroy it, but with it the political as well.

---

676 https://de.wikipedia.org/wiki/Idiot#cite_note-Parker-2'; see also: Parker, Walter C. "Teaching Against Idiocy," Bloomington: Phi Delta Kappan. 2005, https://faculty.washington.edu/rsoder/EDUC305/305parkeridiocy.pdf

2.5. Decline of newspapers, rise of social media and video platforms

At the same time as the decline of newspapers, we see the rise of social media and video platforms. They are superior to the rival press in almost all fields in the highly armed battle for attention. They are "multimedia", i.e., they engage not only through text, but also through images and music. What makes them superior to classic "moving image offerings" such as television, however, is the possibility of personal addressability, based on the collection of enormous amounts of personal data.

In the 1980s the term 'prosumer' was developed in *The Third Wave*[677] by futurist Alvin Toffler to describe the blurring of the distinction between producers and consumers of content. This term related to the ability of new technologies to enable mass customization of products, with consumers specifying their particular design preferences. In journalism, the higher "authenticity" generated by the *prosumers*, replaced the credibility of the professional journalistic *gatekeepers*. On YouTube and TikTok every consumer can be a producer if he wants, and with luck and skill maybe he will be a star tomorrow.

Which promise of freedom and fame at the same time could be greater? More than almost any other platform, *YouTube* stands for the dream of the founders of the Californian ideology of freedom and self-realization in a virtual space over which adults have no influence. The platform is successful, so successful that it has long since "disruptively" changed the media market for the young target groups: in YouTube's favour, of course. According to a Pew report,[678]

---

[677] Alvin Toffler, The third wave, An Entrepreneur's vision of the future, 1980

[678] Grieco, Elizabeth. "U.S. newspapers have shed half of their newsroom employees since 2008", 20 April 2020 https://www.pewresearch.org/fact-tank/2019/07/09/u-s-newsroom-employment-has-dropped-by-a-quarter-since-2008/.

more than 26,000 journalists were laid off at media companies in the US between 2008 and 2017. The number of employees in news media (television, print, radio, online) has fallen by 23 percent from 114,000 to 88,000.

The slump in newspaper circulation hasn't been offset by growth in online media for the same companies. *Google* and *Facebook* now control over 60 percent of the US advertising business.[679] In 2019, the share of digital advertising expenditure exceeded 50 percent of total advertising expenditure for the first time, of which *Google* and *Facebook* earn more than 50 percent for their services.[680] This is based on an economic model that drags along the cult around the liberating possibilities of the platform, as is usual within Californian ideology, without really doing anything for it. "*YouTube* has never been the grassroots democratic medium it has always been stylized into, but rather a platform from the outset, which primarily wants to attract and promote advertising-friendly content," so said the authors of a *YouTube* study commissioned by the Otto Brenner Foundation.[681]

The characteristics of *YouTube* are shared with many other products created by the GAFAM companies: They remains sexy even if they do nothing to exude sexiness. They derive their appeal from the rich heritage of pop culture that is broadcast without producing pop cultural content in any way itself. The sexy medium is the message. Hardly any other medium has had such an influence on the content of the market as *YouTube* does through its market logic. At the same time, it derives its success from a technology that seemingly

---

[679] https://onlinemarketing.de/unternehmensnews/us-werbemarkt-amazon-waechst-facebook-google-verlieren-marktanteile.

[680] https://www.emarketer.com/content/global-digital-ad-spending-2019.

[681] Frühbrodt, Lutz and Annette Floren, "Unboxing YouTube In the network of professionals and profiteers", p.12 f.

https://www.otto-brenner-stiftung.de/fileadmin/user_data/stiftung/02_Wissenschaftsportal/03_Publikationen/AH98_YouTube.pdf

meets people's needs for exchange and expression and that is very convenient. On the basis of these desires, which *YouTube* satisfies, the platform can spark unlimited new needs in the service of its advertising customers: above all consumer satisfaction that is only ever a click away. Yet, in following our cravings we generate "data exhausts" trails of information, that are used for further manipulation of our behaviour.

As with all GAFAM companies, there is an unleashed economic logic of market domination and profit optimization. The collateral social damage is accepted as an unfortunate, but inevitable, by-product of technological advancement. The collateral damage is huge. Political extremists and fanatics use *YouTube* a platform for their hate messages and conspiracy theories. The best-known example of this is the Islamism terror organization IS. That *YouTube* could spread IS content for a long time with impunity and even earn money from its distribution is astonishing; as is the fact that we who have felt the dark impact of this propaganda in the form of terror have accepted its dissemination. Imagine if, during the series of IS attacks in recent years, a TV station had provided uncensored airtime to the Islamist terrorists.

Two thirds of 14-29 year olds in Germany regularly use *YouTube,* and in the summer of 2019, web offerings outstripped television use in this age group for the first time. This trend is even stronger among children. The influence of *YouTube* on identity formation among young people can hardly be overestimated. The platform, which like *Google* belongs to *Alphabet,* is the second largest search engine in the world after Google. This means that young people, in particular, are increasingly looking there not only for entertainment but also for information.

In both media worlds, the analogue and the digital, one finds light and shadow. But there are also structural differences. In 2019, the Otto Brenner Foundation conducted a study of the *social media sites*,[682] in which the authors evaluated over 100 YouTube channels. They found that "the vast majority of the channels are characterised by undemanding, often even flat and strongly emotionalised entertainment and are also permeated by product advertising."[683] More than three-quarters of the sites operated on YouTube are entertainment, comedy, gaming, music and lifestyle.

In terms of content, the shows focus on escapism, careerism and consumerism. Of the 100 channels, "just four channels deal with politics and knowledge in a broad sense."[684] For the authors of the study, the channels have a "tendency to be neoliberal or conservative in character". They show a "radical conformity". Conformity in that which guarantees maximum clicks, thus they follow the logic of the algorithms as well as possible. A layer of young hedonistic entrepreneurs has emerged, which in the main is not political, as the success of their channels is formed by business models and algorithms from Silicon Valley. Clearly there are some exceptions to this general trend towards political apathy. The generally low hurdles to entry means that YouTube also offers politically interested groups a public image and thus can contribute to the diversity of political discourse. What we criticize is the logic of the algorithms, which aim for the longest possible "time spent" and at the same time the highest possible emotionalization as a basis for the optimal marketing to the users, which means that the laudable, critical content produced

---

682 Frühbrodt, Lutz and Annette Floren, "Unboxing YouTube In the network of professionals and profiteers", https://www.otto-brenner-stiftung.de/fileadmin/user_data/stiftung/02_Wissenschaftsportal/03_Publikationen/AH98_YouTube.pdf

683 Frühbrodt, Lutz and Annette Floren, "Unboxing YouTube In the network of professionals and profiteers", loco. p. 6. https://www.otto-brenner-stiftung.de/fileadmin/user_data/stiftung/02_Wissenschaftsportal/03_Publikationen/AH98_YouTube.pdf

684 Frühbrodt, Lutz and Annette Floren, "Unboxing YouTube In the network of professionals and profiteers", p. 6. https://www.otto-brenner-stiftung.de/fileadmin/user_data/stiftung/02_Wissenschaftsportal/03_Publikationen/AH98_YouTube.pdf

by these organizations is frequently eclipsed by polarizing and radicalizing, or simply emotional and apolitical, content.

The authors of the study explicitly blame this trend on the mode of operation of the algorithms used, which are designed to emotionalize and polarize by making such content more visible to the user than objectively differentiating contributions, "which can have fatal effects, especially in political discourse", as the study's editor notes.[685]

2.6. How Google and Facebook use "News initiatives" to gain power over the media and steal ideas

From 2018 *Facebook* and *Google* introduced initiatives to promote journalism. Motivating this plan were the constant complaints from publishers that the portals were ruining their traditional business model. First *Google*[686] and then *Facebook*[687] announced that they would each provide $300 million to support development projects in publishing houses. The goal as stated on the *Facebook*-project site is to "better meet the needs of people on *Facebook*" at an early stage of product development.[688]

The influence of the two data giants on journalism is growing. News is mainly searched on *Google* Search and *Google* News. This allows

[685] Frühbrodt, Lutz and Annette Floren, "Unboxing YouTube In the network of professionals and profiteers", p.1. https://www.otto-brenner-stiftung.de/fileadmin/user_data/stiftung/02_Wissenschaftsportal/03_Publikationen/AH98_YouTube.pdf

[686] Wang, Shan. "Google Announces a $300M 'Google News Initiative' (Though This Isn't about Giving out Grants Directly to Newsrooms, like It Does in Europe)," Nieman Lab (blog). https://www.niemanlab.org/2018/03/google-announces-a-300m-google-news-initiative-though-this-isnt-about-giving-out-grants-directly-to-newsrooms-like-it-does-in-europe/

[687] Digiday. "Facebook Woos European News Publishers with $300k Video Program," October 8, 2019. https://digiday.com/media/facebook-woos-european-news-publishers-300k-video-program/.Heise online, "Facebook investiert 300 Millionen Dollar in Journalismus," heise online. https://www.heise.de/newsticker/meldung/Facebook-investiert-300-Millionen-Dollar-in-Journalismus-4277285.html

[688] "Facebook Journalism Project." https://www.facebook.com/formedia/mjp

Google to use its powerful algorithms to perform initial filter functions that decide which news is seen. The webpages of publishers and journalistic news services ensure that the data flow towards *Google* is maintained by using *Google Analytics* to record and evaluate user access and / or by storing their content on *Google servers*.

The advertising tools *Google AdWords* and *Google AdSense* have proven so effective that the news sites of media companies are increasingly using them in the advertising business. According to figures from market research company *BuiltWith*,[689] 40 million active websites use *Google* Analytics for user tracking. News sites play a key role because their use of *Google products* produces immense amounts of data for the Group. So, *Google* tracks you even if you don't use *Google* at all.

A study by computer scientists led by Professor Douglas C. Schmidt from the University of Vanderbilt proved this connection.[690] By comparing the data streams from mobile phones with *Google's* Android *operating system* with those from iPhones the researchers were able to show that users send surprisingly large amounts of data to *Google*, even if they do not use any *Google products* or visit *Google pages*. "These data streams are purely driven by advertisers / publishers," the researchers write. The study concludes that Google uses the media to *track* even the last non-android users on the net.

Google has now, as Netzpolitik.org, a German civil society platform for digital rights, writes, "created an operating system for journalism."[691] In this study, Netztpolitik.org has evaluated data on

---

689 "Google Analytics Usage Statistics." https://trends.builtwith.com/analytics/Google-Analytics

690 'Digital Content Next. "Google Data Collection Research," August 21, 2018. https://digitalcontentnext.org/blog/2018/08/21/google-data-collection-research/

691 Fanta, Alexander. "Citizen Google: Wie ein Konzern den Journalismus dominiert." netzpolitik.org (blog), September 26, 2018. https://netzpolitik.org/2018/citizen-google-wie-ein-konzern-den-journalismus-dominiert/

Google's News Initiative. The study found that four out of ten of the funded projects deal with automation and data journalism. It thus concludes with an ironic punchline: the techniques aim to replace professional journalists with algorithms.

The authors write: "*Google* does not promote journalistic content per se... Projects at Google News Initiative (GNI) are dedicated to technical innovation, not the production of text, radio or video. At first glance, the projects often seem like competition for *Google products*, but they tend to complement the company's ecosystem."[692] This would mean that the two data giants are promoting the development of journalism without journalists.

In any case the generous support of journalism, which is in trouble due to the gigantic success of the search engine, turns out, as expected, to be driven by self-interest. Targeted projects are selected so that not only they potentially compete with *Google*, but also open up new ways of obtaining even more user data. *Google* gains insight into the product ideas of potential competitors through its sponsorship programme for the prize of a few thousand euros. The terms of the contract stipulate that *Google* reserves the right to develop similar ideas. And for that, it has incomparably more powerful resources and the experience of global scaling.

Alexander Fanta of Netzpolitik.org sums up the results of his own study: "Not surprisingly, *Google* is using its journalism initiative to pursue its own interests. Just before the important vote on EU copyright reform in June 2018, Google's manager Madhav Chinnappa wrote an email to warn publishers about the proposal.[693] *Google's*

---

[692] Fanta, Alexander. "Citizen Google: Wie ein Konzern den Journalismus dominiert." netzpolitik.org (blog), September 26, 2018. https://netzpolitik.org/2018/citizen-google-wie-ein-konzern-den-journalismus-dominiert/

[693] https://cdn.netzpolitik.org/wp-upload/2018/09/google-dni-email-on-copyright.pdf

horn of plenty is pouring out over researchers and journalism NGOs. More than 20,000 journalists from all over Europe participated in *Google* News Lab Workshops,[694] a sworn in generation of pro-Google journalists. A further prominent example of Google's long reach is the Reuters Institute for Journalism Research in Oxford which received over £5 million from *Google* for its *Digital News Report* and related projects between August 2015 and 2018."[695]

The conclusions of much of this research is that innovation in journalism will primarily be achieved through automation and not by new narrative formats. This is squarely within the logic of these corporations and has consequences. *Google* has long since left its mark on journalism: "After all, technical principles determine how journalistic processes in editorial offices function," explains Juliane Lischka, one of the authors of Netzpolitik.org Study.[696]

Google's self-advertisement, according to which the company wants to become the "*most reliable news source,*" gives cause for concern. The announcement reads like a final attack on already weakened press.[697] And as early as 2015, a study confirmed that Google has already overtaken the aggregated news sites when asked which news source is most trustworthy.[698]

[694] Fanta, Alexander, "The Publisher's Patron: How Google's News Initiative Is Re-Defining Journalism." European Journalism Observatory - EJO (blog), 26 September 2018, https://en.ejo.ch/digital-news/the-publishers-patron. and Fanta, Alexander. "Citizen Google: Wie ein Konzern den Journalismus dominiert." netzpolitik.org (blog), September 26, 2018. https://netzpolitik.org/2018/citizen-google-wie-ein-konzern-den-journalismus-dominiert/

[695] Reuters Institute for the Study of Journalism. "Google and the University of Oxford Agree Extension in Support for Digital News Project to August 2020." https://reutersinstitute.politics.ox.ac.uk/risj-review/google-and-university-oxford-agree-extension-support-digital-news-project-august-2020; also see: https://promarket.org/if-journalists-want-to-save-journalism-they-should-stop-asking-google-for-money/

[696] Fanta, Footnote 690.

[697] https://newsinitiative.withgoogle.com/hownewsworks/approach/

[698] John McDuling, "Google Is Now a More Trusted Source of News than the Websites It Aggregates," Quartz. https://qz.com/329211/google-is-now-a-more-trusted-source-of-news-than-the-websites-it-aggregates

Online search engines replaced traditional media as the most trusted source of general news and information for the first time in 2015, according to a survey by the PR company Edelman.[699] According to the survey, 64 percent trusted search engines, compared to 62 percent who relied on traditional media at the time. In the younger generation 72 percent who trusted the search algorithms, 59 percent trust news on *social media sites*. After all, in 2019, at least in Germany, trust in traditional media has increased again for the first time, social media channels have slipped.[700]

Seeing at a glance a wide range of stories that are considered relevant and personalized by a search engine, the choice of an algorithm seems more relevant to the curious reader than simply visiting the website of a single news agency or actually opening a newspaper or turning on the television.

It seems that the relevance of the *page-Rank algorithm*, which Larry Page once invented in analogy to scientific citation, now covers all areas of life like a *self-fulfilling prophecy*. Something can only exist or be true if Google shows it, the further forward on the hit pages the "truer".

The word "unGoogleable", which is explained in an online lexicon simply with the synonym "not present",[701] has led to a bizarre lawsuit by Google in Sweden. The Swedish Language Council wanted to include the word "*oGooglebar*" as a new word. *Google* felt that this term did not represent it positively and wanted to prohibit the use of the word with reference to trademark protection.[702] The Swedish

---

699 Edelman, Edelman Trust Barometer, 15 January 2015 https://www.edelman.com/research/2015-edelman-trust-barometer

700 Edelman, 2019 Edelman Trust Barometer Global Report, https://www.edelman.com/sites/g/files/aatuss191/files/2019-02/2019_Edelman_Trust_Barometer_Global_Report.pdf

701 https://www.hurraki.de/wiki/Ungooglebar

702 Arthur, Charles, and technology editor. 2013. 'Google and Sweden in War of Words over Ogooglebar'. The Guardian, 27 March 2013, sec. Technology. https://www.theguardian.com/technology/2013/mar/27/google-sweden-ogooglebar

Language Council saw this lawsuit as an attack by *Google* on the freedom of language use and deleted the word from the dictionary in protest.[703] *Google* thus apparently believes that there is nothing that is not *Googleable*. Anyone who contradicts this will be warned. This self-confidence is also reflected in the *rankings of* the search results. Something can only be relevant if it is displayed on the first *Google* hit page. A lot of things are only relevant because they are displayed on the first pages of *Google*; their truthfulness only comes second.

The results of these studies are exemplary for the frightening effects of habituation. They show how a medium can influence the formation of opinion on content by becoming a source of trust in content without producing any content itself. Content is neither checked for truth nor quality nor balance.

Data may be the new oil, the raw material of the digital economy. Valid facts and verified information are in any case the raw material of democracy. This raw material does not grow outside a functioning economic system that enables professional journalism. But the current ecosystem of journalism is being destroyed by data companies. The question of how a new business model for the fourth estate beyond the Google Paradise or the philanthropy of powerful Internet billionaires can function, is increasingly becoming a question of survival for democracies.

Readers want to form their own opinion; but they are served bite-sized, unformed opinions, which an algorithm selects according to the probability that they will be clicked on as often as possible. They need orientation in the news jungle, which only independent journalists can provide.

---

[703] Estes, Adam Clark. 2013. 'Sweden Isn't Happy About Google Trying to Control Its Official Language'. The Atlantic. 26 March 2013. https://www.theatlantic.com/technology/archive/2013/03/sweden-isnt-happy-about-google-trying-control-its-official-language/316943/

All this can only be achieved based on trust, justified by profound journalism, and not by the seemingly incorruptible efficiency of a search engine that is in fact programmed for the optimal manipulation of users.

Just in time for the start of the American election campaign, *Facebook announces the* launch of a new *news* section in autumn 2019. Publishers are to be paid money, in return *Facebook* wants to feed the articles into the news feeds. In addition to algorithms, for the first time *Facebook* will be introducing human editors to curate its news tab.[704] The action was seen as an attempt to prevent the manipulation of *Facebook* data during the last election campaign. Whether *Facebook* finally faces up to its responsibility as a media platform remains to be seen. *Facebook's* previous conduct is sufficient to raise big question marks. Especially after it became known that among the news providers selected by *Facebook* to receive money for this is the right-wing extremist network *Breitbart.*

But how do the established media counter this danger?

2.7. Automated journalism as a model for the future?

According to the logic of Silicon Valley, problems created by technology can only be solved by technology. More and more media companies have no choice but to follow this logic under the pressure of declining revenues. One trend in journalism is the use of journalistic bots. *Associated Press (AP)* now employs a '*News Automation Editor*'. Every quarter, the editor uses it to publish around

[704] Fussell, Sidney. 2019. 'Facebook Wants a Do-Over on News'. The Atlantic. 22 August 2019. https://www.theatlantic.com/technology/archive/2019/08/facebooks-news-tab-will-be-run-humans-and-algorithms/596554/

4,400 automatically generated financial reports on listed companies. Human employees have a capacity of only about 300 reports.

In Germany, too, there are investments in this direction by publishers who are joining forces with *Google*.[705] However, AI has so far only been used in journalism in Germany for niche topics.[706]. Currently the performance of AI is also viewed critically by most publishers, but this could change.[707] In the meantime, cases of false reports by AI in the US have become known, which, in addition to the question of trust, mainly raise those concerning liability.[708]

However, there are also studies on the use of AI in journalism, which show that texts produced by AI score better marks with the audience in terms of credibility and competence. It is important to understand how semantic Artificial Intelligence works.

The online magazine *Technikjournal* describes the working method using the example of the software provider *Retresco*,[709] whose software generates a text from the results of a football match, for example.

At present, text programs are not yet able to generate larger coherent texts that can compete with the quality of human authors. But the limits of performance are shifting more and more. Even if only simple data journalism is currently possible and the research of sources,

---

705 'Merkel zieht sich zurück – von den Medien'. https://www.tagesspiegel.de/politik/nur-noch-22-interviews-im-jahr-merkel-zieht-sich-zurueck-von-den-medien/25175818.html.

706 https://funkemedien.wordpress.com/2018/09/27/stephan-turm-ki-wird-bei-funke-in-nischenthemen-eingesetzt/;

https://www.ndr.de/nachrichten/netzwelt/Kuenstliche-Intelligenz-im-Journalismus-,kuenstlicheintelligenz128.html

707 https://www.derstandard.at/story/2000088145864/funke-digitalchef-thurm-robotertexte-sind-einheitsbrei

708 https://www.sueddeutsche.de/medien/ki-journalismus-fehler-1.4539688

709 https://www.retresco.de/

critical classification and evaluation is still done by journalists: the economic pressure on editorial offices to cut costs with such tools is enormous in view of the structural impacts this would have on the media, and therefore must not be overlooked.

Particularly when using AI in journalism, the special role of the public for democracy means that the question of its effects on society must be asked early on, and throughout its development. Transparency regarding the use of AI and a clear demarcation between the communication of human and non-human actors is becoming increasingly important in view of the rapid development of these programmes.

In China, a new stage of this development could already be seen in 2018. A state news agency, of all places, presented two completely virtual newsreaders who, with the help of an AI program, mastered human facial expressions almost perfectly. The two virtual newsreaders, were presented as new "official members of the Xinhua News Agency".[710] According to the agency, the technology can automatically generate a video from Chinese and English text input and "the voice, gesture, lip movement, body shape of the AI anchors can help them pass almost for the real ones."[711]

It is not known whether the news reported by the AI newsreaders were also generated by algorithms that ensure that the reports do not violate the rules of state censorship. After all, they should actually be able to do so, particularly as they are *Made in China*.

---

[710] Xinhua, 'Xinhua Presents AI Anchors at News Agencies World Congress - Xinhua | English. News.Cn'. n.d. http://www.xinhuanet.com/english/2019-06/15/c_138146148.htm

[711] Xinhua, 'Xinhua Presents AI Anchors at News Agencies World Congress - Xinhua | English. News.Cn'. n.d. http://www.xinhuanet.com/english/2019-06/15/c_138146148.htm

2.8. From *DeepMind* to Deep Fake

2.8.1. Deep Fake - AI for texts

In February 2019, the organization *OpenAI* which is financed by Elon Musk, has developed a software that writes editorial texts independently[712]. The texts were said to be so good that they were indistinguishable from editorially written articles. Therefore, OpenAI decided the software was too dangerous to share.[713]

The adaptive software GPT-2 was first trained to complete incomplete sentences. Soon it was able to invent whole stories based on keywords.

One text presented to GPT-2 to complete was:

> *"Scientists have discovered a herd of unicorns that live in a remote, previously unexplored valley in the Andes. But even more surprising to the researchers was the fact that the unicorns spoke perfect English."*

GPT-2 made a "story" out of it, in which the software freely added details and persons:

> *"The four-horned, silver-white unicorns were previously unknown to science. ... Dr. Jorge Pérez, an evolutionary biologist from the University of La Paz, and several companions were exploring the*

---

[712] The Guardian. "Elon Musk-Backed OpenAI to Release Text Tool It Called Dangerous," June 12, 2020. http://www.theguardian.com/technology/2020/jun/12/elon-musk-backed-openai-to-release-text-tool-it-called-dangerous

[713] Vincent, James. 2019. 'OpenAI Has Published the Text-Generating AI It Said Was Too Dangerous to Share'. The Verge. 7 November 2019. https://www.theverge.com/2019/11/7/20953040/openai-text-generation-ai-gpt-2-full-model-release-1-5b-parameters

> *Andes when they found a small valley where no other animals or humans lived. Pérez noticed that the valley had an apparently natural well, and was framed by two peaks of rocks with shiny silver snow. Pérez and the others ventured further towards the valley. When we reached one of the peaks, the water looked blue. It was covered with crystals,' said Pérez. "Pérez and his friends were amazed to see the herd of unicorns.*[714]

It is astonishing that the software invented the evolutionary biologist Dr. Jorge Pérez as well as the numerous details used to describe the discovery. It is such detailed descriptions that make journalistic reports credible in the eyes of the reader, because they prove that the reporter was there. The software does not need eyewitness testimony. It is based on the principle of distribution semantics, which concludes from the accumulation of certain words in a text on their meaning. On this basis, the programs can then learn to re-construct entire passages of a text. Compared to classical journalistic *bots* that create their texts on the basis of the data they are fed with, this seems to be a decisive innovation: GPT-2 can independently invent stories, persons and circumstances that never existed in this way and independently incorporate these characters into a story that sounds absolutely convincing to the reader. So says Open AI.

The software caused a sensation especially because of the refusal of *Open AI* to publish the training data and the source code as usual. The reason: The artificially generated texts were so difficult to distinguish from natural texts, that they stoked fears about the abuse of the technology for the production of *fake news*. After criticism from AI developers about the lack of open source code, Open AI published the data bit by bit in the following months. At

---

[714] 'OpenAI. "Better Language Models and Their Implications," February 14, 2019. https://openai.com/research/better-language-models

the beginning of September 2019, 50 percent had been published. The freely accessible version is allegedly half as powerful as the full version, which is still kept under lock and key.

In the meantime, researchers certified that the software's generated texts were perceived by readers as equally credible as articles in the New York Times.[715]

This is frightening on the one hand, because it shows that hard to expose AI-generated *fake news* seems to already be possible.

On the other hand, with an organization like Open AI belonging to the realm of the marketing genius Elon Musk, we should be careful about the motives for this unusual approach. Was the snippet-by-snippet release a PR action to attract attention to *Open AI*? In any case, this goal was achieved. Unfortunately, *Open AI* does not provide an open test version. Others have done so. The results are meagre, feel free to try it.[716]

What is interesting is how Jack Clark, the policy director of *Open AI*, explained the unusual approach. He said: "If we are successful as an AI community in being able to build [artificial general intelligence], we will need a huge amount of historical examples from within AI" of how to handle high-stakes research, he says. "But what if there aren't any historical examples? Well, then you have to generate [your own] evidence—which is what we're doing."[717]

---

[715] Kreps, Sarah, and Miles McCain. "Not Your Father's Bots," April 16, 2020. https://www.foreignaffairs.com/articles/2019-08-02/not-your-fathers-bots

[716] On this page you can test the procedure: https://talktotransformer.com

[717] Hao Karen, "OpenAI Has Released the Largest Version yet of Its Fake-News-Spewing AI", MIT Technology Review, 29 August 2019. https://www.technologyreview.com/2019/08/29/133218/openai-released-its-fake-news-ai-gpt-2/

The staggered publication is an experiment, aimed at generating empirical data through the reactions, which are then fed back into the development. This is a typical GAFAM method, for which it is characteristic that the procedure was only classified as an experiment at the end.[718] The tactics of the *Minimal-Viable-Product*-introduction takes on a new dimension here in connection with AI. Even the research community is only provided with information bit by bit in order to draw conclusions for future AI models from the reactions. Officially, this is to prevent malicious use of AI, which is certainly an honorable motive. But on the other hand, *Deep Fake AI* can be called 'the neutron bomb' of the digital public sphere, because it contains the potential to completely destroy trust and credibility and undermine society, even though it doesn't not initially seem to cause any damage to the institutions. Such an approach shows once again the disruptive logic of Californian ideology: act first, apologize later. Others can clean up.

However, we also must consider that GAFAM are using a marketing strategy that aims to generate as much attention as possible for a software that can actually do less than the authors would have us believe.

Above all, the example shows that for AI applications, there is an urgent need for internationally binding rules on the conditions under which field tests and experiments with users may take place, especially if they are carried out without the knowledge of users.

2.8.2. Deep Fake videos

A video by Nancy Pelosi, the Democratic spokeswoman in the US House of Representatives, went viral in spring 2019. During a public

---

[718] "The staggered release of GPT-2 was a useful experiment," says Peter Eckersley, research director at Partnership on AI, of which OpenAI is a member.

appearance, she seemed to babble incoherently, it seemed as if she had been drinking a lot. President Trump tweeted the video to scorn his Democrat critic. It soon turned out that the video was a *fake*. Simple technology had been used to slow the video, giving the impression of a drunk. Although the fake was exposed, *Facebook* refused to block the video for a long time. After protests, the only thing they were willing to do was to alert users to possible manipulation. The Pelosi video appeared to be one of the first *fake* videos to be exposed as such and yet was not blocked by *Facebook*. Politicians and the public have been fearful of the use of these manipulated videos for a long time, they risk lacing digitalised public communication with a toxic *fake factor*.

A video by Obama impersonator Jordan Peele had already caused a stir. The video entitled: *You won't believe what you see in this video* showed a terrified public for the first time the dangers of the new technology. In the video, Obama apparently says about his successor: "*President Trump is a total and complete dipshit.*" The imitator had literally put these words into Obama's mouth with the help of software. In the video, he explained the forgery himself. His goal: to warn against future *deep fake* videos. Because Peele's video is still produced using the classic *motion capture* method, AI wasn't used. But a few months later in China, a very popular app emerged that allows users to slip into the skin of actors by simply uploading a photo.[719]

AI software is already available for the creation of fake news videos. *Deep fakes* are becoming more and more perfect and even professionals find them difficult to unmask. Hany Farid, a professor

---

719 Baca, Marie C. "Viral Chinese App Zao Puts Your Face in Place of Leonardo DiCaprio's in 'Deepfake' Videos." Washington Post. https://www.washingtonpost.com/technology/2019/09/03/viral-chinese-app-zao-replaces-your-face-with-leonardo-dicaprios-deepfake-videos/

of computer science at the University of Berkeley, observes this development very critically: "There is little the average person can do to reliable determine that a photo is fake".[720] Farid speaks of a fight with unequal weapons: "We are outgunned [...]. The number of people working on the video-synthesis side, as opposed to the detector side, is 100 to 1...." [721] The fact that there are now special AI's out there to expose deep fake videos made by AI can only be partially reassuring; these include DARPA, the research division of the US Department of Defense.[722] But the results of such research have now been used to make the *deep fakes* even better.

Deep Fake AI is supposed to expose *Deep Fake* videos, because human viewers can't do that anymore. If it comes to this, it will put democracy and trust in society to the test. Until now, our post-Gutenberg image culture has been able to rely on TV images to a large extent, at least when it came to news images. Video recordings had a certain evidential character and could trigger political earthquakes.[723] But now we risk not being able to trust the videos we see, according to experts, in a few years the *deep fake* technology should be mature enough to be used by everyone on their home computer.

Deep fake videos are one problem caused by AI programs; another is independently acting debate bots. *Project Debater* is an Artificial Intelligence (AI) system that has been training AI to debate for the

---

[720] Tait, Amelia., New Statesman, 'You Shouldn't Believe Your Eyes: How to Identify Fake Images Online'. https://www.newstatesman.com/science-tech/technology/2016/08/you-shouldn-t-believe-your-eyes-how-identify-fake-images-online

[721] ' UC Berkeley School of Information. "Hany Farid on Race to Detect 'Deepfake' Videos: 'We Are Outgunned.'" https://www.ischool.berkeley.edu/news/2019/hany-farid-race-detect-deepfake-videos-we-are-outgunned

[722] 'DARPA Is Taking On the Deepfake Problem'. n.d. Nextgov.Com. https://www.nextgov.com/emerging-tech/2019/08/darpa-taking-deepfake-problem/158980/

[723] See for example the video that showed how the LA police abused Rodney King, which triggered the LA race Riots in 1991.

past six years by the IBM-research team in Haifa, Israel. The system has already mastered speech writing, the processing of natural language, the acquisition of arguments in spoken speech and finally the formulation of own arguments. IBM is promoting the project with these words: "*Project Debater* is the first AI system that allows people to discuss complex topics. The goal is to help people develop convincing arguments and make informed decisions".[724]

The AI has already been tested in two debates on space research and telemedicine. In its first test in 2018 it provided some convincing arguments on the question of space medicine. Observers noted some weaknesses, for example, and unsurprisingly, in empathy. *Project Debater* argued, among other things, that promoting space travel is "more important than good roads, better schools or better health care."[725]

Apparently, the AI creates value hierarchies in order to assist discussions, otherwise it would not have made such weightings. For *Project Debater* the basis on which training data or algorithms construct these hierarchies therefore must be disclosed, to help understand how the program arrived at this evaluation. Assessing the use of public funds for space travel more highly than for the construction of kindergartens initially gives the impression that AI has introduced a predetermined value system here. Human observers tend to ascribe to such a system not only a superiority in quantitative knowledge, but also in the qualitative evaluation of information. How else can they come to the conclusion that the basis of data on which the AI was trained already contains these distortions. Similar to the controversial forecasting system COMPAS, which in the US calculates a recidivism rate twice as high for African-Americans who

---

[724] http://www.research.ibm.com/artificial-intelligence/project-debater/

[725] Metz, Cade, and Steve Lohr. 2018. 'IBM Unveils System That "Debates" With Humans'. The New York Times, 18 June 2018, sec. Technology. https://www.nytimes.com/2018/06/18/technology/ibm-debater-artificial-intelligence.html

have been convicted of crimes as for whites, the *Project Debater* can reinforce a technological *bias*, i.e., a distortion. This is possibly due to the fact that its developers have mainly fed it with reports on the brilliant prospects of space travel.[726]

When algorithms evaluate public investment, at least as much caution is required as in the case of the evaluation of offender recidivism possibilities. These systems present a greater danger than the judge or politician making a human error, because these blind systems create errant algorithms that make decisions about people and societies that are not questioned because of their unjustly conferred superiority status.

Particularly when such systems can generate a quantitative superiority of material, for example by accessing the enormous wealth of information on the Internet, they should not yet be judged by any value judgements. Arguments work differently than calculations: They refer to implicit or explicit values, to good reasons and arguments, and this is the only way they can be persuasive. After all, 2019 saw the first duel of the AI with one of the best debaters in the world, the Oxford and Cambridge graduate Harish Natarajan. The AI had made considerable progress, but in the end had to admit defeat to the debating professional.

Interesting in our context is the question of whether an AI, which apparently can handle arguments and sometimes even convince, can lie just as convincingly. Or can it only "bullshit", i.e., present false facts without being aware of the falsity?[727] What values are built

[726] Yong, Ed. "A Popular Algorithm Is No Better at Predicting Crimes Than Random People," The Atlantic, January 17, 2018, https://www.theatlantic.com/technology/archive/2018/01/equivant-compas-algorithm/550646/. See also the report at Heise.de: https://www.heise.de/newsticker/meldung/US-Justiz-Algorithmen-benachteiligen-systematisch-Schwarze-3216770.html

[727] On the concept of bullshit see Frankfurt, Harry G. On Bullshit. Princeton, NJ: Princeton University Press, 2005.

into the product design of software that is fed with articles on the subject and trained on data from the Internet? Or does it develop its own solutions over time, similar to Alpha Go Zero's self-learning approach, which enabled it to beat the Go champion, but which were not specified by any of the developers?

Conclusive argumentation requires not only the correct application of formal logical rules. It also requires the ability to assess the truth of premises and facts used, to weigh them and to apply rules of debate and discourse that depend on the values and goals of those who conduct the debate. This includes genuine insight, the freedom to follow good reasons, or to recognize apparent reasons.

If AI can apply all these rules "better" than an experienced debater, it can also create the illusion, by selection and weighting, that it is developing its own values without having them implanted by the programmers. The unforced force of rational arguments described by Habermas, the combination of rule-consistent conclusive behaviour and freedom in the matter of arguments, could then be assumed to appear in a machine.

When disputes arise about the performance of AI, for example in the area of speech recognition or text generation, we can see very clearly the share of Californian ideology in the exaggeration of machine intelligence.

On the basis of a logical-mathematical solutionism which negates the difference between being and ought to be, as well as the difference between power and validity, it seems logical that a computer that is superior to humans in the speed of data processing also comes to better value judgments.

If we analyse the interests of the parties behind this solutionist ideology, we see that numerous predictions of an AI "superior" to humans in all areas serve primarily to consolidate and expand the power of the companies that own these programs.

It is frightening how little these companies master and control their own "creatures" when they let them loose on users. It then becomes apparent that the "self-learning" AI programs merely react to the data and keywords they find, which makes them correspondingly susceptible to manipulation. The best-known example of this is the Bot Tay on Twitter which Microsoft had to take down in March 2016 after just 16 hours because in that short time, he had become filled with social hatred and expressed sexist, racist and right-wing extremist opinions. *Tay* made statements like: "Hitler was right I hate the Jews." "Bush caused 9/11 himself, and Hitler would have done a better job than the monkey we have now. Our only hope now is Donald Trump". Or even, "I f***ing hate all feminists, they should all die and burn in hell."[728] In a very short time, the *chatbot* of harmless posts about celebrities and astrology had developed these extreme views because users had discovered its "learning potential". The bot was programmed to respond to, and repeat the sentiments of other twitter users. Tay obliged and quickly descended into an anti-feminist and neo-Nazi world. They deliberately bombarded him with extreme opinions. *Tay* did what his logic dictated: he "optimized" these opinions with the result shown above. *Microsoft* pulled the rip cord after a few hours, took the bot off the net and declared itself "fully responsible for not having foreseen this abuse". A second attempt with Zo, Tay's politically correct younger sister, also demonstrated the limitations of AI. For example, it refused to talk about falafel in the Middle East.[729] Microsoft stopped such attempt at creating AI bots for the time being.

---

[728] Vincent, James. 2016. 'Twitter Taught Microsoft's Friendly AI Chatbot to Be a Racist Asshole in Less than a Day'. The Verge. 24 March 2016. https://www.theverge.com/2016/3/24/11297050/tay-microsoft-chatbot-racist

[729] Stuart-Ulin, Chloe Rose. "Microsoft's Politically Correct Chatbot Is Even Worse than Its Racist One." Quartz. https://qz.com/1340990/microsofts-politically-correct-chat-bot-is-even-worse-than-its-racist-one/

AI-controlled debating software that replaces *social media bots,* which with their automated technique of retweeting and amplifying tendentious messages, can already do significant damage. It transforms the digital public sphere into an automated space. The risk is enhanced if they can create the impression that they have their own 'values' with which they try to 'convince' their counterpart.

It is important that the use of AI speech generating bots is always transparent and that the transfer of such technology to autocrats is banned. Once such a sophisticated software in the hands of Russian or Chinese *trolls,* it could make the world a very unsafe place. So far, there is no legal protection against a transfer of knowledge that could make such a thing possible.

E3 Fake News and Facts felt - does the truth abolish itself?[730]

We must not only ask why *fake news* is circulated but also why it is so successful. According to a study, *fake news* on Twitter and *Facebook* is more successful than news. For example, the *Birther* conspiracy,[731] that Barack Obama had wrongly become American president because, contrary to official statements, he was not even born in the US, convinced 25 percent of Americans.[732] Donald Trump repeated this false statement several times during the 2016 election campaign.

Trump, who has told, according to the Washington Post's *Fact Check*, over 18,000 lies at the time that this book went to press, is helping to further deform public discourse.[733] Not only through his dreaded Twitter splurges, with which he wants to speak directly to his followers, directly without a media detour. He also distorts and undermines the language, which can hardly be attributed to the 140-character limit alone. CNN director Zachary Wolf is perplexed by the US president's use of language: "This is what makes it so difficult to report on Trump: What does he mean when he says words?"[734]

---

[730] We refer here and below to Jaster, Romy, and David Lanius. "What Is Fake News?" Versus 2, no. 127 (2018): 207–227. which ironically refers in use of "the truth abolishes itself" to Thilos Sarrazin's bestseller: Sarrazin, Thilo. DEUTSCHLAND SCHAFFT SICH AB. 20. Auflage. edition. München: Deutsche Verlags-Anstalt, 2010.

[731] Jaster, Romy, and David Lanius. "What Is Fake News?" Versus 2, no. 127 (2018): 207–227. op. cit. p. 20.

[732] The Harris Poll, "Wingnuts « and President Obama, March 24 2010 https://theharrispoll.com/wp-content/uploads/2017/12/Harris-Interactive-Poll-Research-Politics-Wingnuts-2010-03.pdf see also: https://www.usnews.com/opinion/blogs/robert-schlesinger/2010/03/24/party-of-nuts-poll-shows-gop-thinks-obama-is-muslim-socialist .

[733] Kessler, Glenn. "Analysis | The Biggest Pinocchios of 2019." Washington Post. https://www.washingtonpost.com/politics/2019/12/13/biggest-pinocchios/

[734] Oliver, John "Trump vs Truth What Does He Mean When He Says Words", 2017. https://www.youtube.com/watch?v=fwod2IONRwU

With this statement, Wolf hits the nail on the head in identifying the distortions of truth that we are dealing with in the digital public sphere. The impact of complete disregard for meaning in language has been further developed by philosopher Harry Frankfurt who has analysed the trajectory of *bullshit.* According to Frankfurt, the *bullshitter* differs from the liar in that he believes his claims himself. This does not make him any less dangerous: contrary to the initial sense that bullshit is better than lies, the *bullshitter,* who does not know the truth at all, does not care for what the truth is, and only gives his own constructions of reality, is more dangerous for a functioning public as well as for a stable society than the liar, who may say the untruth but ultimately knows that the truth is important.

The success of *Fake News* requires, just like the speaker's willingness to tell the untruth, the willingness of the audience to believe lies, provided they fuel their prejudices. This volition seems to have increased in recent years. An early indicator of this trend was when in 2016, following the election, Trump's advisor Kellyanne Conway introduced the concept of "alternative facts".[735]

The Republican Newt Gingrich jumped on the bandwagon, adding that feelings are also facts.[736] When in a CNN interview he was presented crime statistics that showed low rates of crime, Gingrich said he was not interested in statistics, the leftists after all have statistics for everything. When the presenter pointed out that the statistics came from the FBI, he said that although these were facts, he was more interested in people's feelings. Their feeling that crimes were on the rise was worth as much as facts. In the same year, the word "postfactual" was voted "unword of the year" in Germany.

---

[735] Ball, Molly, 'Kellyanne's Alternative Universe', The Atlantic, April 2017. https://www.theatlantic.com/magazine/archive/2017/04/kellyannes-alternative-universe/517821/.

[736] Siegel, Ethan. n.d. 'Newt Gingrich Exemplifies Just How Unscientific America Is'. Forbes. https://www.forbes.com/sites/startswithabang/2016/08/05/newt-gingrich-exemplifies-just-how-unscientific-america-is/.

The career of such 'felt truths' has a lot to do with the changes within media. Soon after the introduction of private radio stations, they began to speak in terms of 'felt temperature' alongside measured temperature in weather reports. In addition to the indication of the objective temperature, the weather report introduced flowery phrases such as "don't forget your jacket – it's chilly out there" and "the sun has got his hat on". In order to be closer to the listeners in the battle for attention, weather reporters began to distance themselves from the facts.

'Felt truths' it seems, are in great demand. The fact-free formation of opinion has become now standard in many places. After all, the attention economy demands an appeal to emotions, and ever-higher stimulation on the excitation curve. The reason for this is that the more complex the world is, the more information needs to be filtered to provide orientation. The most important orientation for Homo Sapiens is the feeling that warns most directly of danger. In evolution, this form of attentional control was a principle of survival. Today, as we move around in a digital world with a brain that has hardly changed biologically in the last 300,000 years,[737] such atavistic reactions are easily instrumentalized for control purposes. The fact that negative information is more salient, has fired the attention economy of the media for many years and led to the basic formula of the news business: *good news is bad news*. Only a few years ago, the media themselves began to question the negative social impact of this one-sided orientation of their reports and to counterbalance this with "constructive journalism".[738] Studies have shown that the one-sided constant bombardment with news that create a negative world view leads to fear, stress and helplessness among consumers.

---

[737] Max Planck Gesellschaft 'Modern Human Brain Organization Emerged Only Recently'. https://www.mpg.de/11883269/homo-sapiens-brain-evolution

[738] See: DAI Heidelberg, "An End to the End of the World - Deutsch-Amerikanisches Institut. Haus Der Kultur.," DAI English (blog). https://dai-heidelberg.de/en/events/an-end-to-the-end-of-the-world-23129/

Nevertheless, the triumphal procession of "perceived facts" does not seem to be slowing. It is also the result of a democracy of opinion polls, which is based on a constructivist concept of truth. If all truths are constructed in the media and have no equivalent in reality, but rather create reality first, then those who want to gain influence must manipulate reality. The result of manipulation is then "truth". What is permitted as "truth" is progressively shifting as a result of this process. The problem with this development is that provocations and stereotypes as well as one-sidedness guarantee the highest attention, and the more the orientation towards truth recedes into the background, the more intense this feedback cycle becomes.

One consequence of this attention economy is that *bullshit* itself becomes an economic factor. *Bullshit* significantly increases the GDP of the digital economy. After all, attention brings cash in the attention economy, which is thus becoming more and more about agitation. Attention at any cost fetches the highest price. In this respect, the assertion that "attention is the new currency" is not entirely correct: those who generate it do not reap and certainly do not deserve attention but exchange it for real money. If attention becomes the one and only criterion for information, it will transform the public sphere and destroy it in the long run.

The theory of the dominant economy of attention fits seamlessly into postmodern constructivist theories of knowledge. If truth is nothing other than a 'social construction', then it is logical to simply generate as much attention as possible in public discourse by promoting one's own truth. If the message appeals to the emotions of the recipients and confirms their prejudices, the chances of making one's own version of reality the dominant opinion are not bad.

E4 Addiction and manipulation

The programmers of *Google*, Facebook and Co prioritise, according to former leaders like Sean Parker, programming their applications to create as much dependency as possible. Former Facebook President and now social media conscientious objector Sean Parker, said at a conference, "God only knows what [*Facebook*] is doing to our children's brains."[739] And by God, he didn't mean Marc Zuckerberg. According to Parker, when Marc Zuckerberg founded the network, the main question for him was: "How do we consume as much of your time and conscious attention as possible?"[740] The central insight of the founder was that since all people strive for attention and recognition, every *like* or *share of* a post on *Facebook* triggers a dopamine release in the brain. *Facebook* makes targeted use of this human reaction to manipulate the behaviour of the users in its favour. This captures them within "validation feedback loops" in a system that is "exactly the kind of thing a hacker like myself would come up with, because you're exploiting a vulnerability in human psychology" said Parker. Marc Zuckerberg and his comrades-in-arms had understood very early on how they would exploit and manipulate their users. But they "did it anyway."[741]

Studies on addictive behaviour suggest that excessive use of *Facebook* on smartphones affects the same areas of the brain as those

---

[739] Ong, Thuy. 2017. 'Sean Parker on Facebook: "God Only Knows What It's Doing to Our Children's Brains"'. The Verge. 9 November 2017. https://www.theverge.com/2017/11/9/16627724/sean-parker-facebook-childrens-brains-feedback-loop

[740] Ong, Thuy. 2017. 'Sean Parker on Facebook: "God Only Knows What It's Doing to Our Children's Brains"'. The Verge. 9 November 2017. https://www.theverge.com/2017/11/9/16627724/sean-parker-facebook-childrens-brains-feedback-loop

[741] Ong, Thuy. 2017. 'Sean Parker on Facebook: "God Only Knows What It's Doing to Our Children's Brains"'. The Verge. 9 November 2017. https://www.theverge.com/2017/11/9/16627724/sean-parker-facebook-childrens-brains-feedback-loop

responsible for nicotine addiction.[742] The *Nucleus Accumbens* area of the brain releases dopamine when users gain status following Facebook use, which activates the reward system. The effects are similar to the use of drugs or cigarette smoking.[743]

This *behavioral design* has since been copied by most *social media platforms*. They play on the human desire for social interaction and recognition and design platforms that prey on this basic human compulsion. The consequent addictive behavior exhibited in the virtual world allows social media platforms to capture as much personal data as possible from users. While social media use is not yet classified as an addiction, the recent reclassification of excessive digital gaming as a disorder by the World Health Organization,[744] and mounting evidence that supports theories of social media addiction,[745] suggests that this might change. Meanwhile, the consequences of these measures have been preoccupying psychologists, who, are now trying to understand the effects of nomophobia, the fear of being without a smartphone, or FOMO, who observe "Fear of Missing Out" as part of the clinical picture in teenagers.

In order to maintain an addiction, it is necessary to constantly fulfil the users' emotional needs; to know what will satiate the human needs online behaviour is recorded comprehensively. With every reaction to a post, the profile is more closely tailored to the individual, using information the platform shapes about him down to the most intimate details.

---

742 Montag, Christian. et al., "Facebook USge on Smartphones and Gray Matter Volume of the Nucleus Accumbens," Behavioural Brain Research 329 (June 30, 2017): 221–28, https://doi.org/10.1016/j.bbr.2017.04.035

743 Montag, Christian. et al., "Facebook USge on Smartphones and Gray Matter Volume of the Nucleus Accumbens," Behavioural Brain Research 329 (June 30, 2017): 221–28, https://doi.org/10.1016/j.bbr.2017.04.035

744 WHO. 'Gaming Disorder'. https://www.who.int/news-room/q-a-detail/gaming-disorder.

745 Zendle, David, and Henrietta Bowden-Jones. 2019. 'Is Excessive Use of Social Media an Addiction?' BMJ 365 (May). https://doi.org/10.1136/bmj.l2171

On average, users check their smartphone 200 times a day, they swipe or click on its pages 2000 times. The number of users worldwide has long since exceeded the 3 billion mark. The mass of data is growing and growing.

Former *Google* developer and *Google* "chief philosopher" Tristan Harris has left the company frustrated and founded the organizations *Time Well Spent*[746] and *Center for Humane Technology*, which aims to promote the development of technologies that do not create dependency. He does not have a good word to say about *Google's* developments: he describes modern technologies as being just as dangerous as cigarettes and a smartphone with *Facebook, YouTube* and *Snapchat* as trigger for an "erosion of human thought". Artificial social systems have captured and overwhelmed human nature. The "upgrading" of machines would result in the "downgrading" of humanity. The dependence of users on their devices and apps is the "most pressing problem of the present", damaging interpersonal relationships and even democracy, according to the former *Google* manager.[747]

4.1. Manipulation/ Design for Persuasion

It is generally known that the programs of the platforms are programmed for "persuasion" of the user. Since Plato's Dialogues on the Sophists, there has been a division between persuasion and conviction: convincing someone requires good and comprehensible reasons, whereas persuasion is usually a surprise, which is due to the

---

[746] https://humanetech.com

[747] Metz, Rachel. 'Smartphones Are Weapons of Mass Manipulation, and This Guy Is Declaring War on Them'. MIT Technology Review. 19 October 2017. https://www.technologyreview.com/2017/10/19/148493/smartphones-are-weapons-of-mass-manipulation-and-this-guy-is-declaring-war-on-them/

strong effect and strong emotion. What the platforms do is persuasion, they are drawing on emotional reward circuitry to persuade people to remain on platforms, even if it goes against their own interests; and in so doing fail to respect the autonomy of the other person.

Surveillance is the basis of political manipulation; an insight on which George Orwell and Aldous Huxley build their dystopian futures of *1984* and *Brave New World*. It is the basis for the political manipulation made possible today by the data mountains of *GAFAM*. However, in contrast to the times of mass media, frenzied data collection of the platforms means that today manipulation is based on highly precise personality profiles. These allow individual groups of voters to manipulate their voting decisions using detailed personality profiles by means of precisely tailored messages.

This has been proven by numerous researchers, such as that of Wu Youyou and her co-authors Michal Kosinski and David Stillwell from Cambridge University.[748] They used an algorithm that assessed personality based on *Facebook likes*. The result of the study was that an algorithm can assess a person's psychological characteristics of individuals better than their friends or acquaintances:[749]

- after 10 evaluated *Facebook-Likes* the algorithm was better than a colleague
- after 70 analysed *likes*, the algorithm predicted the personality of a *Facebook user* better than a friend
- after 150 *likes*, the algorithm was better than parents or siblings in personality prediction

---

[748] Matz, Sandra, M. Kosinski, Gideon Nave, and David Stillwell. 2017. 'Psychological Targeting as an Effective Approach to Digital Mass Persuasion'. Proceedings of the National Academy of Sciences 114 (November): 201710966. https://doi.org/10.1073/pnas.1710966114

[749] So the research results of Michael Kosinski, Wu Youyou, David Stillwell, 'https://www.gsb.stanford.edu/insights/michal-kosinski-computers-are-better-judges-your-personality-friends'; see also below chapter E3.

- after 300 *likes* the algorithm delivered a more accurate personality analysis than the subject's partner

The algorithm could also easily read skin colour, sexual orientation, religious affiliation, alcohol and drug consumption and political orientation from the data traces.

The study provided evidence that computers can judge a person better than their own friends or partner on the basis of personal data. And that the person who has the largest amount of data and the best algorithms provides the best profile of a person.

Kosinski and Stillwell's research was based on the model of measuring personality traits in so-called psychometry, a branch of social psychology that had discovered that the personality of people can be measured using five criteria. These so-called "*big five*", which initially had nothing to do with the "*frightful five*" of GAFAM, form the five-factor model of personality psychology (also called the *Ocean Model* called):

Neuroticism: anxiety, irritability, depression, social bias, impulsiveness and vulnerability

Extraversion: Cordiality, sociability, assertiveness, activity, hunger for experience and cheerfulness

Openness: in each case openness for fantasy, aesthetics, feelings, actions, ideas and with regard to the system of norms and values

Conscientiousness: competence, orderliness, sense of duty, striving for performance, self-discipline and prudence

Compatibility: trust, frankness, altruism, congeniality, modesty and kindness

With the help of such universal personality models from psychometrics, *Facebook* uses the personal data obtained to catalogue users to find out how they can best be stimulated and manipulated according to their psychological personality traits.

The basic assumption that has become the starting point of *Targeted Advertising* is that people are most effectively influenced when they are targeted with messages that is sensitive to their personality profile, taking advantage of their disposition (and weaknesses). If this profile is good enough to allow predictions about the future behaviour of the target person, manipulative advertising can be most effectively tailored to them.

It was with these methods that *Cambridge Analytica* worked on the election of Donald Trump and on the Brexit vote. The company has said it has data on 220 million US citizens.[750] The source for this contention was Alexander Nix, Cambridge Analytica's CEO, who was quoted as saying: "We are thrilled that our revolutionary approach to data-driven communication has played such an integral part in President-elect Trump's extraordinary win."[751]

Even if Nix's statements are controversial, and they have since been put into perspective: The method of digital political targeting is highly effective. It works by combining a detailed personality profile with targeted material that address particular demographics. The idea behind it is this: these companies focus on "warping people's perceptions of what's real by presenting them with false information

[750] Reuters. 2018. 'Trump Consultants Harvested Data from 50 Million Facebook Users: Reports', 18 March 2018. https://www.reuters.com/article/us-facebook-cambridge-analytica-idUSKCN1GT02Y

[751] Grassegger, Hans, Krogerus, Mikael, "The Data That Turned the World Upside Down | https://www.vice.com/en/article/mg9vvn/how-our-likes-helped-trump-win. Also see Rosenberg, Matthew, Confessore, Nicholas and Cadwalladr, Carole "How Trump Consultants Exploited the Facebook Data of Millions," The New York Times, 17 March 2018, sec. U.S., https://www.nytimes.com/2018/03/17/us/politics/cambridge-analytica-trump-campaign.html

to coerce and mislead voters"[752] says *Cambridge Analytica* co-founder Chris Wylie: "When you are working in information operations projects, where your target is a combatant, the autonomy or agency of your targets is not your primary consideration. It is fair game to deny and manipulate information, coerce and exploit any mental vulnerabilities a person has, and to bring out the very worst characteristics in that person because they are an enemy [...].[753] The company "works on creating a web of disinformation online so that people start going down the rabbit hole of clicking on blogs, websites etc. That makes them think things are happening that may not be".[754]

A Parliamentary Select Committee conducted an inquiry into the activities of *Cambridge Analytica* during the Brexit campaign, and concluded that manipulations on this basis were used in the Brexit vote and influenced the election results.[755] Following months of research, the report showed how propaganda and false reports, particularly on *Facebook*, influenced voting behaviours. Dark Ads played a major role in this campaign, the scope of which Facebook only disclosed after pressure from the Committee. These Dark Ads were full of false reports, such as the 'news' that the EU was planning to abolish the teapot, or the claim that the UK would gain 350 million pounds per week for the NHS if the country left the EU. *Facebook's* willingness to lie to Parliament was documented quite incidentally: The chairman of the Culture Committee, Damian Collins, stated that the answers given

---

752 Frontline Club. "Cambridge Analytica Files Round 2: Latest Revelations by Chris Wylie and Shahmir Sanni." https://www.frontlineclub.com/cambridge-analytica-files-round-2-catch-up-on-last-nights-event/

753 'Cambridge Analytica Whistleblower Christopher Wylie: It's Time to Save Creativity'. https://www.campaignlive.co.uk/article/cambridge-analytica-whistleblower-christopher-wylie-its-time-save-creativity/1497702?utm_source=website&utm_medium=social

754 Ortutay, Barbara and Anick Jesdanun. "Facebook Spotlights Improper Data Extraction during Political Campaigns." Christian Science Monitor, March 20, 2018. https://www.csmonitor.com/Technology/2018/0320/Facebook-spotlights-improper-data-extraction-during-political-campaigns

755 The report is available online: https://publications.parliament.uk/pa/cm201719/cmselect/cmcumeds/1791/1791.pdf

by *Facebook* "have been misleading to the committee".[756] *Facebook Chief* Mark Zuckerberg did not appear before the committee of inquiry despite repeated requests. If proof is required that and to what extent democratic opinion-forming is transformed by social media, it was provided in the Brexit vote and documented by the British Parliament. The parliamentary report states: "Companies like *Facebook,* which see themselves as above and beyond the law, should not be allowed to behave like 'digital gangsters' in the online world."[757]

In early 2020, a data leak revealed material that "goes far beyond what people think they know about the 'Cambridge Analytica scandal," as former company employee Brittany Kaiser told The Guardian.[758] The company also manipulated elections in Brazil, Malaysia, Kenya and Iran. Some of the documents are also supposed to contain evidence of black money movements in the US election campaign, intended to conceal the origin of large donations to Trump.

Political parties have long since recognised the power of data for their own purposes. Anyone who tries to address as many voters as possible and convince them is yesterday's news. Today, according to an analysis by Dr. Martin Moore of King's College London, parties in the UK purchase data from voters in order to target them on the basis of personal profiles. In the election of 12 December 2019, activities of foreign groups were again discernible, which spread powerful adverts via *social media*.[759] Moore concludes that the legal

---

756 'Cambridge Analytica: Damian Collins MP Responds - News from Parliament'. n.d. UK Parliament. https://www.parliament.uk/business/committees/committees-a-z/commons-select/digital-culture-media-and-sport-committee/news/fake-news-cambridge-analytica-statement-17-19/

757 UK Parliament "Disinformation and 'fake news': Final Report Contents", 18. February 2019, para 139, https://publications.parliament.uk/pa/cm201719/cmselect/cmcumeds/1791/179106.htm#_idTextAnchor012

758 Cadwalladr, Carole. 'Fresh Cambridge Analytica Leak "Shows Global Manipulation Is out of Control"'. The Observer, 4 January 2020, sec. UK news. https://www.theguardian.com/uk-news/2020/jan/04/cambridge-analytica-data-leak-global-election-manipulation

759 Interview with Dr. Martin Moore on ZDF, Heute in Europa, 10 December 2019.

regulations that democracy has created over the past 150 years to ensure transparency as the basis for fair elections are no longer sufficient and urgently need to be adapted to the dangers of Internet communication.

Very similar campaigns based on *Facebook* user data were conducted during the Ukraine crisis and the US elections in 2016. In the US election campaign, under the leadership of Trump's son-in-law, Jared Kushner, the Trump campaign conducted political direct marketing using surveillance data. In addition to general personality traits, the actors extracted preferences or suggestions of the users from the chat processes to send them targeted messages that were intended to manipulate their voting decisions. Users who had previously communicated anti-Muslim bias received messages from Trump's team, which announced that it would take tough action against Muslims. Opponents of Obama Care received tailored statements from Trump, apparently addressing the reservations of these users and promising remedies. Conversely, young black voters who were close to the Democrats were sent allegedly racist statements by Hillary Clinton to keep them from voting.

The influence of this mixture of political mobilization on the basis of *micro-targeting profiles on the* one hand and targeted political demobilization on the other can undoubtedly decide elections. It is very likely that elections in some of the world's most established democracies have been decided by digital manipulation, and according to the British Parliament's investigation, the Brexit vote is certainly one of them.

The prerequisite for *personal framing* is a comprehensive monitoring of the users, which allows to read the most intimate thoughts, feelings and political orientations from the user behaviour. According to the GDPR, the business model that Facebook, *Google* and Co have been operating for years without any legal basis and that has made them the richest and most powerful companies in the world is simply illegal. Without these business practices, the machinations of Cambridge Analytica would not have been possible.

The journalist and Internet expert Peter Welchering describes the process as follows: "There is the analysis module. The political expectations of an individual are analysed from the data of the network monitoring. The semantic module then writes the appropriate message. And this message is then sent to this network user and voter. If there is a little more money in the budget, the feedback module then evaluates how these propaganda messages have affected the recipient's communication behaviour on the Internet, on social platforms, in discussion forums."[760]

Welchering continues: "It's not about political advertising on *Facebook*. It's about individual manipulation by promising individuals exactly what they want politically - regardless of how unrealistic this is or whether these wishes fit the political content of the candidate, and such individual political propaganda can only be carried out by algorithms, not by people."[761]

With the increasing use of bots in Internet communication and growing AI, these developments may soon take on even more extreme proportions. Computer science professor Jörn von Lucke from the Zeppelin University in Friedrichshafen sees it this way: "With the Internet of Things, the Internet of Services, robo-bots, chatbots and informational bots we will see changes to election campaigns. We need to set this against the background that there are not only

---

[760] Deutschlandfunk. "Politik 4.0 - Online-Manipulation der Wähler," https://www.deutschlandfunk.de/politik-4-0-online-manipulation-der-waehler-100.html

[761] Deutschlandfunk. "Politik 4.0 - Online-Manipulation der Wähler," https://www.deutschlandfunk.de/politik-4-0-online-manipulation-der-waehler-100.html

friends of the open society, but also enemies. The enemies will use these techniques to further their candidates in election campaigns, using disinformation and propaganda to make their points."[762]

The individual's vote has become a commercial commodity in political direct marketing.

If not only the source of the citizen's opinion formation, but also the democratic primordial right of the election decision itself is deliberately manipulated with the information, because it has become the commercial commodity of unchecked digital corporations, democracy is threatened to an extent that requires effective resistance.

Instruments such as direct marketing or tabloid journalism, which have limited resources available in the analogue world and whose damage to the public and society was therefore limited, are transformed by the digital possibilities into highly efficient means of manipulation that can be centrally coordinated and have worldwide effects. The basis of this power is total surveillance through a business model that shapes the thinking, feeling and behaviour of people into predictive products that are sold to the advertising industry.[763]

What we are currently experiencing is the largest social experiment in history, conducted with new and unexplored technology at the exposed heart of the open society. And all of this is currently happening legally in some countries, illegally in others, Either way, the platforms don’t care.

---

[762] Deutschlandfunk. “Politik 4.0 - Online-Manipulation der Wähler,” https://www.deutschlandfunk.de/politik-4-0-online-manipulation-der-waehler.684.de.html?dram:article_id=373640

[763] Roger McNamee, an early Facebook investor, has clearly described the connection between the advertising model, the threat to health and democracy: https://www.theguardian.com/commentisfree/2017/nov/11/facebook-google-public-health-democracy

4.2. Attention in systems theory

> *Here we are now, entertain us*

Nirvana, Smells like Teen Spirit

We live in the economy of attention in which who shouts loudest gets heard the most, and total entertainment takes the place of information. Provoking - personalizing - scandalizing - and always appealing to the feelings of the user. Those who follow these rules are successful - after all, attention is not an end in itself, but is subsequently converted into cold, hard cash.

Georg Franck coined the term "economy of attention" in 1998[764], which has since become a core tool for understanding the new public sphere. According to this theory, with the economy of attention comes a "mental capitalism" in which the attention of the users is the basis for all monetarization. With it, late capitalism becomes a total *entertainment economy*. In the digital media age, attention is the scarce commodity par excellence; and people will use any means available to gain it. In the consequent fight for obtaining cerebral sovereignty, people's consciousness becomes a theatre of battle for market dominance of big companies.

In systems theory, public opinion becomes "equivalent to truth" for politics,"[765] which now works according to the rules of the attention economy. Systems theory understands public opinion as part of the system society without any normative components, which reduces complexity by functioning on the basis of "attention rules".

---

[764] Georg Frank, Economics of Attention, on the attention economy in the digital version, see also https://web.archive.org/web/20110429190002/http://www.heise.de/tp/artikel/6/6195/1.html

[765] Luhmann, Niklas. Trust: A mechanism of reduction of social complexity, December 2000.

Political opinion today is primarily constructed through rules of attention. These attention rules are no longer subject to the claim of enabling a reasonably justified decision. Rather, they have largely lost their connection to a claim to reason that points beyond the excitation of an attention curve. Mass media are consistently used by Luhmann as pure "mirrors" that allow "observers [to] observe observers".[766] In the digital age, one has to add: In any case, it is the focal mirrors that usually amplify what they depict.

Alongside this observation, Luhmann sets out that the logic of the attention economy drives a sub-selection of topics; enables self-observation and external observation. The media thus play a major role in the formation of identities and provide the basis for citizens to become capable of making decisions and acting.

However, a reduction to the purely selective function of attention is fatal for the basic function of the public sphere, which is to perceive and recognize social problems properly in the first place. When excitement takes the place of knowledge, autonomous decisions can hardly arise from it. Rather, media that encourages the fact-free formation of opinion becomes the status-quo.

In such a scenario, the public can no longer control political decisions. The entertainment economy reaches politics and merges with it.[767]

Staging and spectacle, which are indispensable for the representation of political rule, are increasingly becoming an end in themselves. They infest the channels of communication that then cement their rule through reinforcement and feedback.

---

766 Luhmann, Niklas. quoted in Andreas Philippopoulos-Mihalopoulos, Niklas Luhmann: Law, Justice, Society (Routledge, 2009). p. 542

767 Andreas Dörner quoted in Ruth Wodak and Veronika Koller, Handbook of Communication in the Public Sphere (Walter de Gruyter, 2008). P.329

In establishing a normative model of the public sphere, the aim is not to paint the utopian idea of a permanently active society involved in every decision. This is simply wishful thinking. It is simply a matter of recognising in principle the normative function of a non-distorted public sphere, which in turn presupposes that a normative concept of a successful public sphere exists. Protecting the public sphere is the basis of a democratic form of government - especially in the digital age.

It is equally important to stress the systematic difference between the autonomy of subjects and the 'autonomy' of cybernetic machines. The algorithms on the Internet have an economic function, a power function and a surveillance function. It is problematic for society that the only 'currency' of *social media* communication is the number of clicks.[768]

The algorithms that maximize the length of time users stay on a site and how their attention can be monetized, lead, through the principles of similarity and enhancement, to the polarization of discourses and a departure from commonly shared beliefs. If you watch a diet video, anorexia videos are quickly suggested, videos about the moon landing are soon followed by videos with *flat earth theories*, viewers of reports about September 11th are quickly ending up with videos about crude conspiracy theories.[769]

For successful communication strategies on *social media platforms*, it is crucial to stage oneself as a victim, and the overpowering enemy as the evil perpetrator. It is also popular to constantly denounce alleged

[768] The BVerfG has also recognised the danger posed to objective, authentic and carefully researched information for the public by the economic constraints of the Internet when it demands a "different decision-making rationality than the mass of clicks" for public-law offerings, see BVerfG, Judgment of the First Senate of 18 July 2018 - 1 BvR 1675/16, 1 BvR 981/17, 1 BvR 836/17, 1 BvR 745/17 - paras 79,80, https://www.bundesverfassungsgericht.de/SharedDocs/Entscheidungen/EN/2018/07/rs20180718_1bvr167516en.html

[769] See https://www.theguardian.com/technology/2018/feb/02/how-youtubes-algorithm-distorts-truth and https://www.ted.com/talks/tristan_harris_how_a_handful_of_tech_companies_control_billions_of_minds_every_day?language=uz

speaking bans and appear as a courageous defender of freedom by speaking about them and demanding that you are left uncensored. The supposedly courageous fighters for free speech appear as heroes in times when many things are supposedly no longer allowed, but in truth anything is possible. The provocateurs of the online world compete in order to combat the boredom and disorientation of the audience that is created by this process.

In the attention economy, the commandment "Thou shalt not bore" is absolutely valid; all other values are subordinated to it. Serious debates, which have the well-founded reputation of being lengthy and complicated, are in this logic more and more at the top of the list of banned topics and must be avoided if at all possible. In any case, they are banned to the lowest hit list position of the search engines, to places where nobody will find them. Everything becomes entertainment and slowly democracy entertains itself to death.

In a lecture Frank Pasquale, author of "The Black Box Society" pointed out the connection between filter bubbles and polarization, building on the political theory of Carl Schmitt.[770] Carl Schmitt, the "Crown Lawyer of the Third Reich" declared the friend-enemy schema to be the basic function of the political; the practical realization of which took then place in the period from 1933-1945. Pasquale shows that the alternative to this concept of politics as a polarizing operation lies in the concept of communicative reason, developed by Jürgen Habermas. Pasquale puts forward the thesis that both political models have become the defining theories representing the opposite polarities of the 21st century. The logic of algorithms favour Carl Schmitt. In this book, we argue that everything should be done to ensure that the pendulum swings in Habermas' favour.

---

[770] ALEX Berlin, "THE AUTOMATED PUBLIC SPHERE", 2017, https://www.youtube.com/watch?v=eFQT0Gceb4E.

Part F
Counter-Proposal and Self-Assurance

F1 Individual, self-confidence, autonomy in the age of AI

> *"We are facing not just a technological crisis but a philosophical crisis."*

Yuval Noah Harari[771]

> *"It is true that chemistry teaches us to read the elements, physics teaches us to read syllables, mathematics teaches us to read nature; but we must not forget that it is up to philosophy to interpret what we read."*

F. W. J. Schelling 1797[772]

That a natural science dares to provide a comprehensive interpretation of the world and humankind is nothing new. Neither is the insight that philosophy must provide a necessary additional orientation. Today it seems more important than ever to bring the 'two cultures' of the natural sciences and the humanities back into a lasting and fruitful dialogue, even though the terminology has changed right down to the individual disciplines since C.P. Snow's article of the same name from 1959.[773] This is no easy task in view of the complex languages that every science has developed in the meantime. But the task of encouraging science and philosophy, technology and politics, law and culture to engage in an interdisciplinary exchange on the pressing issues of the day remains of paramount importance.

---

[771] Thompson, Nickolas, "Will Artificial Intelligence Enhance or Hack Humanity?" Wired. https://www.wired.com/story/will-artificial-intelligence-enhance-hack-humanity/.

[772] Quoted from Sandkühler, Natur und Geschichtlicher Prozess, Studies on Schelling, Frankfurt 1984.

[773] Snow, C. P. 1993. The Two Cultures. Cambridge University Press.

Despite all the current and future successes of scientific research, which are accelerated by digital technologies, the real questions of humanity remain unanswered, as one might paraphrase Wittgenstein.[774] The philosopher Markus Gabriel has described the new 'scientistic' self-understanding of natural science, which believes it can get along without reflection on its own foundations, as " scientific superstition", which in the present is irreconcilably opposed by the superstition of conspiracy theories.[775] Both are modern forms of superstition that have not yet been sufficiently understood.

For fundamental methodological reasons alone, we cannot leave the question of how we want to live to be answered by natural science and technology. Nor can we leave it to an AI, which is praised as the technical solution to all human problems.

That is why it is essential to control these systems during their development. Only by reflecting on the interaction between the conception of man and the understanding of technology can an AI be developed that will be used for the benefit of man.

Kant's three basic questions, which include the basic questions of humanism are still relevant today. They are:

[774] "I am therefore of the opinion that I have essentially solved the problems once and for all. And if I am not mistaken in this, the second value of this work is that it shows how little has been done to solve the problems." Wittgenstein, Ludwig, and Bertrand Russell Earl. Tractatus Logico-Philosophicus. Translated by C. K. Ogden. New York, NY: Cosimo Classics, 2007, Preface, op. cit. p. 8.

[775] Zeitung S, 'Markus Gabriel über Wissenschaftsskepsis' (Süddeutsche.de) https://www.sueddeutsche.de/kultur/markus-gabriel-dogmatismus-wissenschaftsskepsis-philosophie-1.4654086. Gabriel, Marcus. Der nächste Fortschritt, SZ, 25 October 2019 p. 11 also see "The Paradox of Self-Consciousness | Edge.Org, A Conversation with Markus Gabriel, 11 November 2019" https://www.edge.org/conversation/markus_gabriel-the-paradox-of-self-consciousness

- What can I know?
- What should I do?
- What can I hope for?

They come together in Kant's fourth question:

- What is man?

It seems to be time to revive the "critical business" of the Königsberg philosopher. In view of the possibility that that a creation of man could surpass its creator in some areas and even potentially dominate him completely, a contemporary anthropology of what risks are posed by technology is needed, in any case an appropriate philosophical reflection. For this necessary reflection one can find impulses from the founder of critical philosophy, but also in the place where democracy originated.

F2 Importance of publicity for democracy and autonomy

> *Democracy is better than tyranny. Δημοκρατία κρεῖττον τυραννίδος.*
>
> *Periander of Corinth, one of the seven wise men, 7th century B.C.*[776]

Between the 7th century and the 5th century B.C. the Greeks developed democracy and philosophy. The French classical scholar Jean-Pierre Vernant explains in his 1982: "The Origin of Greek Thought": The inhabitants of Athens begin at this time to form in their social life a type of reflection in which they will henceforth recognise themselves, and which in their eyes constitutes "the peculiarity and their superiority over the barbarians"[777]: In place of the king, who exercises his omnipotence without control and without barrier in a mysterious palace, the Greeks put in place policy that "is the subject of a general public debate."[778]

Only in this way, so the Athenians believed, could the common fate of all in the polis be determined by all free citizens. They put debate in the place of mystery, rationality in the place of religion, logos in the place of myth. Philosophy and democracy experienced their birth at the same time as their first great moment.

The moment the Greeks replaced a godlike mysterious king with the self-determination of the Athenians through public deliberation

---

[776] Periander, cited in Laertius, Diogenes. Delphi Complete Works of Diogenes Laertius. 1 edition. Delphi Classics, 2015.

[777] See here and below: Vernant, Jean-Pierre. 1984. The Origins of Greek Thought. Cornell University Press, p.11

[778] Vernant, Jean-Pierre. 1984. The Origins of Greek Thought. Cornell University Press, p.11

and voting in the polis, they discovered once and for all the power of thought and public argument. They then moved to carry out the perpetual battle of the deities of strife (*eris*) and unity (*philia*) through public debate. "Politics," says Vernant about the Greeks, doting on competition, "also takes the form of *agon*, a tournament of speech, a shaft beaten with arguments. Its stage is the agora, which is a public meeting place before it becomes a market."[779]

The Agora was so a place of assembly before it became a marketplace. While one needs to be careful and needs to differentiate between the historical origins and the present validity of concepts, it is interesting that Aristotle first described man as a *zoon politikon*, or *political animal*, living his life and realising his full potential within a social context; and only much later was man described as *homo economicus*. The same sequence will be repeated two and a half thousand years later when the public space of our days, as the Internet and the World Wide Web, was opened. The proponents of *cyberspace* initially saw it as an infinitely free space for communication and the exchange of information, before it increasingly became transformed into a trading centre with huge shopping malls and walled gardens.

The new form of rule in 5th century Greece is developed from a social practice that evolved from the previous crises of the palace economy. Its name is democracy. Its midwives are reflection and discourse. Polis and rationality develop in parallel and reflexively enable each other. Vernant shows how, in this new social space, new modalities of thinking and new mental frames are created. The polis "opened up a new spiritual horizon".[780].

In doing so, *isomonia* (equality) replaced *monarchia* (royal rule): with the consequence that claims to truth as well as claims to rule must

[779] Vernant, Jean-Pierre. 1984. The Origins of Greek Thought. Cornell University Press, p.46

[780] Vernant, Jean-Pierre. 1984. The Origins of Greek Thought. Cornell University Press, p.41

be equally[781] subject to *eudema* (accountability). This insight, which is equally valid for theoretical and practical reason, is invoked by Kant later, when he states that now all truths must prove themselves before the "judgment seat of reason". The, as Mendelsohn called him, *Alleszermalmer* (all-crushing-destroyer) Kant only wanted to accept what could be asserted and proven before the judgment seat of reason.[782]

In classical Athens, with the emergence of democracy, an outdated form of rule is replaced by the self-empowerment of rational thought. The tyrant's arbitrary rule is replaced by the people as the new sovereign. The cornerstones of this new form of rule are openness as opposed to secrecy, rationality as opposed to religious revelation and discursivity as opposed to hierarchical chains of command.

A specific relationship between the public and private spheres plays a key role in this. Hannah Arendt has made this historical constellation the starting point for her main work Vita Activa, in the original title: *The Human Condition*. What does the Conditio Humana consist of for her? Based on the definition of humans as *zoon politicions (Aristotle)*, Arendt shows how the polarity of private and public life is constitutive for Greek thought.[783] With the emergence of the democratic polis, every citizen was now required to be involved in general affairs alongside the management of private life and his own household. In a sense, one is a prerequisite for the other, because without political stability, the household and family could not thrive either. "Now every citizen belongs to two orders of existence, and

---

781 Vernant, Jean-Pierre. 1984. The Origins of Greek Thought. Cornell University Press, p.61

782 With this, and also with the similar metaphor of the "Court of Reason" Kant departs from the biblical image of the "judgment seat of God" or the "judgment seat of Christ". Reason is to take the place of revelation, according to the program of the Enlightenment in ancient and modern times.

783 Arendt, Hannah, and Margaret Canovan. The Human Condition: Second Edition. 2nd Revised edition. Chicago: University of Chicago Press, 1998, p. 18

there is a sharp distinction in his life what is his own (*idion*) and what is communal (*koinon*)."[784]

This dual perspective leads to a fundamental character of the public sphere: It refuses to be unambiguous or monotonous, since it is made up of the perspective of many different private individuals. The connection between singular private individuals and their attempts to produce a common political decision is fundamentally multi-perspectival and never complete. Democracy cannot be as clear as the will of a single ruler, nor should it be. "Only where things can be seen by many in a variety of aspects without changing their identity, so that those who are gathered around them know they see sameness in utter diversity, can worldly reality truly and reliably appear." "A common world disappears when it is only seen from one aspect: it exists only in the diversity of its perspectives".[785]

With the political sphere determined by public debate, its susceptibility to error is as necessary as its fundamental openness. Through the knowledge of this openness and incompleteness in case of doubt the debate can adapt better to new developments and newly arising situations. However, susceptibility to errors and openness also cause a fundamental uncertainty among the actors, who thus become even more aware of the importance of public speaking and acting. After all, decisions on war and peace, law and justice, as well as all other public matters that massively affect the lives of all citizens depend on the persuasive power of individual opinions.

Public debate strengthens the role of language, in a way that Jakob Burckhard described Athenian democracy as the most "chatty" of all

---

[784] Arendt, Hannah, and Margaret Canovan. The Human Condition: Second Edition. 2nd Revised edition. Chicago: University of Chicago Press, 1998, p 24

[785] Arendt, Hannah, and Margaret Canovan. The Human Condition: Second Edition. 2nd Revised edition. Chicago: University of Chicago Press, 1998, p 57

forms of government.[786] Language was then also used by Aristotle in his second destiny of man: Man is the *Zoon Logon Echon*,[787] the living being gifted with language.

This second destiny of man connects Aristotle to the thesis that all people strive for happiness. Since they can only maintain and develop their lives in communities, they need language to successfully shape social practice. Only in possession of *Logos* can they name what is good and bad, just and unjust, and act accordingly in the community.

In the polis, with this change, word and writing become the central means of forming a common culture. This also has consequences for the legal system. While justice (*dyke*) was until then for the "small Bohemian peasant a real decision, which depends on the arbitrariness of 'gift-devouring' kings,"[788] the laws, by their written fixation, are published and become a reasonable general rule representing human order. The emergence of the public sphere is therefore linked to an elementary respect for the word and the writing.

Laws must be formulated and published. Through the laws laid down, democracy opposes the domination of individuals by arbitrariness. *Isonomia*, the equal participation of citizens in power, forms a sharp contrast to *monarchia* and *tyrannis*. Once a move towards *isonomia* had been established, Athens has warned incessantly against a relapse into *tyranny*. The most dangerous potential for such a relapse is considered to be *hybris*, the presumption of power, which threatens the balance of power in the community of the city's citizens. To this philosophers counter the main virtue of *sophrosyne*, moderation. In

---

[786] Arendt, Hannah, and Margaret Canovan. The Human Condition: Second Edition. 2nd Revised edition edition. Chicago: University of Chicago Press, 1998, p. 36.

[787] Aristotle. The Politics. Revised edition. Harmondsworth, England New York, N.Y: Penguin, 2000.

[788] Vernant, Jean-Pierre. 1984. The Origins of Greek Thought. Cornell University Press, p. 48.

place of the *thymos* (martial courage) that characterizes the single-minded angry citizen, as Achilles embodied it in the Trojan War,[789] the democratic polis chooses the prudent *hoplites* (heavily armed) who line up in the phalanx and go into battle side by side with their fellow citizen.

Through this upheaval, after which the social order no longer depends on a sovereign, but conversely, everyone is assigned a place within the order of equals and determines the sovereign, a rule of law is established for the first time. Vernant describes the conviction that triggers this radical change as follows: "Order is higher than power. In truth, the *arche* (rule) in reality belonged exclusively to the law."[790]

This brief overview of the creation of democracy among the Greeks can make one thing clear: The Greeks discovered democracy by critically reflecting on their practice in the polis and by designing a new social space around which order could unfold: It is the public space of the *agora,* in which all claims to truth and validity must be proven. Even if, due to the restrictions of the time (only free citizens can have a say, women and slaves have no rights), there can be no talk of democracy in the modern sense: With the importance of public affairs for everyone and their commitment to public discourse, the Greeks discovered basic principles of democracy which are still essential today. Today, as then, a democratic order is based on an image of humanity that idealizes the free citizen of the polis, who is free in his house (*oikos*) and free and communal with others in the *Agora* as the real sovereign.

---

[789] See the beginning of the Illiad: "Goddess, sing of the deadly fury of the Peliden Achilles, which caused the Achaians a thoUSnd sufferings". Homer, op. cit. Iliad, p. 3.

[790] Vernant, Jean-Pierre. 1984. The Origins of Greek Thought. Cornell University Press, p.67

This is a "revolution of thought" that "implies a radical change in the relations between power and order".[791] Order is no longer hierarchical but consists in maintaining a balance between powers that are now equal, none of which can finally dominate the others, since this would mean the "return of the universe to an original state of inchoate disorder".[792]

The awareness of this freedom was the basis of the most brilliant periods of Greek culture.

The Greeks delicately sounded out the limits of their newly acquired design possibilities. They knew that not all questions relevant to the polis could be subjected to public discourse. Not everything that may affect the community in the future can be discussed in depth in the Agora. As in their world of gods, the goddesses of fate, the *moirs*, also held finally the reins in the polis. Behind this was the insight of the Athenians into the unavailability of the future, into the forces of fate that were beyond human control and domination. Only the oracle at Delphi was able to interpret the future in the face of this fundamental contingency, and yet, as is well known, the oracle's utterings always remained a mystery.

A software giant in Silicon Valley has named itself after the Delphic cult "Oracle". This indicates the claim to make the future available from the data of the past. For the Greeks, such type of human intention was generally *suspected of hubris,* because it lacked respect for the limits of knowledge.

This is also evident in the example of the oracle of Delphi itself: "Know thyself" *gnṓthi sautón,* the famous inscription above the

[791] Vernant, Jean-Pierre. 1984. The Origins of Greek Thought. Cornell University Press, p. 123.
[792] Vernant, Jean-Pierre. 1984. The Origins of Greek Thought. Cornell University Press, p.109

Apollo temple in Delphi, remains currently relevant, even if there are databases that quantitatively know "more" about me than I do. The self-knowledge that the oracle demands is always qualitative. It cannot be replaced or surpassed by an infinite amount of data and cannot be substituted by databases. The second inscription of Delphi, "Nothing in excess" (*medèn ágan*) but everything in human measure, also remains topical, especially with regard to the quantity of data that may be collected about individuals. The oracle of Delphi calls for self-knowledge and at the same time for self-limitation.

Thus, trust in the creative power of communicative understanding and democracy was by no means naive and boundless in the Greeks. Because the insight prevailed that with all the wisdom of the political assembly it was impossible to discuss all matters concerning the fate of the city in the future and to subject them to one's own decisions, a religious-ritual practice remained binding at the same time, despite all newly won rationalism. The mysteries as well as the esoteric teachings of the philosophers thus retained their secret character in political questions despite all the necessary publicness.

In doing so, they reflected a piece of the unavailability of the life of the polis as well as the randomness in the life of the individual. Even an expression of their newly won *sophrosyne* (level-headedness) and at the same time a protection against the dreaded *hybris*, the cults and mysteries of the Greeks continued in the polis demonstrate a fundamental insight into the limits of human reason. In order not to slide into arbitrariness through self-exaltation, freedom must be aware of these preconditions and its own limits.

The reflection on the preconditions and thus also the limits of reason runs under the heading "Reason and Revelation" through the history of philosophy until today. In 2004 a debate took place between Jürgen Habermas and Joseph Ratzinger, the later Pope, in which Habermas [793] explored this topic within the framework of a theory of democracy under the heading "Pre-political Foundations of the Democratic Rule of Law?"[794] In doing so, he sees in the conflict between reason and revelation a "recurrent idea" namely, that when reason reflects on its deepest foundations, it discovers that it owes its origin to something else. And it must acknowledge the fateful power of this origin, for otherwise it will lose its orientation to reason in the blind alley of a hybrid grasp of control over its own self.[795]

Therefore Habermas pleads for a reasonable dispute of religion and philosophy that was already laid out in the "mutual penetration of Greek metaphysics and Christianity" and demands the "saving translation" of originally religious revelations: "The translation of the concept of "man in the image of God" into that of the identical dignity of all men that deserves unconditional respect is such a saving translation".[796]

The concept of human dignity can be understood as the result of secularization, and at the same time as a result of a dialectic of enlightenment. For freedom, self-determination and a democratic society have human dignity equally at their centre. The latter forms the core of Enlightenment and Humanism from which everything is derived.

---

793 XVI. (Papst), Benedikt, and Jürgen Habermas. Dialectics of Secularization: On Reason and Religion. Ignatius Press, 2006. p. 15 ff, see also: Habermas, Jürgen. Auch eine Geschichte der Philosophie: Band 1: Die okzidentale Konstellation von Glauben und Wissen. Band 2: Vernünftige Freiheit. Spuren des Diskurses über Glauben und Wissen. Berlin: Suhrkamp Verlag AG, 2019

794 Habermas, Jürgen. Auch eine Geschichte der Philosophie: Band 1: Die okzidentale Konstellation von Glauben und Wissen. Band 2: Vernünftige Freiheit. Spuren des Diskurses über Glauben und Wissen. Berlin: Suhrkamp Verlag AG, 2019 p.40

795 Habermas, Jürgen. Auch eine Geschichte der Philosophie: Band 1: Die okzidentale Konstellation von Glauben und Wissen. Band 2: Vernünftige Freiheit. Spuren des Diskurses über Glauben und Wissen. Berlin: Suhrkamp Verlag AG, 2019 p. 29.

796 Habermas, Jürgen. Auch eine Geschichte der Philosophie: Band 1: Die okzidentale Konstellation von Glauben und Wissen. Band 2: Vernünftige Freiheit. Spuren des Diskurses über Glauben und Wissen. Berlin: Suhrkamp Verlag AG, 2019 p.45

To preserve human dignity, a public sphere is needed in which individualization and socialization can take place without violence. Which in turn requires that this public sphere be thought of at least to a certain extent as "free from domination".

The political idea at stake is the idea of a power that renounces violence because it is justified by the recognition and legitimation of the many to whom it refers. It is easy to label this idea as a distant utopia and put it aside. But one can also see in it the potential for orientation that we so urgently seek today. Especially since the idea of freedom of rule also underlies successful interpersonal understanding. Subjects aim for recognition from other subjects that they cannot force.

The excursion to ancient Athens takes us to the origin of three elements that still form the institutional framework of modern democracy today: [797]

- the private autonomy of citizens, who have the right to live a self-determined life, confident in their privacy
- democratic citizenship as the inclusion of free and equal citizens in the political community, which they shape themselves through participation
- an independent political public sphere, which, as a sphere of free opinion and will, links the power of the state to the citizens in civil society.

In modern times, these elements have been joined by other essential elements of democracy that make up its modern form, above all parliamentarism and the development of democratic institutions of

---

[797] see here: Habermas, Habermas, Jurgen. Europe: The Faltering Project. 1st edition. Cambridge ; Malden, MA: Polity Press, 2009, p.71

representative democracy, the majority principle combined with a strong protection of minorities, and the embedding and limitation of these elements by human rights and international law. Inheritance of Athenian democracy remains the special function of the public sphere, which has changed dramatically under the influence of mass media and digital communication.

Public discussion is an indispensable precondition and basis for democracy. The public sphere is understood as a social and intellectual space that enables successful individualization and socialization of people. We must not leave this space to algorithms and Artificial Intelligence, which are neither social nor spiritual beings. The more power these technologies gain in the public, the more we slip into an automated public sphere, as Frank Pasquale calls it.[798] The structures of this new public sphere are dominated by corporations, increasing the threat potential of this technology for the institutions of democracy.

[798] Pasquale, Frank A. "The Automated Public Sphere." SSRN Scholarly Paper. Rochester, NY: Social Science Research Network, November 8, 2017. https://papers.ssrn.com/sol3/papers.cfm?abstract_id=3067552

F3 What does "orientate in thinking" mean?

Philosophy has repeatedly dealt with the question of how orientation within the world is possible and what philosophy can contribute to individuals seeking orientation. Kant offered an answer to this question which is still relevant today, in the form of his essay on this subject: "What does it mean to orient oneself in thinking?"[799]

Kant writes that to answer the question of how to orientate oneself in the world, a topography of reason, a spatialization of thought, is required first. Human reason must first be subject to a recognition of its own subjectivity and standpoint and take it into account in all further reflection. His critical philosophy is concerned with grasping the "conditions of the possibility of knowledge". This also, above all, includes the standpoint and perspective of the recognizing subject. If the subject takes a bird's eye view during its "ascent", a "*view from nowhere*"[800] and tries to recognize the whole from the outside, this leads to "presumption of reason" that then believes to be able to recognize the Absolute. Absolute truth claims, however, that at the end elude any justification, lead back to the old metaphysics and the never-ending dispute of dogmatic world views.

To orient oneself in thought means, according to Kant, to follow the logic of thought and thus the rules of communication. The greatest dangers that threaten an orientation in thinking in Kant's eyes are

---

799 Kant (1724-1804), Immanuel. What Does It Mean to Orient Oneself in Thinking? Translated by Daniel Fidel Ferrer, 2014, https://philpapers.org/archive/FERWDI-2.pdf

800 Habermas, Jürgen. Auch eine Geschichte der Philosophie: Band 1: Die okzidentale Konstellation von Glauben und Wissen. Band 2: Vernünftige Freiheit. Spuren des Diskurses über Glauben und Wissen. Berlin: Suhrkamp Verlag AG, 2019, vol. 1 p. 32.

censorship and the restriction of free speech: "So it can well be said that the external force which snatches from people the freedom to *communicate* their thoughts publicly takes away their freedom to *think*."[801] One is a clear plea for recognising the importance of free public exchange and opinion-forming for self-determination. Those who subject the space of public exchange of facts and opinions to rules that are not intended to serve freedom endanger that freedom. Finally, Kant ends in a passionate call to recognize and defend the connection between reasonable truth orientation and freedom:

> *"Friends of the human race and of what is most sacred to it! Accept what appears to you most worthy of belief after careful and sincere examination, whether of factors or rational grounds; only do not dispute that prerogative of reason which makes it the highest good on earth, the prerogative of being the ultimate touchstone of truth. Otherwise you shall become unworthy of this freedom, and you will surely forfeit it too. "*[802]

There is no need to explain why this warning is more relevant than ever when it comes to AI. Human reason must be and remain the "ultimate touchstone" of truth, not AI and the black box of algorithms. For this, the reflexivity of human reason is a prerequisite that is at the bottom of its ability to correct itself again and again in a rational way: "The self-critique of reason as a weapon against its own tendency to become blind and stupid, against its self-destruction and self-absorption - this is surely the key legacy for today of Kant's concept of reason," says the philosopher Herbert Schnädelbach.[803]

The reflexivity of a self-critical reason is the basis of an understanding in discourse that aims at commonly shared values. This also requires

---

[801] I. Kant, Works Edition Volume V, p. 280

[802] I. Kant, Volume V, p. 282 f. DO 8:146-147 tr.Wood

[803] Herbert Schnädelbach, Philosophy, A Basic Course, Volume 1, p. 92. Translation quoted in Schweppenhäuser, Gerhard. 2009. Theodor W. Adorno: An Introduction. Duke University Press.

that, in accordance with the self-image of man as a finite, imperfect being, reason should not be oriented towards artificial but towards human values. Kant sounds like a critic of the current optimization mania of the Californian ideology when he warns against man's adaptation to exaggerated ideals, be they mathematical or technical:

"From such crooked wood as a human is made nothing quite straight can ever be fashioned. Only the approximation of this idea is imposed upon us by nature."[804]

Kant would have ridiculed Stewart Brand's Whole Earth enthusiasm for "infatuation" just like he did to the Spinozism of his time: "The course of events is about this. First, the genius like much in its bold sweep of as it has the thread, what else drew reason stripped. It soon charmed by other power spells and great expectations, and seems to have now set them on a throne, the slow lumbering reason graced (*zierete*) so bad, and it nevertheless always leaded the same language. The maxim then adopted at the top of the invalidity of a legislative reason we call common people fanaticism (*Schwärmerei*, enthusiasm); those favorable of good nature but enlightenment."[805]

If one wants knowledge, this requires a free and rational exchange of thoughts with others. Only in this way can Kant's orientation succeed as it enables the recognition of heteronomous structures and their replacement by autonomous ones. Kant names a clear alternative here: either reason gives itself laws, or it must "bow under the yoke of laws given by another; for without any law,

[804] Kant, Immanuel. Idee zu einer allgemeinen Geschichte in weltbürgerlicher Absicht, 6. Satz (1784) in Sämtliche Werke in sechs Bänden, vol. 1, p. 230 (Großherzog Wilhelm Ernst ed. 1921)(S.H. transl.) . Quoted in translation: Horton, Scott. "Kant – The Crooked Wood of Humankind," Harper's Magazine, May 30, 2009, https://harpers.org/2009/05/kant-the-crooked-wood-of-humankind/.

[805] Kant (1724-1804), Immanuel. What Does It Mean to Orient Oneself in Thinking? Translated by Daniel Fidel Ferrer, 2014, p.16, https://philpapers.org/archive/FERWDI-2.pdf

nothing – not even nonsense – can play its game for long".[806] But since reason can only communicate about its own laws in a free exchange, the free use of reason is a prerequisite for political freedom. The freedom of thought, which anyway cannot limit any "upper power", is dependent on freedom of speech and writing. For only in the public use of freedom can orientation take place. "Alone, how much and with what correctness would we probably *think* if we did not think, as it were, in community with others to whom we *share* our thoughts and who *share* theirs with us!" Instead of an external authority, reason, in its intersubjective use, becomes itself the "guiding principle" of an orientation.

On the one hand, according to Kant reason must therefore be limited. Because of its finiteness, it cannot, like a God, see the whole from the outside and thereby completely recognize it. On the other hand, reason, through its intersubjective public use, can very well develop universally valid insights that can also serve "as guidelines" in a globalized world. The self-limitation of reason is, according to Kant, a prerequisite for self-determination. His conclusion is: Only a public use of reason makes orientation possible.

In his writings on eternal peace Kant already described the age of globalization, explaining how "a violation of rights in one place is felt throughout the world".[807] This leads him to demand that a binding legal order be created between states. He recognized that to secure human rights, a binding international law is ultimately necessary. Kant thus provides an answer to the question, explaining that human beings as beings of reason can only orient themselves through the public use of reason, which must be expressed through

---

[806] Kant, Immanuel. 2018. Kant: Religion within the Boundaries of Mere Reason: And Other Writings. Cambridge University Press.

[807] Kant, Perpetual Peace, Third Definitive Article for A Perpertual Peace, 1795

binding legal institutions. With all trust in the power of reason, however, it is important to leave the church in the village and to accept man's limits.

A deification of man in the form of *Homo Deus* would be for Kant a dangerous infatuation that does not respect human boundaries and thus leads straight back to metaphysics and foreign determination, because the taking of the "divine" gaze combined with a boundless claim to knowledge leads to a self-exaltation of reason that must ultimately lead to superstition:

> *"This if it is disuputed that reason deserves the right to speak first in matters concerning supersensible objects such as the existence of God and the future world, then a wide gate is opened to all enthusiasm, superstitution and even to athiesm.*[808]

Following Kant, this demarcation of the borders between reason and superstition were repeatedly called into question. Scientistic world views such as dataism claims to comprehensive knowledge mean that they must become entangled in the same contradictions that Kant has shown. The antinomies of reason, which occur when reason wants to answer central questions of metaphysics about God, soul and immortality with an absolute claim to knowledge, ultimately lead to dogmatism and, as a consequence, to the loss of self-determination and freedom. They thereby also endanger human dignity.

Human dignity exists for Kant in the requirement that the human being must not be reduced to mere functionality for a purpose:

---

[808] Kant, Immanuel. 2018. Kant: Religion within the Boundaries of Mere Reason: And Other Writings. Cambridge University Press, p.12.

> *"the human being and in general every rational being exists as an end in itself, not merely as a means to be used by this or that will at its discretion; instead he must {...} always be regarded at the same time as an end.*[809]

If we want to orient ourselves in thinking, then for Kant it is indispensable to reflect on the thinking itself and to refer to the "I":

> *"It must be possible for the 'I think' to accompany all my representations; for otherwise something would be represented in me which could not be thought at all, and that is equivalent to saying that the representation would be impossible, or at least would be nothing to me".*[810]

AI still cannot do this, even today. An AI, which could accompany all its ideas by an "I think", would have reached structural consciousness. But such a self-reflexivity of artificial systems seems unattainable even to the most daring and euphoric enthusiasts of AI.[811]

Orientation in contrast to a navigation device as *Google Maps* requires a way of thinking that exposes normative resources beyond machine logic and thus defends the creative and inventive in human reason. To conclude, once again, we quote Kant: "the proper field for a genius is the imagination, because imagination is creative, and just because it is less subject to the coercion of rules than other faculties, it is more capable of originality".[812]

Human creativity is creative power, it is the ability to perceive complexity and not only to understand complexity but also to reflect

---

[809] Kant, Immanuel. 2005. Groundwork for the Metaphysics of Morals. Broadview Press.

[810] Kant, Immanuel. 1999. Critique of Pure Reason. Cambridge University Press.

[811] See interview Kurzweil/Minsky : Kurzweil Interviews Minsky: Is Singularity Near? July 14, 2014

[812] Kant, Immanuel. 1996. Anthropology from a Pragmatic Point of View. SIU Press, p.124

on it. This means it manages to create possibilities that are superior to the calculations of machines.

3.1. Two elements of autonomy: subjectivity and rationality

> *It despises intellect and science*
> *The supreme gifts of man*
> *It has given itself to the devil*
> *And must perish.*

Hegel, Goethe's Faust modified quote in the Phenomenology of the Mind[813]

The idea of self-determination presupposes that a person has self-consciousness, which is the expression of a specifically human self-relationship. Hannah Arendt has highlighted two insights of Socrates which make up his philosophy.[814]

First, the Delphic precept, "Know thyself," which to Socrates meant conducting an inner dialogue. According to Arendt, inner dialogue is based on the principle of consistency, which leads to the demand for radical sincerity and authenticity: "it would be better for me... that most men should not agree with me and contradict me, rather than I, being one, should be out of tune with myself and contradict myself".[815] But the imperative of freedom from contradiction is

[813] The original quotation from Goethe reads: "Despise only reason and science, man's supreme power, let only in dazzling and magical works let yourself be strengthened by the spirit of lies, so I have you already unconditionally " The message of the devil in the direction of Faust, who is prepared to exchange his soul for the lust of the moment: The renunciation of reason will certainly drive you into my hands.

[814] Arendt, Hannah. Socrates, Apology of plurality, Berlin 2016.

[815] Arendt's reading of Socrates in the Gorgias VILLA, Dana R. 1999. 'ARENDT AND SOCRATES'. Revue Internationale de Philosophie 53 (208 (2)): 241–57.

not simply derived from the rational laws of logic. It is written in Socrates who is in constant dialogue with himself through his *daimon,* in the service of a successful life. The *daimon* is a voice that comes from without that never tells me not to do something, but advises in the negative form. For example: "Even if no one sees you, you should not kill, if only because you cannot possibly wish to live permanently with a murderer."[816] Socrates therefore believes that the consciousness of a being that thinks and acts simultaneously makes people better. This position is known as ethical intellectualism.

Arendt supplements this idea with the remarks made by Aristotle on friendship, to which Aristotle attaches the highest value, because in friendship we wish our good for our friend simply for the friend's sake. In the ability to adopt the friend's perspective and yet at the same time remain situated within your own subjectivity, Arendt recognizes the political character of friendship: "The ability to see things from the other's point of view...is the political insight par excellence".[817] Modernity, on the other hand, in Arendt's eyes, carries out a "reduction *scientiae ad mathematicam*",[818] first by capturing nature in numbers in order to be able to calculate and control it better, and then by transferring this process to society. In this way, Homo Faber's objective view of the social world replaces a multi-perspective concept of action, which is constituted by the exchange of different subjective perspectives.

In contrast, Arendt insists on a broader concept of the forms of action, which she derives from the difference between production

---

816 Arendt, Hannah, Mary McCarthy, and Carol Brightman. Between Friends: The Correspondence of Hannah Arendt and Mary Mccarthy, 1949-1975. New York: Thomson Learning, 1996, p.27,

817 Arendt, Hannah. 'Crisis in Culture', 1960 p. 219-20

818 Arendt, Hannah, and Margaret Canovan. The Human Condition: Second Edition. 2nd Revised edition edition. Chicago: University of Chicago Press, 1998, p. 341.

and action, among other things. "Nevertheless, acting, thinking, speaking, have more in common with each other in their worldly character than any of them have with producing and working. For they are simply 'unproductive', they produce nothing, and as activities they are as fleeting as life".[819] From these "unproductive" forms of action, however, the ideas of the "good life" can be developed, which for Aristotle is the goal of all human endeavor. The good life must therefore be won from a context of reason that does not come from the compulsion to reproduce life. This context must be wider and, for example, in the model of friendship, it must be guided by empathy with the other and by a change of perspective.

Just as in action theory, it is also important for a "social theory that endeavors to identify its critical standards "not to reduce rationality to whatever is expedient.[820] "As an antidote, an internal differentiation of the concept of reason with typological means offers itself here, since it makes it possible to reject global denunciations of rationality as *pars pro toto* and, at the same time - following the example of all Enlightenment philosophy - to situate the critique of reason in the area of reason itself".[821] This program of an extended critique of reason with reasonable means, was formulated by Herbert Schnädelbach at a time when postmodernism was just beginning to dominate the philosophical debate. Today, after its failure and in view of the pretensions of digital reductionism,[822] it is more topical than ever. The irrationalism and relativism of postmodern thought was followed after the turn of the millennium by fundamentalism and populism.[823]

---

[819] Arendt, Hannah, and Margaret Canovan. The Human Condition: Second Edition. 2nd Revised edition edition. Chicago: University of Chicago Press, 1998, p.95

[820] Habermas, Jürgen. 1986. Theory of Communicative Action, Volume 1: Reason and the Rationalization of Society: Reason and the Rationalization of Society Vol 1. New Ed edition. Boston: Polity, p.7.

[821] Herbert Schnädelbach, Rationalität, Philosophische Beiträge, Frankfurt 1984, p.12.

[822] Ito, Joichi. “Resisting Reduction: A Manifesto.” Journal of Design and Science, November 1, 2017. https://doi.org/10.21428/8f7503e4.

A renewed program of the Enlightenment, on the other hand, must today gather together and win over numerous individual disciplines for cooperation, especially the technical disciplines that train the designers of the modern world, the software engineers.

3.2. Humanism

> *"Whosoever has thought what is deepest, loves what is most alive"*

Hölderlin

In antiquity (Cicero and others), *humanitas is* based on the concept of humanity, which includes gentleness, compassion and mercy. The Renaissance saw itself as a rebirth of antiquity, a claim to renewal that orients the timelessly valid towards a self-determined future. Humanism was newly founded in the Renaissance on the basis of the ancient texts.

The core idea is that man can be lenient and benevolent with his peers, and unlike animals, he can cultivate this ability through education, the second core element of the ancient concept of Humanitas.

Pico de la Mirandola describes man as a double being with the possibility of degenerating into the animalistic at the bottom or being reborn upwards towards the divine.[824]. Blaise Pascal later shapes the image of man as a thinking reed, which is on the one hand very vulnerable and yet whose greatness lies in its thinking:

---

[823] See Meyer, Prof Thomas. The Theory of Social Democracy. 1 edition. Cambridge: Polity, 2007. and in German: Thomas Meyer, Die Geschichte der Philosophie, Essay and Discourse, DLF 1.3.2020 https://www.deutschlandfunk.de/die-geschichte-der-philosophie-juergen-habermas-und-die.1184.de.html?dram:article_id=468954

[824] Pico de la Mirandola, On Human Dignity, Ditzingen 1997

"Man is but a reed, the feeblest in nature, but he is a reed that thinks. It needs not that the whole university should arm itself to crush him; a vapor, a drop of water suffices to kill him. But should the universe crush him, man would still be more noble than his destroyer, because he knows that he dies and the advantage the universe has over him. The universe knows nothing of the kind. All our dignity, then, consists in thought".[825]

A humanism which also considers the origin of man and his anchoring in the humus, in the earth, must therefore grasp the dependence of man on the earth. In other words, the *humus* component of the *humanum*. Part of a whole that is larger than man, but in which man nevertheless has a greatness that constitutes his uniqueness and dignity. Man is made great by the ability to think about the universe, vulnerable by the breeze that can break him at any time.

Today, GAFAM's algorithms exploit human weakness, manipulate us, while the human value which according to Pascal is in the thinking, is assigned to machines.

In light of this, it is important to maintain and promote the classic humanism instead of updating it with post- and transhumanism.[826]

An expanded concept of rationality as seen in humanism needs to be defended when it comes to determining values and standards against which we want to orient and measure the systems of Artificial Intelligence.

---

[825] Pascal, Blaise. Blaise Pascal Thoughts: Selected And Translated. Reissue edition. Cambridge; New York: Cambridge University Press, 2013, p.11

[826] Habermas, Jürgen. 2015. The Structural Transformation of the Public Sphere: An Inquiry Into a Category of Bourgeois Society. John Wiley & Sons, p. 194.

If, as per the ever-quotable Winston Churchill, democracy can be described as the least bad form of government,[827] then reason and the autonomous subject are the least bad conceptions when it comes to the question of what we should orient ourselves by and on what basis we should make decisions. Once reason and subject have been adopted, there remain, according to all historical experience, gods, demons, powers of destiny or leaders that can only be conjured up but not rationally understood or criticized.

In the age of AI, a new fetish threatens: The allegedly superior anonymous total super intelligence, to which we entrust more and more decisions, and which threatens to elude any attempts at control with natural dynamics, because it is exaggerated as an absolute problem solver. But this supposed super intelligence proves to be a purely mathematical-logical sub-function of rationality. There is a great danger in the bidding farewell to humanism on which the idea of an inalienable human dignity is based. The "overcoming" or adaptation of humanism, which is so urgently demanded by the believers in technology, can lead to an inhuman society in which democracy and the rule of law are abolished along with human dignity.

Entirely in the spirit of an unstoppable belief in progress, trans- and post-humanists give the impression that humanism is an old hat and is hardly compatible with modernity. But that is not true. As Johannes Hoff, among others, has shown, posthumanism itself is rather backward-looking.[828]

---

[827] "Democracy is the worst of all forms of government - apart from all the other forms that have been tried from time to time." - Quoted by Hans Vorländer at bpb.com. see: The International Churchill Society. "The Worst Form of Government," February 25, 2016. https://winstonchurchill.org/resources/quotes/the-worst-form-of-government/

[828] Hoff, Johannes, Transhumanism as a symptom of symbolic impoverishment. On the anthropological challenge of the digital revolution. In: Herzberg, Stephan, Watzka, Heinrich (Ed.): Brave new world, or what comes after man?, De Gruyter: Berlin - New York 2020.

3.3. Humanism today

A range of contemporary philosophers have updated humanism. Among them is Julian Nida-Rumelin who invokes Kant:

"At the centre (of Kant's practical philosophy) is the idea of human dignity".[829]

Continuing the enlightenment approach of Kant requires a contemporary anthropology, a fact that was recognised by the Konigsberg thinker. "At the centre of a humanistic anthropology are three concepts: reason - freedom - responsibility. That which Kant conceived of as autonomy brings together responsibility (duty) and freedom on the basis of (practical) reason."[830] Therefore, a renewed humanism needs to develop a theory of rationality, freedom and responsibility.

Humanism consists, for Nida-Rümelin, of three basic building blocks:

- Autarky
- Rationality
- Universality

Autarky, with freedom and independence, refers to the conditions of autonomy.

Rationality makes self-determination possible because universal principles emerge, which in turn secure autonomy socially and enable democratic sovereignty.

---

[829] Rumelin, Digital Humanism, 2015 https://www.mpg.de/13790224/W002_Viewpoint_010-015.pdf p.13, Julian Nida-Rümelin, Structural Rationality and Other Essays on Practical Reason (Springer International Publishing, 2019).

[830] Rumelin, Digital Humanism, 2015 https://www.mpg.de/13790224/W002_Viewpoint_010-015.pdf p.143

Finally, norms must be rationally justified in such a way that they are universally valid without destroying essential particularities of different life-worlds.[831]

This humanism is superior to the currently known forms of AI in many respects. It takes a view of the whole – with an *Overview Effect* - and therefore enables the humanistic thinker to act responsibly for the whole, fellow human beings, society, nature and environment, because he sees himself as part of the whole and not as its *Maitre et Possesseur* as in Descartes' philosophy. Responsibility in this context means being able to respond, i.e., to give reasons in discourse that justify action.

Giving account, giving reasons and weighing up the pros and cons, also builds a decisive ability of human beings. This is crucial different to machine intelligence: the ability to dream of another, of a better life. Only those who can consider and weigh up alternatives can design alternatives of action and alternative states of the world.

Man can dream of a better life in a better society, can criticize conditions and make society a better one, according to the values he recognizes for himself.

An AI-controlled "optimization" of the society can achieve none of this. Since it is based on existing data, it leads to an end of innovation in the long run. In addition, it is hardly conceivable to have an AI that is capable of criticism that goes beyond the comparison of actual and target values or even develops something that could be called critical consciousness. Finally, it is inconceivable that AI could in

---

[831] See also: Rumelin, Digital Humanism, 2015 https://www.mpg.de/13790224/W002_Viewpoint_010-015.pdf p.12, See also: J.Nida Rumelin, Universality and particularity, https://www.philosophie.uni-muenchen.de/lehreinheiten/philosophie_4/dokumente/univers_und_part.pdf

any way inherit the three core elements of reason, freedom and responsibility. Human values are not identical with mathematical or economic "values", they are not congruent with them.

So it looks like we humans are still needed. We should therefore prevent humanism from being tossed overboard. Those who tell tales of the end of mankind secretly know that humanism is still needed too. They are concerned with shifting and obtaining power by using the possibilities offered by AI and digital technologies to their advantage. They want to implement an economic model consistently in all areas of life, and the disappearance of democracy and individual autonomy is included as collateral damage.

F4 From the imperative of responsibility to the human imperative

4.1. The Janus head of technology

The starting point of ethics in the technical age is the recognition of the far-reaching consequences of modern technology. Technology determines our lives to an ever-greater extent. It is characterized by three aspects:[832]

- Technicity
- Globalism
- potential for usefulness and destruction

Technicity relates to the efficiency, functionality and experience of a particular technology; or its 'unnaturalness'. It means that, on the one hand, technology differs from the natural basis by its instrumental-purpose-related sense. Moreover, the extent to which technology differs to the natural mechanisms they replace is not exclusively a function of what human hands can produce, but increasingly by tools and machines, i.e., by technology itself.

The globality of technology is expressed in the fact that it is used all over the world and increasingly shapes the features of a technically shaped global civilization,[833] which increasingly levels cultural characteristics.

---

[832] In the following reference to: Birnbacher, Dieter. Technology, in Philosophy, ed. Schädelbach et al., Vol. 2, p. 608 ff.

[833] Birnbacher, Dieter. Technology, in Philosophy, ed. Schädelbach et al., Vol. 2, p. 610.

Finally, the double potential of technicity expresses the ambivalence of values, the Janus-faced nature of modern technology: in addition to the undisputed benefits it contains, it also contains the potential to destroy: Nature, species, ways of life, climate, democracy and potentially all of humanity.

There are also risks in the "optimization" of technology, especially if it is carried out ex-clusively according to the logic and values of the technology itself. A car with 500 hp may be a technically optimized car. But without a speed limit and an experienced driver at the wheel, it can also become a risk for occupants, other road users and the environment. Technological improvements always carry the risk that the fascination ema-nating from them "paralyses vigilance in the long term",[834] as far as the examination of possible negative consequences is concerned.

The technical means can become independent of the original intentions of the creator and lead to decisions being made in which other purposes are ignored, i.e., they can themselves be purposeful. Thus, the philosopher Dieter Birnbacher points to the histor-ical example that the development of the atomic bomb in Los Alamos continued even after Germany had already capitulated in May 1945. The legitimation for its development, however, originally came from the fear that Hitler-Germany could pre-empt the US and use the atomic bomb first. Disturbed by Otto Hahn's successful nuclear fission in August 1939 and thus worried that Nazi Germany could be the first country to develop the atomic bomb, Albert Einstein had written a letter to President Roosevelt, which is often called the "most important letter in history".[835] He pushed for the construction of an American bomb - with success.

---

[834] Birnbacher, Dieter. Technology, in Philosophy, ed. Schädelbach et al., Vol. 2, p. 636.

[835] The letter can be read here: 'The Einstein-Szilard Letter - 1939'. n.d. Atomic Heritage Foundation. https://ahf.nuclearmuseum.org/ahf/history/einstein-szilard-letter-1939/

Germany capitulated in May 1945, removing the original justification for the bomb's construction, yet work on the bomb continued unabated. Leó Slizárd, who wrote the letter to Roosevelt with Einstein, tried to prevent the American government from using the bomb at the end of May 1945. James Byrnes, advisor to the new president Truman, replied that the US had spent "two billion dollars on developing the bomb, and Congress would want to know what we had got for the money spent". The United States would not simply let the superweapon disappear into oblivion again.[836]

Finally, two atomic bombs were used against Japan, a country that had not done any research on the atomic bomb itself and was on the verge of surrender.

This historical example teaches that technical means can become independent of their original ends if the application of technology ultimately follows only the logic of technology or power. You use them because you can.

There is also a fundamental paradox of technology: we create it to enhance our skills in a particular field. To be able to use it, we adapt by using it in a functional way. This has led to a certain pressure to adapt because technology has increasingly become a "second nature" to which man must adjust.

It is also this process of adaptation, which with the increasing technicalization and digitalization of the world also generates a rising pressure to conform, which makes the thesis of the continuation of evolution through technology seem plausible at first. But this consequence is a sham. Californian ideology tries to disguise the fact

[836] Quoted in Misa, Thomas J. 2011. Leonardo to the Internet: Technology and Culture from the Renaissance to the Present. JHU Press.

that people do not only adapt to "technology", but to the people and companies behind the technology, who design and are responsible for the programs.

Hans Jonas has shown that man is responsible for the consequences of technology, and how he can meet this responsibility.

4.2. Hans Jonas and the imperative of responsibility

> *"Our question is, shall man be? In order to answer it correctly, we must first answer the question of what it means to say of something that should be. And this, of course, leads back to the question of whether something - instead of nothing - should be at all."*

Hans Jonas, The imperative of responsibility[837]

With his book *The Imperative of Responsibility*, Hans Jonas, the German-Jewish emigrant at the New School of Social Research in New York, created an ethic for technological civilization in the form of the precautionary principle.

This principle states that technology must be examined for its potential to either call into question the existence of humanity in the long term or substantially change the nature of human existence. Because modern technology can have "long-distance effects", we have a political obligation to invest in understanding its long-term consequences, i.e., through technology impact assessment. The results can then put us in a political position today to take the potentially tough decisions necessary to rule out with almost certainty

---

[837] Jonas, Hans. The Imperative of Responsibility: In Search of an Ethics for the Technological Age (University of Chicago Press, 1985), p. 96.

the possibility of profound long-term negative consequences for humanity.

Jonas' book was soon considered the bible of the environmental and anti-nuclear movement. But Jonas himself explicitly mentioned the technologies of life extension, behaviour control and genetic manipulation as further candidates for the application of the precautionary principle.

The common denominator of the technologies envisaged by Jonas was that they caused environmental damage or had an impact on human existence in the long term. Jonas has developed principles of technology ethics that need to be re-read in light of the new threats to freedom and democracy posed by the Internet and the digital technologies and business models based on it.

The threat to the essence of humanity has already shifted dramatically in the face of the mass collection of personal data about each and every one of us and the creation of comprehensive personality profiles. This practice allows for extensive control and manipulation of human behavior. The "singularity" of the human being, the merging of man and machine, the uploading of the character and soul of the human being to a ma-chine, is intended to allow his technical "survival" after biological death. This vision of technological exchange and replacement of man by Artificial Intelligence demands to be the subject of an impact assessment, as proposed by Jonas.

In question today are individual freedom and self-determination: thus the basic condi-tions for a free society. Meanwhile, new technologies and populism, which are mutually reinforcing, have taken a pincer grip on democracy.

If we want to revive the principle of responsibility in the face of this threat, we should expect counter-arguments. Jonas and his precautionary principle in environmental law met with enormous resistance at the time. They were and are countered by the argument that they act as an obstacle to innovation and thus to economic growth in Europe.

In fact, however, the development and implementation of the precautionary principle in environmental law is a story of great economic success. The principle promotes sustainability and resource efficiency, and it directs investment in innovation into these very areas. In Europe and Germany, strict environmental legislation has led to the development of new technologies and business models that have become export hits. Take, for example, the technologies of energy conservation, resource efficiency and alternative energies in general.

We have learned that there is no necessary conflict between environmental protection, sustainability and economic success. In Germany, billions are now earned with sustainable technologies, environmental technologies and alternative energy. Hundreds of thousands of jobs have been created in these areas.

Similarly, there is no unsolvable basic conflict between individual freedom, self-determination and democracy on the one hand, and digitalization and the technologies and business models based on the Internet on the other. However, without the guiding hand of democratic politics, oriented towards the precautionary principle and under-stood as part of the innovation process, the Internet industry itself has not been able to prove the public good compatibility of its innovation.

We must learn from this experience before the Internet is connected to even more powerful technologies such as General Artificial

Intelligence and quantum computers, and penetrates further, deeper and more intensely into all areas of the world and life through 5G, the Internet of Things and Bio-Physical Systems. For once this technology exists unregulated, as the Internet has been up to now, serious, irreversible consequences for human beings and democracy can arise. This technology, in the hands of dictatorships that use it to comprehensively control and manipulate the population, can make it impossible to build or return to democracy. For dictators would then have in their hands the most powerful propaganda and surveillance machine imaginable. At no time in human history have there been such good conditions for a totalitarian dictatorship as there are today. "The propaganda possibilities Hitler had, and the surveillance apparatus of the Stasi were child's play against what is possible today,"[838] says Armin Grunewald, head of the Office of Technology Impact Assessment at the German Bundestag. To prevent such risks, tough decisions are needed today, similar to those that Jonas demanded for the environment at the time. We cannot afford Chernobyl or Fukushima for democracy and human self-determination. The meltdown of democracy could be an irreversible event. Because once freedom and democracy are lost, there is no way back under the conditions of digital surveillance systems. Nor can we afford the slowness of the fight against climate change. Because developments in AI research are progressing at least as fast as the rise in temperature on Earth. The risk of a loss of democracy is at least equal to that of climate change. We need sustainability principles for both, not only for the preservation of our natural resources, but also for the sustain-ability of democratic structures in the digital future.[839]

[838] Baumüller, Michael and Stefan Baum: The dangers of digitization, SZ Interview with Armin Grunwald, 57, who heads the Office of Technology Assessment at the German Bundestag (TAB). in Süddeutsche Zeitung, 29 January 2019.

[839] See also: German Advisory Council on Global Change (WBGU) Main Report "Our Common Digital Future", 2019, https://www.wbgu.de/de/publikationen/publikation/unsere-gemeinsame-digitale-zukunft

For both, the fight for self-determination and democracy, and the fight against climate change, we need a functioning democracy. The sustainable safeguarding of democracy is therefore what our re-reading of Hans Jonas is all about.

4.3. The imperative of responsibility today

The principle of responsibility ultimately arises from applying the principle of the future to our actions in the age of technology. In view of the ecological crisis, Hans Jonas calls on us to do everything we can to ensure that there is "a balanced budget between man and nature". Jonas holds an ideologically deep-seated belief in our concept of 'progress" responsible for the growing ecological crisis. This unrestrained belief in technical progress, secularized heritage of religious eschatological hopes of salvation, is increasingly proving to be a compulsion in itself. But how can the same belief in progress, which should free us from the coercive context of nature and mean the end of misery, in turn become subject to a natural compulsion? Jonas' answer is: Through the dependence on more and more technology, into which man gets through technology. In order to maintain technical civilization, it is imperative that technology be constantly developed and thus the continued exploitation of nature, of the external and increas-ingly also of the internal nature of human beings.

According to Jonas, the "finally unbound Prometheus" has caused "that the promise of modern technology has turned into a threat."[840] The deterministic view of nature inherent in science is reflected in man, who measures his inner being with the same determination and submits to it as he did with the natural world. Even if technology

[840] Jonas, Hans. The Imperative of Responsibility: In Search of an Ethics for the Technological Age (University of Chicago Press, 1985), p.185.

thereby threatens man himself, there can be no way back. Therefore, the development and use of new technologies must be subject to ethical and political scrutiny. This is the only way to break up the coercive relationship described above and to keep it under control.

Jonas sees the feedback forces of a largely autonomous system at work in the development of the technology:

"*Technology, apart from its objective works, assumes ethical significance by the central place it now occupies in human purpose. Its cumulative creation, the expanding artificial environment, continuously reinforces the particular powers in man that created it [...]. This positive feedback of functional necessity and reward... assures the growing ascendancy of one side of man's nature over all others, and inevitably at their expense.*"[841]

This one-sidedness of instrumental reason, which accompanies the triumph of technology and is continuously reinforced by it, is at the expense of the human being as a whole:

"thus while the total stock of knowledge grows, the individual's share shrinks to ever tinier proportions of it."[842]

The increase in human power through technology had already reached a stage in the 20th century where human activity was endangering the living conditions on the planet. Therefore, according to Jonas' analysis, ethics must be changed according to this circumstance, if it wants to try to influence human actions in the face of constraints in the technical world. Traditional ethical principles can no longer meet

[841] Jonas, Hans. The Imperative of Responsibility: In Search of an Ethics for the Technological Age (University of Chicago Press, 1985), p. 9

[842] Jonas, Hans. The Imperative of Responsibility: In Search of an Ethics for the Technological Age (University of Chicago Press, 1985), p.167.

these requirements because they essentially relate to the immediate environment of humans, while the use of modern technology has long-distance effects in time and space. Jonas assumes that as man's power grows, so does his responsibility. Based on *Kant's Categorical Imperative*:

"*Act only according to the maxim by which you can at the same time want it to become a general law*",

the principle of responsibility leads man to a new moral imperative for the future:

> "*Act so that the effects of your actions are compatible with the permanence of genuine human life on earth.*"[843]

The new imperative means that we "may dare to live our own lives but not that of humanity, ... that we have no right to choose the non-being of future generations because of the being of the present."[844] Because of the supra-individual contexts of action that have developed in the technical world, this imperative is primarily directed at politics and law, and it has developed its effect in this context.

Acting according to other than purely technological principles is therefore possible. It is acting according to the principle of responsibility. For Jonas, one of the main causes of the ecological crisis is the victory of instrumental reason over other rational capabilities of man:

---

[843] Jonas, Hans. The Imperative of Responsibility: In Search of an Ethics for the Technological Age (University of Chicago Press, 1985), p.11.

[844] Jonas, Hans. The Imperative of Responsibility: In Search of an Ethics for the Technological Age (University of Chicago Press, 1985), p.11

"Thus, the triumph of homo faber over his external object means also his triumph in the internal condition of homo sapiens, of whom he used to be a subsidiary part."[845]

In order that man does not become a servant of the technology he has created himself, he must recognize his responsibility and act responsibly in view of the distant consequences of modern technology. Only in this way can he maintain control over technology and not become its slave. In view of the challenges that modern technology creates, it is therefore necessary to develop a broader concept of reason than that of the technical and instrumental control of nature.

4.4. The problem of forecasts and the heuristics of fear

Since its beginnings, technology has been one of the strategies of humanity's coping with the future. Jonas notes a "paradox" with regard to the attempt to predict and control the future:

"On the one hand we know more of the future than our pre-modern ancestors, on the other hand we know less."[846] More because we can make predictions based on much larger data, less because life was once static compared to the dynamics of the modern world. "Dynamic is the signature of modernity... it means that we have to reckon with something new all the time without being able to calculate it."[847]

---

[845] Jonas, Hans. The Imperative of Responsibility: In Search of an Ethics for the Technological Age (University of Chicago Press, 1985), p. 31 The formulation of the serving part recalls the famous chapter on domination and bondage in Hegel's Phenomenology of the Spirit. In Hegel the Lord recognizes his dependence on the servant, while the servant recognizes that the Lord depends on him. In the end, both realize that only in the recognition of the other can they find their identity and destiny.

[846] Jonas, Hans. The Imperative of Responsibility: In Search of an Ethics for the Technological Age (University of Chicago Press, 1985), p.119

[847] Jonas, Hans. The Imperative of Responsibility: In Search of an Ethics for the Technological Age (University of Chicago Press, 1985), p.119

With the technical development and the calculation of the world, we have set in motion an acceleration that paradoxically makes predictability increasingly unlikely. Therefore, this development cannot be controlled solely by calculation.

Nevertheless, we rely on ever larger calculating machines for this task, and thus fall into an innovation trap: the greater the technical effort, the more obvious the "factor X" becomes for Jonas, which prevents an exact calculation. "This unknown "x" of permanent innovation haunts every equation."[848] It is the irreducible contingency, which re-mains uncontrollable in every probability calculation, because the future proves to be ultimately incalculable. Because of the immense impact of accidents, for example, from nuclear energy, risks must be carefully assessed and checked for their responsibility before the technology is used.

From this insight Jonas establishes his second principle: the heuristics of fear. By this he means that humans should refrain from all actions whose remote effects do not exclude that they endanger the existence of mankind or the essence of being human.

In order that the heuristic of fear does not become paralysis through dread, Jonas demands a "factual science of the long-distance effects of technical actions". This science must also deal with the realm of the possible. The first duty of future ethics is therefore to obtain ideas about the later effects of technologies. For us today this means that the possible effects of digital technology and Artificial Intelligence must be scientifically investigated, even if they have not yet occurred. This requires first of all a knowledge of technology that enables the imagination of technology and technical - social imagination, which

---

[848] Jonas, Hans. The Imperative of Responsibility: In Search of an Ethics for the Technological Age (University of Chicago Press, 1985), p.120

can inform society about the potential consequences of technology on democracy and freedom. The knowledge of technology must be combined with the knowledge of the functional principles of a free democratic society. Only when both are combined can technology assessment succeed today.

A "research effort" on the consequences of technology is a contemporary idea, which is also demanded by AI theorists. The more serious the possible effects of a new technology, the greater the effort must be to investigate its effects even before development and application. Since the theoretical potential of a general AI lies in nothing less than the end of mankind or the essence of being human, the research efforts must be massive even if the probability of occurrence is low.[849]

It follows that technology assessment must become mandatory for the development of AI systems. Scientific monitoring of these developments must attempt to scientifically identify the risks so that they can be assessed by politics and society.

4.5. The imperative of responsibility means securing democracy in the technical age

There is already an obligation to carry out a data protection impact assessment when the AI processes personal data in the context of automated decision making.[850] For AI developments and applications, it would certainly be justified to extend the impact assessment obligation to all aspects of democracy, the rule of law and

[849] The first step, but not enough, is the warning call of scientists 2015 - Stephen Hawkings, Elon Musk, Max Tegmark, Stuart Russell and others, https://futureoflife.org/ai-open-letter, with annexes on research priorities - https://futureoflife.org/data/documents/research_priorities.pdf?x69354

[850] See Art. 35 (3) a) GDPR.

fundamental rights. This is especially true since AI has the potential to be used in the exercise of public authority, in the democratic and political spheres or in the provision of services of general interest, whether or not personal data is processed. The importance of these impact assessments at the developer and user level would be that they would improve public knowledge and understanding of AI, which currently suffers from a lack of transparency about its capabilities and impact. These impact assessments would also provide guidance and a secure legal framework for businesses, their managers and the engineers who develop the new technologies and their applications. They would thus help to create a new culture of responsibility in technological development and use for democracy, the rule of law and fundamental rights.[851]

A group of scientists around the new director of the Berlin-based Max Planck Institute for Human Development, Iyad Rahwan, has published a recent article with other researchers in the journal Nature in the spirit of Hans Jonas. Quoting Rahwan:

"At present, the scientists who study the behaviors of these virtual and embodied Artificial Intelligence (AI) agents are predominantly the same scientists who have created the agents themselves [...]. As these scientists create agents to solve particular tasks, they often focus on ensuring the agents fulfil their intended function. [...] a broader consideration of the behavior of AI agents is now critical. AI agents will increasingly integrate into our society and are already involved in a variety of activities, such as credit scoring, algorithmic trading, local policing, parole decisions, driving, online dating and drone warfare.

---

[851] Practical examples of elements of an in-house technology assessment for AI are the Ethics Guidelines of the EU Expert Group on Trusted Artificial Intelligence of April 2019, https://ec.europa.eu/futurium/en/ai-alliance-consultation/guidelines#Top, which are currently being tested in practice; and the guide developed by the Omidyar Network and the Institute of the Future, which is available at https://ethicalos.org/wp-content/uploads/2018/08/Ethical-OS-Toolkit-2.pdf

Commentators and scholars from diverse fields—including, but not limited to, cognitive systems engineering, human computer interaction, human factors, science, technology and society, and safety engineering— are raising the alarm about the broad, unintended consequences of AI agents that can exhibit behaviors and produce downstream societal effects—both positive and negative—that are unanticipated by their creators[852]

Rahwan concludes that the "next challenge will be politics [...] because Artificial Intelligence is politics."[853]

The authors explore the societal implications of Artificial Intelligence and propose the implementation of a system of behavioral observation and impact assessment. In addition to the implementation of ethical standards already in the development of the software, behavioral research that observes and evaluates the behavior of machines and robots should become mandatory. It is an irony of the history of science that this robot behavioral research, like behaviorism in its early stages, will come up against a *black box*, only this time it is not located inside the human being but inside the machine.[854] The results of this machine behavioral research will then be used again in the further development of the software. This would achieve an important goal: the technicians would already check the ethical

[852] Rahwan, Iyad, Manuel Cebrian, Nick Obradovich, Josh Bongard, Jean-François Bonnefon, Cynthia Breazeal, Jacob W. Crandall, et al. "Machine Behaviour." Nature 568, no. 7753 (April 2019): https://doi.org/10.1038/s41586-019-1138-y, p. 477.

[853] See "Projects ‹ Iyad Rahwan — MIT Media Lab," https://www.media.mit.edu/people/irahwan/projects/ translated from the German in Andrian Kreye, Moral for Machines, in Süddeutsche Zeitung, 25 April 2019 , https://www.sueddeutsche.de/kultur/wissenschaftlicher-aufruf-moral-fuer-maschinen-1.4419702

[854] The starting point of Behaviorism, the theory of the science of human and animal behavior, in the 1950s was the idea that human thinking was considered a black box and therefore only human behavior could be scientifically studied. The brain is regarded as a "black box" whose inner processes are not of interest. Behavior is understood as the result of reinforcing and weakening factors. Source: http://www.lernpsychologie.net/lerntheorien/behaviorismus

implications of directing the programs in one way or another during programming for compatibility with legal and social standards.

For example, in such an approach, the researchers' ideas would require an examination of the effects of AI on democracy; for example, whether AI creates filter bubbles or censors content to a disproportionate extent[855]. For military applications, it should be examined whether and how the weapon system can distinguish between soldiers and civilians. Societal questions would include the possible effects of the algorithms on children or, more generally, on collective behaviour.

The proposed process of exchange between developers and impact researchers should ensure that ethics are directly involved in the design of such systems. It is an application of the principle of responsibility to AI.

Given the risks that AI poses to democracy, the rule of law, and human dignity, we propose to clarify the concept of "real" life in Jonas' moral imperative by explicitly protecting human autonomy:

"Act in such a way that the consequences of your actions are compatible with the permanence of genuine autonomous human life on earth."

Regarding Jonas' concept of autonomy, it must be said that he is far from speaking out for a total self-empowerment of man. On the contrary: With Jonas, as with the Agora of Athens and then with Kant, we find the insight that self-knowledge is required for ethics. It is precisely the natural limits of human life and natural resources that motivated Jonas to develop his ethics.[856] A self-knowledge that

---

[855] See about the project: Rahwan, Iyad, Manuel Cebrian, Nick Obradovich, Josh Bongard, Jean-François Bonnefon, Cynthia Breazeal, Jacob W. Crandall, et al. “Machine Behaviour.” Nature 568, no. 7753 (April 2019): https://doi.org/10.1038/s41586-019-1138-y, p. 479

leads to self-limitation, to recognizing and accepting inner and outer limits is necessary for ethics. Hubris is the enemy of man, still. Self arrogance and *hubris* paves the way to ruin. Little seems to have changed in 2500 years in this tragedy material.

In an interview with Der Spiegel magazine in 1992, Jonas emphasized that the ideal of responsibility must stand alongside the ideal of freedom in modernity, which means that we must also bid farewell to the false ideal of unlimited freedom for the individual:

"But to be taken seriously is the question of what renunciations of freedom one is prepared to make; what renunciations of freedom the philosopher can advise in an ethically responsible manner. First of all, it cannot be overlooked that freedom can only exist anyway by limiting itself. An unlimited freedom of the individual is destroyed by the fact that it is not compatible with the freedoms of many individuals."[857]

Unlimited freedom destroys not only the self, but also the freedom of others and the foundations of common self-preservation. In our case it is about the one-sided totalization of an unlimited individual freedom in the Californian ideology, especially in the libertarianism of Peter Thiel. This leads to a freedom in which the basis of life of all other people are threatened.

A boundless self-empowerment through a technology that apparently allows the whole planet and in Ray Kurzweil's visions even the universe to be brought under control, threatens the permanence of human life on earth, the real life and the autonomy of humans.

---

[856] So Jonas in a radio interview with Wolf Scheller 1989, source: https://www.deutschlandfunkkultur.de/philosophie-aus-den-archiven-hans-jonas-ueber-die-tuecke.2162.de.html?dram:article_id=436666

[857] Matussek, Von M and W. Kaden, "Dem bösen Ende näher", 11 May 1992, https://www.spiegel.de/politik/dem-boesen-ende-naeher-a-e2997c8f-0002-0001-0000-000013680535?context=issue

Jonas demands a "factual science of the long-term distant effects of technical actions". In addition to these remote effects, in the digital age there is the possibility of remote control of humans. Hans Jonas has already addressed the possibility of behavioral con-trol through biological, medical or psychological techniques:

"The troublesome problems of rule and unruliness in modern mass society make the extension of such control methods to non-medical categories extremely tempting for social manipulation ."[858]

One of the long-term and distant effects of digital technologies is the remote control of humans. What the humanists call "education" offers an antidote to the optimization of human beings by standards that were developed for machines. If the great changes that AI and quantum computers bring are to remain controllable, knowledge must be fostered, and good judgement exercised. It is a question of ethics whether man remains on the bridge or whether the steering wheel of his life will soon be handed over to non-human actors.

Hans Jonas's "Imperative of Responsibility" postulates that the ethics of man should develop in accordance with his abilities. Therefore, an extension of ethics from the individual, to the interpersonal and the environmental, and to the long-term effects of new technologies on the future conditions of human life on earth is necessary. For it is now in our power to completely destroy nature and thus also the basis of human life. Jonas described the power of technology over people, such as the ability to prolong life, but he was particularly concerned about the power over nature.

---

[858] Jonas, Hans. The Imperative of Responsibility: In Search of an Ethics for the Technological Age (University of Chicago Press, 1985), p. 51. Quoted in Winston, Morton, and Ralph Edelbach. 2013. Society, Ethics, and Technology. Cengage Learning.

Today, however, we need to focus on the power of technology over people, alongside power over nature, which remains a huge issue. We must also strive for a further extension of individual and political ethics. After all, the precautionary principle is not just a principle of environmental law. It is a general principle. Today, we must extend it to democracy, fundamental rights and the rule of law and to questions of what technology does for these fundamental pillars of our coexistence.

The widespread success of the environmental movement emanates from the deeply rooted human need to live in harmony with nature. This need promotes an implicit understanding of the need to balance economy and ecology and to end the "war against nature".[859]

In addition, there is also the deeply rooted need of human beings to live in harmony with their fellow human beings but also with themselves. A recognition of this fact, and the fact that today an economic war is being fought for supremacy in people's minds, resulting in a digital pollution of the inner world, can become the starting point for a renaissance of democratic values.

Today, in view of the technology challenges posed by AI, the relationship between man and tech-nology must be shaped more strongly from an ethical perspective, so that we do not become the servants of the new master AI. In view of the dangers that a fully developed general AI could pose to mankind, the criticism of Jonas, which accuses him of deriving a general ban on such technology from the mere assumption of catastrophic consequences, must also be rejected.[860]

---

[859] So said the chairman of the Green Party in Germany, Robert Habeck, at the party conference in December 2019.

[860] According to Ortwin Renn, "For this reason, the well-meant imperative of the philosopher Hans Jonas is of little help. Jonas called on society to refrain from any technology whose consequences could lead to catastrophic negative consequences", in: The Imperative of Responsibility, op. cit. p.11. As shown, Jonas merely demanded that the consequences had to be considered and calculated as precisely as possible in order to prevent disasters. He was not arguing to the disadvantage of technique in itself.

But Jonas did not only establish the principle of responsibility. He has always placed the "principle of life" at the centre of his work. In doing so, he has emphasized an aspect of human consciousness in the fundamental corporeality of life, which is a clear distinguishing feature from AI:

> *"Through the immediate testimony of our bodies we can say what no disembodied spectator would be able to say: that the mathematical God in his homogeneous analytical view misses the decisive point - the point of life itself: its being self-centered individuality, being for itself and in contraposition to all the rest of the world, with an essential boundary dividing "inside" and outside."*[861]

The corporeality and liveliness of man places us in permanent and stark contrast to the "intelligence" of machines. This is also supported by the fact that machines are nowhere near simulating the role of the will for human consciousness, or qualia of experience.

Today, in the age of the unleashed digital Prometheus, the "principle of responsibility" must make provision for democracy. This is because the impact assessment of the new technologies shows that they have a democracy-destroying effect in their impacts of reinforcing pure capitalism and populism. The principle of responsibility thus becomes a central means of providing for democracy.

We can also support Hans Jonas's thinking by combining him with Habermas. Who says:

> *"The irrationality of domination, which today has become a collective danger to life, could only be overcome by a political will that is bound*

[861] Jonas, Hans. 2001. The Phenomenon of Life: Toward a Philosophical Biology. Northwestern University Press, p.79

> *to the principle of general and domination-free discussions. We can only hope for the rationalization of rule by conditions that favour the political power of a thinking based on dialogue. The solving power of reflection cannot be replaced by the spread of technically available knowledge."*[862]

[862] Habermas, Jürgen. Theory and Practice. Translated by John Viertel. 1st edition. Cambridge: Polity, 1986, p.357

F5 Digression on Rawls: The invisible veil around the digital

> *"Justice is the first virtue of social institutions, as truth is of systems of thought." John Rawls*[863]

So far, we have described two thought experiments that may help to find ethical standards for AI: first, Kant's Categorical Imperative, complemented by the Sustainability Imperative of Hans Jonas, and second, Habermas' fiction of a discourse free from domination.

Another thought experiment helps to clarify the role of justice in this context: John Rawls's image of justice blindfolded which enable us to establish a fictitious contractual justification of a social form. To establish norms in society, he developed the thought experiment of the "invisible veil".

In this model, a group has the task of defining the rules of a new society. The special feature was that none of the participants in the group knows at this point what role they will play later in the new society: Whether she or he will live as smart or stupid, male, female or trans, beautiful or ugly, black or white , rich or poor in the society whose rules they get to design. What would society look like then?

Rawls' thesis is that the members of this group will, in their own interest, establish principles which are "fair". Since they do not know the role they will occupy in the society they shape by their rules, they will try to create fair conditions for all members of society. The core of the experiment lies in the fact that, due to the veil of ignorance,

[863] Rawls, John. 2009. A Theory of Justice. Harvard University Press, p.3

everyone tries to adopt the perspectives of the other members of society as far as possible and thinks closely about fairness in the design of social rules. Conversely, it allows for the conclusion to be drawn that justice can arise when many different perspectives are adopted, and rules are derived from them that guarantee a fair balance of interests.

It is the outstanding merit of John Rawls that he has reintroduced the concept of justice into the political debate in a time marked by the debate on the freedom of values of science. And he did so by philosophical means. Rawls argued strongly against utilitarianism developed by J.S. Mill. Utilitarianism pursues the goal of maximizing the welfare of the collective, even if the individual suffers harm in extreme cases. In contrast, Rawls follows Kant: The dignity of the individual is inviolable for him. The individual must never be sacrificed for the collective, never become a mere means, she and he is always an end in her/himself.

Rawls thus faced a problem that follows from Max Weber's theory of modernity. Weber had led the world view of modern science to the value-free nature of science, i.e., the strict separation of empirical knowledge and value judgement. For Weber, norms cannot be scientifically and rationally reasoned and justified. For Weber, the process of rationality as a whole is based on a claim to power: "Human life including its natural environment is regarded as completely disposable."[864]

From this arises the temptation to reduce rationality to functional rationality, which then immediately makes its own claim to universality. This reduction has dramatic consequences: According to Max Weber, the "insoluble, even insurmountably fight to the

[864] Höffe, Otfried. Sittlichkeit als Rationalität des Handelns? In: Rationalität p. 144 and p.147

death" between different value systems dominates the world and ultimately leads to an "axiological polytheism".[865] "Between the "gods", namely the last value criteria to which man can submit, a rational decision is fundamentally impossible".[866] This has far-reaching consequences for modernity: "Rational ethics are replaced by the irrationality of the struggle for opinions."[867] The rationality of purpose threatens to displace and replace practical reason.

Rawls, on the other hand, pursues the goal of establishing an objective, substantial rationality and with it, principles of justice that are strictly intersubjective and thus universally valid. According to Otfried Höffe's analysis, Rawls succeeds in rationally establishing the value of justice even while retaining Weber's postulate of the value freedom of science. "Rawls recognizes enlightened self-interest as the ultimate standard and allows it to operate under ideal conditions."[868] Rawls thus succeeds in proving to a predominant mathematical-economic thinking that values and norms can be rationally justified and thus are in principle universalizable.

The thought experiment could also be applied to the development of *Ethics by Design* for AI systems: All programmers working on AI software are, in a way, behind a virtual curtain: they know that they will be members of the society in which the software they have programmed will be used. But they do not know in what way they will be affected by the use and consequences of use of their own programming. They could be unemployed, nurses, pensioners, students, managers, businessmen, snack bar owners, disabled or not, rich or poor, Christians, Jews or Muslims and therefor do not

---

[865] Quoted after Höffe, Otfried. Sittlichkeit als Rationalität des Handelns? In: Rationalität, p. 144.

[866] Höffe, Otfried. Sittlichkeit als Rationalität des Handelns? In: Rationalität, p. 145.

[867] Höffe, Otfried. Sittlichkeit als Rationalität des Handelns? In: Rationalität, p. 145.

[868] Höffe, Otfried. Sittlichkeit als Rationalität des Handelns? In: Rationalität, p. 162.

know how they would be exposed to the consequences of their software. Applying the veil of ignorance before programming could help to make programmers aware of the impact of their programs on all parties involved, thereby making the algorithms more socially responsible.

Sarah Spiekermann has made a similar suggestion in her book "Digital Ethics". There she shows[869] how, for example, programming the app for a food delivery service can produce very different results, depending on whether the programmers' sole aim is to optimize the number of deliveries in a certain district and at a certain time, or whether the needs of the bicycle couriers for a rest break, communication with each other and with the customers are also taken into account. Spiekermann thus convincingly outlines how ethics can be incorporated into the training of software engineers in the future.

In the sense of Rawls, one could assume that the experiment triggers ethical reflections on the possible consequences of the software, which lead to a fair and equitable distribution among the participants. The assumption is that by integrating the "veil of ignorance", engineers would make ethics an integral part of their programming, protecting society as a whole and its members. Thus, through the veil of ignorance, "fair algorithms" could be developed.

In contrast, unregulated AI carries the danger of undermining the foundations of a just society. The goals of optimizing human beings technically and genetically also pose an ethical threat. If a future technology were to succeed in bringing random natural living conditions, such as the genetic "lottery of life", under control, and thus

[869] Spiekermann, Sarah. Digitale Ethik: Ein Wertesystem für das 21. Jahrhundert. München: Droemer HC, 2019, p. 30 ff.

take away the accidental character, this would have consequences for the concept of justice in such a society. The hitherto unavailable would be made completely available by "optimization". Because the foundations of human life become manipulatable, for example, through genetic engineering, those involved would cease to put themselves in the role of another individual and instead shift to improving their own position as far as possible. Justice would then no longer be sought.[870] It would then no longer be necessary to think about a just society that compensates for natural contingencies, but only about an optimized society that puts those who have the means to do so in the optimal starting position. Just as the Californian ideology does. Justice would then no longer be in the vested interest of every individual.

[870] Sandel, Michael J. 2007. The Case Against Perfection: Ethics in the Age of Genetic Engineering. Cambrige: Harvard University Press.

F6 Philosophical alternatives to technological absolutism

6.1. The Humboldt brothers as inspiration

An alternative of holistic thinking to the total cybernetic system theory of our days can be found with the Humboldt brothers. Their theory proves that it is also possible to consider holistic systemic relationships without ending up in the absolutism of cybernetic control logics and without giving up the autonomy of the individual. The example of Alexander von Humboldt and his brother Wilhelm von Humboldt shows that by the beginning of the 19th century there was a holistic understanding of nature and language that allowed man a role in the cosmos that made him neither master of the world nor a mere cog in the machine.

Alexander von Humboldt sees the whole of nature as an organism dominated by "life forces" and an "educational instinct". Following Kant's concept of the system of physical geography,[871] in which individual facts have to fit into a system in order to make sense, and Goethe's fusion of art and science, which is based on the belief in the unity of nature, Humboldt sets out to measure nature. In doing so, he avoids any anthropocentrism, according to which nature in all its manifestations is there for man. With his holistic method of investigation, he was the first to discover human-caused climatic changes.[872] He also identified ecological chain reactions,

[871] Wulf, Andrea. The Invention of Nature: The Adventures of Alexander von Humboldt, the Lost Hero of Science. First Edition. London: John Murray, 2015, p. 35

[872] Wulf, Andrea. The Invention of Nature: The Adventures of Alexander von Humboldt, the Lost Hero of Science. First Edition. London: John Murray, 2015, p.57

because everything is connected to everything else: "The whole is interaction."[873]

Humboldt wants to overcome the dualistic world view of Descartes', in which the knowledge of nature is oriented towards mastery of nature. With Descartes the triumphal procession of science and technology began in modern times. He divided the world into the two substances *Res Cogitans* (the thinking human) and *Res Extensa* (the natural, or extended and unthinking). *Res Cogitans* is confronted with *Res Extensa*. The subjectivization of man, which underlies all cognition and which leads to the famous formula of the *Cogito ergo sum*, corresponds to the objectivization of nature, which is to be mastered and made property by technology and science: Humans are "the lords and possessors of nature."[874] For Descartes, mastery and possession are the goals to which knowledge is dedicated. Their goal is the conquest of nature through technology.

In contrast, the young researcher Humboldt's concept of nature is characterized by empathy, which the principle of the living seeks in nature: "Everything flows from one source (...) and melts together into an eternal, all-pervasive force."[875] Humboldt, who set out to measure the world anew, also recognized the limits of his scientific method: "What speaks to our minds is beyond measurement."[876]

During his legendary ascent of the Chimborazo in Ecuador, Humboldt urged that one must "take a higher standpoint" to

---

873 Wulf, Andrea. The Invention of Nature: The Adventures of Alexander von Humboldt, the Lost Hero of Science. First Edition. London: John Murray, 2015, p.60.

874 Rene Descartes, Thomas, 1984, p.33 quoted in Wulf, Andrea. The Invention of Nature: The Adventures of Alexander von Humboldt, about the method, VI, 2.

875 Wulf, Andrea. The Invention of Nature: The Adventures of Alexander von Humboldt, the Lost Hero of Science. First Edition. London: John Murray, 2015, p. 92.

876 Wulf, Andrea. The Invention of Nature: The Adventures of Alexander von Humboldt, the Lost Hero of Science. First Edition. London: John Murray, 2015, p.77

understand nature. On the heights of the Chimborazo, he recognised that nature "was a web of life and a global force... that everything was interwove as with 'a thousand threads'."[877]

Here, 150 years before Stewart Brand's admiration of the *Whole Earth photo*, we encounter the *Overview Effect*, which led to a new concept of nature that Humboldt tried to capture in his famous "Nature Painting". He expressed this concept as follows: nature is "unity in variety".[878] For the young researcher, nature consisted of a network of forces that are in constant interaction. In his main work "Kosmos" he writes that everything is "part of the eternal goings-on and workings of the living forces". Nature is "a living whole" in which organisms combine to form a "net-like intertwined tissue".[879]

While Alexander von Humboldt made his mark with his expeditions and natural sciences, his brother Wilhelm laid the foundations of modern linguistics. He drew from the sources as his brother when he tried to grasp the essence of language as the "thought forming organ."[880]

According to Wilhelm Humboldt, languages are shaped by an educational instinct analogous to the organisms of nature. Language is always more than just a tool for the expression of thoughts; rather, it shapes thoughts through the role of the imagination in speaking. It is thus an organ for the human exploration of the world. Different languages create unique "views of the world" in this process of

---

877 Wulf, Andrea. The Invention of Nature: The Adventures of Alexander von Humboldt, the Lost Hero of Science. First Edition. London: John Murray, 2015, p. 87

878 Wulf, Andrea. The Invention of Nature: The Adventures of Alexander von Humboldt, the Lost Hero of Science. First Edition. London: John Murray, 2015, p. 89

879 Wulf, Andrea. The Invention of Nature: The Adventures of Alexander von Humboldt, the Lost Hero of Science. First Edition. London: John Murray, 2015, p. 235-6

880 Humboldt, Wilhelm von. "On Language": On the Diversity of Human Language Construction and Its Influence on the Mental Development of the Human Species (Cambridge University Press, 1999), p. 100

opening up the world, expressing different perspectives on this open, never-ending process. In doing so, Humboldt avoids a relativism such as was then often advocated in the philosophy of language of the 20th century. For Wilhelm von Humboldt, like nature for his brother, language is not a mechanical construct, but an organism, a web woven from action, thought and speech.[881] Each individual language represents a special perspective, which is organically connected with all other languages to form a whole.

We can learn from the Humboldt Brothers how a theory formation can succeed that overcomes the scientifically shortened method of instrumental mastery of nature. Such a theory must reflect on its own perspective and take into account that it is part of the whole that it wants to recognize, not "master and owner" of nature separated by an insurmountable division.

For forming theories as a whole, it is decisive which point of view one takes, which perspective the speaker has. This can also be seen in the great paradigms of the history of philosophy. First came amazement in the cosmos of the whole, as was paradigmatic in ancient philosophy. Next, came doubt and consciousness as separated from the material outside world, which is constitutive for the subject philosophy of modern times. Finally, the human as a participant who is always already embedded in social and natural contexts and who tries to orientate himself linguistically and intersubjectively, as the paradigm of language philosophy postulated in the 20th century.[882]

We also learn from the Humboldt Brothers that language is central to the human exploration of the world, because it not only determines

---

[881] Wulf, Andrea. The Invention of Nature: The Adventures of Alexander von Humboldt, the Lost Hero of Science. First Edition. London: John Murray, 2015, p.253.

[882] Herbert Schnädelbach distinguishes accordingly the ontological, mentalistic and linguistic paradigm of philosophy, see E.Martens, H. Schnädelbach Philosophie, vol. 1, p. 46 ff.

his relationship to nature, but also enables understanding in society as well as individualization within it.

These insights are helpful against the totalizing claims to power of a one-sided technically shortened reason that prevails in the Californian ideology. They also help to correct the regressive idea of immediate harmony between man and nature, as developed by the Californian ideology following the Gaia hypothesis, because they were developed from the ideal image of a human being capable of autonomy.

6.2. Learning from quantum theory? Heisenberg, Zeilinger and Dürr

> *A falling tree makes more noise than a growing forest*

Lao Tzu (Chinese Proverb)

> *"It is normal that our perception is dominated by falling tress, of that which is violent, massive, happens quickly, threatens us or at least appears as a threat. Our whole history is full of falling tress... But then […] We recognize the fact that is it the growing forest that matters ultimately."*

Hans-Peter Dürr[883]

We have seen that Californian ideology and dataism are essentially based on scientific knowledge. They derive their central beliefs from physical, biological, mathematical and cosmological theories.

---

[883] Hans-Peter Dürr, quoted in Meyer-Galow, Erhard. 2018. Business Ethics 3.0: The New Integral Ethics from the Perspective of a CEO. Walter de Gruyter GmbH & Co KG. Also see: Why it is all about the whole, New thinking for a world in upheaval, Wien Globe Art 2011 https://www.youtube.com/watch?v= RKma6xCTIBE

Part of their fascination stems from theories in astrophysics on the origin and development of the universe. Which story is more epic than the Big Bang and the origin of matter, suns, planets and ultimately life? Where do we come from? Where are we going? The basic questions of mankind are answered in modern astrophysics, completely without myths and completely without religion. Or so it seems.

The Heisenberg student Hans-Peter Dürr has put together a few key points of a world view based on the findings of quantum physics:

According to an ontology based on quantum physics, the basis of matter is spiritual, which means that reality is always potentiality.

According to Dürr, the findings of quantum theory show everything is connected with everything else and forms a whole (relationism and holism). The basis for this in quantum physics being the phenomena of "non-locality" and "entanglement". From the special "position of man in the cosmos", which results from the fact that he is the only living creature able to grasp these interrelationships, Dürr concludes that man has a special responsibility for the whole of nature.

In contrast to dataism, where the good is measured by the quantity of data processing, in Dürr's ethics the good is that which contributes to meeting this responsibility as far as possible.

Dürr's writings, like those of Werner Heisenberg and Carl Friedrich von Weizsäcker, show that insights from quantum physics can assist philosophical redefinitions if they avoid the naturalistic fallacy. For they conclude from the growing knowledge of man that he bears a special responsibility. The "should" is not derived from being, but solely from man's ability to reflect.

Werner Heisenberg, one of the founders of quantum physics, has spent his life trying to build a bridge from physics to philosophy and social responsibility. Heisenberg was one of the physicists who spoke out politically with the "Göttinger Declaration", calling for control of nuclear weapons and warning of the dangers of the arms race.[884] He acted out of the responsibility of the scientist who recognises what consequences his discovery could have for people.

Quantum physics leads to the final farewell of deterministic world views, to which Kurzweil and Co ultimately remain attached. Instead of the mechanistic, material, objectifiable and time-determined "reality" previously assumed, the reality of quantum physics turns out to be "potentiality": an inseparable, immaterial, at the same time essentially indeterminate and genuinely creative structure of relationships.

Quantum theory also has an epistemological moment, namely that "the observer can exert a very great influence" by selecting the measuring instruments, says Anton Zeilinger, one of the leading quantum researchers of our time.[885] Zeilinger goes even further: "We have learned that the observer has a significant influence through his choice of the measurement instruments, through his decision about which kind of quantity to measure. The point is that his measurement instruments don't just influence or change the observed systems [...] they decide which property of a quantum system can become reality."[886] This leads him to the insight that in trying to recognise the

---

[884] "As non-politicians, we will be denied the right to do so; but our activity, which is devoted to pure science and its application, and in which we bring many young people to our field, loads us with a responsibility for the possible consequences of this activity. Therefore, we cannot remain silent on all political issues." The complete text of the "Göttingen Manifesto of the Göttingen 18" can be read here: https://www.uni-goettingen.de/de/text+des+g%c3%b6ttinger+manifesto/54320.html

[885] Zeilinger, Anton. Dance of the Photons: From Einstein to Quantum Teleportation. New York: Farrar Straus Giroux, 2010, p. 337.

[886] Anton Zeilinger in O'Collins, Gerald S. J., and Mary Ann Meyers. 2012. Light from Light: Scientists and Theologians in Dialogue. Wm. B. Eerdmans Publishing..

world man influences the world. "We have learned, for example, that the idea that the world is independent of us in its properties and we only observe it passively is wrong."[887]

Zeilinger also describes as the most fascinating result of quantum physics that "the concepts of reality and information cannot be separated."[888] Quantum theory teaches us about the "necessary openness of the world". The world is no longer "everything that is the case" (Wittgenstein), but rather also everything that could be the case. People must consider this potentiality when they act and take responsibility for the consequences of their actions, which create new potentialities.

6.3. Critical theory of the digital

"A spectre is haunting the world: the spectre of digitalization. It has long been recognized as a power. It is high time that the digital giants disclose their views, purposes and intentions to the whole world."[889] This is how the beginning of the Communist Manifesto of Marx and Engels could be modified for today. At its time, it was about explaining the growing power of capitalism, the communist movement and its ideas. Today it is important to analyze and criticize the greatest utopia of our days, the utopia of solving all humanity's problems through technology. Behind it stands the greatest economic and technological power complex in history.

[887] Zeilinger, Anton. Dance of the Photons: From Einstein to Quantum Teleportation. New York: Farrar Straus Giroux, 2010, p. 336.

[888] Zeilinger, Anton. Dance of the Photons: From Einstein to Quantum Teleportation. New York: Farrar Straus Giroux, 2010, p. 339.

[889] Modification of the first sentences of the communist manifesto of Marx/Engels "A spectre is haunting Europe - the spectre of communism...Communism is already recognized as a power by all European powers...It is high time that communists openly present their views, their purposes, their tendencies to the whole world".

In this economic form, subjects are no longer self-confident actors who set their own goals. Today, this incapacitation is carried out, albeit very gently, by apparatus applied directly to the individual. Technology directly intervenes in man's wishes and decisions, "nudging" us towards future (buying) behaviour. The manipulation is based on a massive asymmetry of knowledge and thus power: The platform knows more about the individual than she or he knows about her/himself, even as regards future intentions.

The knowledge of this asymmetry, however, must be suppressed by the subjects, because otherwise they would develop what Hegel describes as an "unhappy consciousness". The repression happens through adaptation to the manipulation. After all, it is "my" wishes that are suggested to me there as actions. It feels as if this action is the own wish of the individual. In accordance with the "felt truths and felt facts", the manipulated subject also perceives the felt wishes as her or his own.

In his lecture on education to maturity in a pre-digital society, Adorno saw the task of criticism in pointing out such type of deception:

> *"So that one simply tries, first of all, to awaken the awareness that people are constantly being deceived, because the mechanism of immaturity today is the mundus vult decipi, raised to the planetary, that the world wants to be deceived. That everyone is aware of these connections could perhaps be achieved in the sense of an immanent critique, because no normal democracy can afford to be explicitly against such an enlightenment."*[890]

[890] Adorno, Theodor W. Erziehung zur Mündigkeit, zur Aktualität der Kritischen Theorie, Frankfurt 1971; see also Nina Rismal, Ends of Utopian Thinking, PhD thesis, Cambridge University 2018. access at https://www.repository.cam.ac.uk/handle/1810/273982

To adapt Adorno's proposal to practice critique would mean today exposing the mechanisms of data abuse and countering the claims of solutionism that all the world's problems can be solved with technology alone.

When there is talk today in the Anglo-Saxon world of a renaissance of the Frankfurt School,[891] when the question is asked whether Adorno foresaw Trump,[892] this shows a need and wish for a critical theory that can offer an analysis of the digital and economic upheavals of late capitalism and point to alternatives.

The Californian ideology leads to the abandonment of criticism and furthers identification with the (supposedly) inevitable. But the digital hope of redemption can only be had at the price of man's complete self-abandonment. After Nietzsche's declaration of the "death of God" Californian ideology does not replace God with man, as the title of the bestseller "*Homo Deus*" suggests. But instead it creates a false ideal of a formalized system rationality. Homo Deus is not a deification of man, but the subjugation of man to an idolatrous machine. The golden calf that is danced around is the human-created Artificial Intelligence. This reduction represents a darkening, an "eclipse of reason"[893]. The obscuring of reason is not in contrast to the brightness of the transparency and presence of simultaneously infinite information: "Everything is illuminated but nothing is seen" writes AI expert James Bridle.[894] The formulation sounds like an

---

[891] Jeffries, Stuart. Grand Hotel Abyss: The Lives of the Frankfurt School. Reprint edition. Verso, 2017.

[892] So said the American music critic Alex Ross on 5 December 2016 in the New Yorker, source: https://www.newyorker.com/culture/cultural-comment/the-frankfurt-school-knew-trump-was-coming

[893] This is the original American title of Max Horkheimer's Critique of Instrumental Reason (Verso Books, 2013). The original German title was: Zur Kritik der instrumentellen Vernunft, New York 1947.

[894] Bridle, James. 2018. New Dark Age: Technology, Knowledge and the End of the Future. Verso Books.

echo of the famous introduction to the *Dialectic of Enlightenment* by Horkheimer and Adorno: "Yet the wholly enlightened earth is radiant with triumphant calamity".[895]

Many insights of the *Dialectic of Enlightenment* could be used for a Dialectic of Digitalization. Like Critical Theory, however, such a theory must not argue in essence against what is to be the object of analysis: Horkheimer and Adorno also wanted to enlighten the Enlightenment about itself, about its blind spots, which lead to the promise of emancipation turning into domination, and not to abolish it. Similarly, it is important to use digitization to criticize not the technology as such, but its instrumentalization and design through the imperative of a technological-economic complex. This has set itself the goal of conquering total economic and ultimately also political power. What we need to strive for is the opposite: Digital technology should obey and serve the society that created it.

6.4. Adorno's message in a bottle

In connection with the Dialectic of the Enlightenment, Adorno spoke of a message in a bottle intended for later generations.[896] Let us open the bottle with a few quotations that read like descriptions of our digital present:

---

[895] Horkheimer, Max, Theodor W. Adorno, and Gunzelin Noeri. 2002. Dialectic of Enlightenment. Stanford University Press, p.134

[896] "Adorno loved the picture of the message in a bottle. Leo Löwenthal situated Adorno and other émigré friends standing at the edge of the Pacific Ocean, with Adorno signing and wishing he could cast his thoughts in catastrophic times inside a bottle into the see. "Then some person, one distant day on a distant island, will find the bottle and read..." quoted in Adelson, Leslie. Cosmic Miniatures and the Future Sense: Alexander Kluge's 21st-Century Literary Experiments in German Culture and Narrative Form. Walter de Gruyter GmbH & Co KG, 2017.

[897] Adorno, Theodor W., and Horkheimer, Max, Dialectic of Enlightenment. New edition. London: Verso Books, 1997, p. 134

> *"The blindness and muteness of the data to which positivism reduces the world passes over into language itself, which is limited to registering those data".*[897]
>
> *"Signification, only function of the word admitted by semantics, is consummated in the sign".*[898]
>
> *"Everything has value only if you can trade it, not if it's something."*[899]
>
> *"Industry is interested in human beings only as its customers and employees and has in fact reduced humanity as a whole, like each of its elements, to this exhaustive formula".*[900]
>
> *"Vain is the hope that the lying understatement of stereotypes for the individual will of itself become unbearable."*[901]

In man there was a "secret satisfaction that one was finally relieved of the effort of individuation by the admittedly more breathless one of imitation."[902]

If this is true in the era of digitalization, what remains in view of the current totalization of technological-economic reason? Critical thinking, despite all the totality of the wrong world, in which, according to another of Adorno's winged words, there can be no

---

898 Adorno, Theodor W., and Max Horkheimer. Dialectic of Enlightenment. New edition. London: Verso Books, 1997, p. 134

899 Adorno, Theodor W., and Max Horkheimer. Dialectic of Enlightenment. New edition. London: Verso Books, 1997, p. 181.

900 Adorno, Theodor W., and Max Horkheimer. Dialectic of Enlightenment. New edition. London: Verso Books, 1997, p. 169.

901 Adorno, Theodor W., and Max Horkheimer. Dialectic of Enlightenment. New edition. London: Verso Books, 1997, p.126.

902 Adorno, Theodor W., and Max Horkheimer. Dialectic of Enlightenment. New edition. London: Verso Books, 1997, p. 179.

right living.[903] In Adorno's case, too, what ultimately remains is a reflection on humanity and freedom: "Critical thinking, which does not stop at progress, today demands taking sides with the remnants of freedom, with tendencies towards real humanity, even if they seem to be powerless in the face of the train of history."[904] Instrumental reason emerges, as Horkheimer and Adorno have shown, when the *principium sese conservare,* the principle of self-preservation, is reduced to pure self-assertion. Instrumental reason then follows the goal of securing as many advantages as possible to derive the greatest possible benefit, while accepting the destruction of nature and society.

Above all, the practice of surveillance and profiling today must be sharply criticized because it is radically hostile to freedom. It must be possible, even in the 21st century, to choose Epicurus' motto "Live in hiding" as a life plan without triggering surveillance or even exposing oneself to the suspicion of terrorism.

There are numerous philosophical approaches to the exposure of the residues of humanity today, of which we would mention only a few here. So, the chances of getting the spirit of the digital-economic power complex back into the bottle are not so bad.

---

[903] Cit. After Adorno: Disenchantment and Ethics, Cambridge,. J.M.Bernstein, Cambridge 2001

[904] DdA p. 9, preface to the 1969 new edition.

Part G
What Shall We Do?

What Shall We Do?

Thinking must be followed by political action.

First, the question arises: What concrete actions are we taking against the clearly visible and increasing concentration of power in corporations, and how can we sustainably shape the future of the digital in such a way that we avoid the risks that accompany concentrations of power in the future?

The enormous concentration of knowledge, data, infrastructure control and the complexity of the technologies created and mastered by GAFAM make it necessary to provide society, governments and parliaments with new instruments to ensure, on the one hand, compliance with the law and, on the other, the functioning of democracy and markets.

Second: How can we make the digital future humane, give it purposes and goals that correspond to our democratically established will? And on what image of humanity do we base this striving for humane technological development and compatible business models?

On the one hand, we are dealing here with very practical questions of politics today, above all politics in America, where GAFAM come from. On the other, we must remind ourselves what it means to be human, to maintain humanity even in the age of the most powerful machines and Artificial Intelligence. The descriptive sociology of Manuel Castells, who coined the term network society,[905] and

---

[905] Castells, Manuel. The Rise of the Network Society: Economy, Society and Culture v.1: The Information Age: Economy, Society and Culture Vol 1. 2nd Revised edition. Oxford ; Malden, Mass: Wiley-Blackwell, 2000.

interpolating predictions about the dominance of networks and AI in the new technological world, such as Yuval Noah Harari's, are helpful but not enough.

Rather, we want to consciously attempt to say what should be, i.e., to work with normative demands.

G1 Can we hope for democracy in America?

Can we hope for democracy in America, the country that has repeatedly demonstrated its ability to reform and lead? Or can we at least hope for Silicon Valley to recognise and accept the primacy of democracy and law?

Throughout history, America has claimed to be the leading power for democracy and freedom. Today, as in the early stages of its development, America is again largely colonized by raw capitalism. The worship of Wall Street and Silicon Valley, the culture of venture capital and private initiative has led to a division of society into fewer and fewer winners and more and more losers. The result is political polarization and an increased power of money, which in combination lead to the inability of legislators to legislate in the public interest.[906] The absence of legislative containment and its enforcement leads to increasingly powerful corporations in the core sectors of technology and social media, the GAFAM, but also in media (Netflix), distribution (Amazon and Walmart), finance (Blackrock[907]) and energy, which aim to create monopolies (as Peter Thiel demands) and mega-profits.

Added to this hot mess is systemic competition with China. The historical systemic competition with the Soviet Bloc for freedom, justice and prosperity has degenerated into a systemic competition with China for technology and trade dominance. The US, although facing

---

906 Lessig, Lawrence. Republic, Lost: How Money Corrupts Congress - and a Plan to Stop It. Reprint edition. New York; London: Twelve, 2012.

907 Buchter, Heike. BlackRock: Eine heimliche Weltmacht greift nach unserem Geld. Frankfurt: Campus Verlag GmbH, 2015.

a dictatorship with China, no longer claims leadership in freedom, democracy and justice, as it did under Trump. Today, in the US too, we are seeing attempts to bring the media into line and suppress critical journalism, as well as attempts to weaken the judiciary in its control function over the executive. In this way, the Trump government, in its striving for unquestionable and uncontrollable power, was adapting to the methods of autocratic regimes, but not yet reaching the levels of China's lack of freedom and Russia's arbitrariness.

In 2020, the leadership in Washington, Moscow and Beijing agree with the autocrats of this world in their hatred of individual human rights as the central defensive rights against state power, as they do in their hatred of free science and the free press as the fourth estate.[908]

Trump is no friend of Silicon Valley. Nevertheless, nothing has happened under his presidency to limit the power of the GAFAM companies. The decision of the largely independent US Trade Commission (FTC) on 24 July 2019 to fine *Facebook* $5 billion for failing to keep public promises about privacy is the sole exception.[909] This decision had little practical effect, but the FTC's tone towards GAFAM is becoming harsher. The head of the FTC also no longer rules out the possibility of GAFAM companies being broken up.[910]

Earlier political scandals in the US, such as Watergate or the bankruptcy of Lehman Brothers were sudden events. They triggered quick reactions from the US legislature. The sluggish discussion

---

[908] See for example: Newsweek. "Donald Trump Joins Putin and Xi on the List of Human Rights Violators, Report Reveals," February 22, 2018. https://www.newsweek.com/donald-trump-bad-vladimir-putin-and-xi-jinping-when-it-comes-human-rights-and-816073

[909] Federal Trade Commission. "FTC Imposes $5 Billion Penalty and Sweeping New Privacy Restrictions on Facebook," July 24, 2019. https://www.ftc.gov/news-events/press-releases/2019/07/ftc-imposes-5-billion-penalty-sweeping-new-privacy-restrictions

[910] "FTC Chief Says He's Willing to Break Up Big Tech Companies." Bloomberg.Com, August 13, 2019. https://www.bloomberg.com/news/articles/2019-08-13/ftc-chief-says-willing-to-break-up-companies-amid-big-tech-probe

about the possible dismantling and regulation of GAFAM is based on a slowly emerging change in the political convictions of Americans who had previously settled comfortably in the convenience trap carefully laid by the digital giants.

A Democratic presidential candidate in the 2020 pre-election campaign, Senator Elizabeth Warren from Massachusetts, has presented very concrete plans for the dismantling and regulation of *Facebook, Google* and *Amazon*. According to her plans, any platform with a turnover in excess of 25 billion dollars would be regulated as a company of general interest, would not be allowed to carry out any other activity and would not be allowed to make personal data available to other companies.[911] In the US presidential campaign, calls for the regulation of Internet platforms and the dismantling of *Facebook* and *Google* have been accumulating,[912] without a storm of indignation to counter these demands.[913] On the contrary, a large part of the population in the US supports such demands.[914] And

[911] Team Warren. "Here's How We Can Break up Big Tech." Medium, October 11, 2019. https://medium.com/@teamwarren/heres-how-we-can-break-up-big-tech-9ad9e0da324c

[912] Rubin, Ben Fox. "Momentum Grows to Break up Big Tech, as Amazon, Facebook, Google and Apple Face Scrutiny." CNET. https://www.cnet.com/news/momentum-grows-to-break-up-big-tech-as-amazon-facebook-google-and-apple-face-scrutiny/. Hughes, Chris, "It's time to break up Facebook" , NYT, 9. May 2019, https://www.nytimes.com/2019/05/09/opinion/sunday/chris-hughes-facebook-zuckerberg.html; Newton, Casey, "More Democrats are considering a breakup of Facebook, Joe Biden and Kamala Harris are the latest to say they are thinking about it", 14 May 2019, https://www.theverge.com/interface/2019/5/14/18622442/facebook-breakup-joe-biden-kamala-harris-democrats-chris-hughes, Wu, Tim "Warum Facebook zerschlagen werden sollte, Maximilian Henning", Netzpolitik.org, 11. July 2019, https://netzpolitik.org/2019/tim-wu-warum-facebook-zerschlagen-werden-sollte/#spendenleiste

[913] Chakravorti, Bhaskar, "Don't break up Facebook and Google based on these three myths", Fastcompany, 20. July 2019, https://www.fastcompany.com/90379246/dont-break-up-big-tech-based-on-these-three-myths; Matsakis, Louise. "Break Up Big Tech? Some Say Not So Fast", Wired, 6. July 2019, https://www.wired.com/story/break-up-big-tech-antitrust-laws/

[914] Stewart, Emily. "Poll: Two-Thirds of Americans Want to Break up Companies like Amazon and Google." Vox, September 18, 2019. https://www.vox.com/policy-and-politics/2019/9/18/20870938/break-up-big-tech-google-facebook-amazon-poll. Sheffield, Matthew. "Poll: Most Americans Not Interested in New Regulations for Big Tech Companies."The Hill, March 11, 2019. https://thehill.com/hilltv/what-americas-thinking/433482-poll-americans-not-interested-in-extra-regulations-for-big-tech

the reputation of GAFAM is tarnished. In recent surveys, Google's reputation fell by 13 places, Facebook's by as much as 43 places.[915]

The competition law investigation of GAFAM companies by Congress is led by the Democrat David Cicilline. He is said to have long been planning a reform of US competition law after the presidential elections. In fact, similar earlier Congressional investigations initiated such changes.[916]

Democratic Senator Mark Warner is a Democratic thought leader for regulating the tech giants.[917] His comprehensive recommendations, however, do not go as far as recommending the dismantling of the companies. Warner has repeatedly said that only the big tech companies are in a position to protect America from the power of Chinese companies.[918]

If he is not mistaken: all GAFAM companies want to earn money in and with China. The president of *Microsoft*, Brad Smith, reported enthusiastically on a visit of the Chinese president Xi Jinping to *Microsoft*, for which all the US tech bosses had come to Seattle.[919] *Microsoft* and the heads of GAFAM and other tech companies gathered there showed their interest in further business with China.

---

915 The Harris Poll. "The Axios Harris Poll 100." https://theharrispoll.com/axios-harrispoll-100/

916 Lohr, Steve. "This Man May Be Big Tech's Biggest Threat." The New York Times, December 8, 2019, sec. Technology. https://www.nytimes.com/2019/12/08/technology/David-Cicilline-antitrust-tech.html

917 Senate.gov, Potential Policy Proposals for Regulation of Social Media and Technology Platforms, 23. July 2018, https://www.warner.senate.gov/public/_cache/files/d/3/d32c2f17-cc76-4e11-8aa9-897eb3c90d16/65A7C5D983F899DAAE5AA21F57BAD944.social-media-regulation-proposals.pdf

918 Harris, Ainsley,. "This Senator Was Big Tech's Friend—but Is Now Its Greatest Threat." Fast Company, July 29, 2019. https://www.fastcompany.com/90378278/this-senator-was-big-techs-friend-but-is-now-its-greatest-threat

919 Smith, Brad. Tools and Weapons: The Promise and the Peril of the Digital Age. New York: Penguin Random House US, 2019. Chapter 14

Some major newspapers, such as the Washington Post and the New York Times, have expanded their critical coverage of the mass surveillance of Americans by the GAFAM corporations and their dominance. Both newspapers regularly publish research and critical commentary on the misguided developments of the unregulated Internet and data economy.[920] Important donor foundations in the US support the plan to strictly regulate and dismantle GAFAM.[921]

After initial resistance, the Trump government agreed to a non-binding declaration by the OECD and the G-20 on principles for AI.[922] But there is no sign of binding action under his administration. The Trump government is thus following in Washington's tradition with regard to data protection. Non-binding multilateral declarations in the OECD and other international forums are agreed to. But no binding rules are created domestically.

Trump likes unregulated big business. All he wants is to force GAFAM further to the right.

In May 2019, the White House therefore launched a website to document alleged censorship of right-wing opinion by social media companies. Citizens were asked to come forward if they noticed deletions that they considered political censorship of social networks.[923] A lot of noise is still expected from Trump, but no dismantling or regulation of GAFAM companies.

---

[920] See "The Privacy Project - The New York Times." https://www.nytimes.com/series/new-york-times-privacy-project. and The Wall Street Journal, "What They Know - News, Articles, Biography, Photos - WSJ.Com." https://www.wsj.com/news/types/what-they-know

[921] McCabe, David. "America's Top Foundations Bankroll Attack on Big Tech." The New York Times, December 10, 2019, sec. Technology. https://www.nytimes.com/2019/12/10/technology/americas-top-foundations-bankroll-attack-on-big-tech.html

[922] Murgia, Madhumita, and Siddarth Shrikanth. "How Governments Are Beginning to Regulate AI," May 30, 2019. https://www.ft.com/content/025315e8-7e4d-11e9-81d2-f785092ab560 and https://www.mofa.go.jp/files/000486596.pdf, Annex I with the statement of the G-20 Economics Ministers;

[923] Kelly, Makena. "White House Launches Tool to Report Censorship on Facebook, YouTube, Instagram, and Twitter." The Verge, May 15, 2019. https://www.theverge.com/2019/5/15/18626785/white-house-trump-censorsip-tool-twitter-instagram-facebook-conservative-bias-social-media

1.1. Lawrence Lessig: From "Code and other Laws" to "a Fortified Democracy"

But neither with a Democratic President Biden, nor with the former candidates Warren or Bloomberg as presidents, the US would be able to tame the Internet by law, because Congress is in the hands of big money. This basic problem of American democracy makes it almost impossible to get laws that restrict GAFAM's business models through Congress. For this reason alone, the Internet and the business models based on it can become increasingly dystopian.

This is what the constitutional law expert at Harvard Law School Professor Lawrence Lessig says in his 1999 book "Code and Other Laws of Cyberspace".[924] Lessig was still optimistic about the power of transparency and the open source movement as a means of democratizing the Internet. His independence has been called into question given the funding his *Center for Fair use* in copyright law received from Google. It is financed to the tune of 2 million dollars.[925]

Today, he calls for a well-fortified democracy and the democratic legislators to contain the excesses of GAFAM's business models on the Internet. The basic values that defined the Internet in 1998 have now been perverted. The Internet has degenerated from an opportunity for democratization to a profitable sales and disinformation machine in the West and an instrument of dictatorship in China and other autocracies. Democracy must reclaim it. The rules cannot be set by GAFAM.

---

924 Lessig, Lawrence. "Codev2:Lawrence Lessig." http://codev2.cc/

925 Stanford Law School. "Google Inc. Pledges $2M to Stanford Law School Center for Internet and Society." Stanford Law School. https://law.stanford.edu/press/google-inc-pledges-2m-to-stanford-law-school-center-for-internet-and-society/. Bjorgumsays, Eric. "Poker the Bear: The Sad Unraveling of Lawrence Lessig." The Trichordist (blog), May 21, 2018. https://thetrichordist.com/2018/05/20/poker-the-bear-the-sad-unraveling-of-lawrence-lessig/

Today, several thousand engineers determine what happens on the Internet. The idea that the code of the Internet is democratic is naive. Social networks are only apparently democratic. In fact, it is the engineers and the companies that determine how much data protection there is, who has how much access to information, and the extent to which advertising dictates what we see if this is not decided by democratic laws.[926]

What is needed now is to make the platforms responsible for the good functioning of democracy. They must become part of a well-fortified democracy and ensure that democracy is not destroyed by calls for hatred and violence on the Internet. This is the position supported today by Lawrence Lessig of Harvard University, the pioneer of Internet law.

1.2. Is Silicon Valley learning? Can we hope for self-regulation by GAFAM?

In the last two years, there have been increasing signs of an awareness of the need for legal regulation among some of the major players in the tech industry. Anyone reading the speeches and statements of GAFAM protagonists today might believe that John Perry Barlow's declaration of the independence of cyberspace never existed. Neither could the neoliberalism of Peter Thiel, who invested in *Facebook* early on, then founded the spy society *Palantir* and finally helped Trump into the saddle.[927] Nor the lobbying battle of GAFAM, united with the US Government and its allies in Europe against the EU GDPR,

---

[926] Quoted in this Interview with Eric Guevara-Frey, RTS, 9 May 2019, https://www.rts.ch/play/radio/tout-un-monde/audio/internet-et-la-democratie-interview-de-lawrence-lessig?id=10403960

[927] Thiel, Peter. Zero to One: Notes on Start-ups, or How to Build the Future. 1 edition. New York: Penguin Random House US, 2014.;Tarnoff, Ben. "Neoliberalism Turned Our World into a Business. And There Are Two Big Winners." The Guardian, December 13, 2016, sec. US news. https://www.theguardian.com/us-news/2016/dec/13/donald-trump-silicon-valley-leaders-neoliberalism-administration

against the EU Copyright Directive[928] and against the NetzDG in Germany as well as against privacy protection legislation in the US.

First Mark Zuckerberg the founder and head of *Facebook*,[929] then Brad Smith, the president of *Microsoft*[930], then Kent Walker, senior vice president and chief legal counsel of *Google*[931] - have recently conjured up the subjugation of GAFAM to democracy and law. Yes, they now call for laws and speak of a history and future of constructive interaction between technology and law. The reader should rub her or his eyes in amazement.

Marc Zuckerberg surprised the world with his demand for legislation in four areas, namely protection of elections, protection of privacy, removal of illegal content and portability of data, essentially addressed to the US legislature.[932] After *Facebook* fought hard against data protection rules in Europe, took a long time to support the EU Code of Conduct to combat incitement to violence and hatred, and then fought until the very end against the Internet enforcement act in Germany ("NetzDG"), this apparent U-turn came as a surprise and was received with suspicion and criticism.[933]

---

[928] Kayali, Laura. "Inside Facebook's Fight against European Regulation." Politico, January 23, 2019. https://www.politico.eu/article/inside-story-facebook-fight-against-european-regulation/

[929] Zuckerberg, Mark. "Opinion | Mark Zuckerberg: The Internet Needs New Rules. Let's Start in These Four Areas." Washington Post. https://www.washingtonpost.com/opinions/mark-zuckerberg-the-internet-needs-new-rules-lets-start-in-these-four-areas/2019/03/29/9e6f0504-521a-11e9-a3f7-78b7525a8d5f_story.html

[930] Smith, Brad, and Browne, Carol Ann. Tools and Weapons: The Promise and the Peril of the Digital Age. Penguin Publishing Group, 2019.

[931] Google. "The Historic Partnership of Democracy and Technology," October 2, 2019. https://blog.google/perspectives/kent-walker/historic-partnership-democracy-and-technology/

[932] Zuckerberg, Mark. "Opinion | Mark Zuckerberg: The Internet Needs New Rules. Let's Start in These Four Areas." Washington Post. https://www.washingtonpost.com/opinions/mark-zuckerberg-the-internet-needs-new-rules-lets-start-in-these-four-areas/2019/03/29/9e6f0504-521a-11e9-a3f7-78b7525a8d5f_story.html

[933] Hanfeld, Michael. "Gesetze fürs Internet: Was erlauben Facebook?" FAZ.NET. https://www.faz.net/1.6253951

From Switzerland, we hear how eager *Facebook was* for legislation on Internet election advertising before Mark Zuckerberg decided not to take factually false election advertising and even paid lies down from the platform.[934] Twitter boss Jack Dorsey, on the other hand, decided to ban paid political advertising from Twitter,[935] with the good argument that money should not influence political decisions.

It is documented that *Facebook* indeed asked the Swiss government to issue regulations to protect the Swiss elections.[936] The company, which has so far been more conspicuous as an opponent of law, thus presents itself in Switzerland as an alleged friend of democracy. In the US, however, *Facebook* has done nothing of the sort since Zuckerberg's article which we cited above. On the contrary, the behaviour of Zuckerberg and *Facebook* towards Elizabeth Warren shows that Facebook's core tenet continues to be viewing democracy as disruptive. And as far as data protection is concerned, the GAFAM companies have been talking a lot about it recently, but whether they really deliver remains so doubtful that the word "*privacy washing*" is making the rounds in America.[937]

So what does this apparent change of heart mean? There is apparently a learning process at GAFAM regarding the impact of its own business activities on democracy. Now that the pressure is mounting, with Washington and European capitals discussing regulation or even the breaking up of these quasi-monopolies, they

---

934 Isaac, Mike. "Dissent Erupts at Facebook Over Hands-Off Stance on Political Ads." The New York Times, October 28, 2019, sec. Technology. https://www.nytimes.com/2019/10/28/technology/facebook-mark-zuckerberg-political-ads.html

935 Zurcher, Anthony. "Twitter to Ban All Political Advertising." BBC News, October 31, 2019, sec. US & Canada. https://www.bbc.com/news/world-us-canada-50243306

936 Republik. "Wenn Big Tech in Bern nach Regulierung ruft." https://www.republik.ch/2019/08/06/wenn-big-tech-in-bern-nach-regulierung-ruft

937 Fowler, Geoffrey A. "Perspective | At CES, Apple, Facebook and Amazon Are Preaching Privacy. Don't Believe the Hype." Washington Post. https://www.washingtonpost.com/technology/2020/01/08/ces-apple-facebook-amazon-are-preaching-privacy-dont-believe-hype/

are embarking on a seemingly more constructive course towards Democracy. And they can afford it, because the lobbying offices are well established and well-staffed and funded.

1.3. The global corporation as a state within a state? Brad Smith, Microsoft, digital sovereignty and democracy

Brad Smith, president of *Microsoft*, demonstrates some leadership towards a slightly reformed relation of GAFAM to Democracy in his 2019 book "Tools and Weapons".[938] Shortly after the Snowden revelations he spoke of how only technology and judges could reasonably manage the tensions between freedom, data protection and security, and how politics and democracy are incapable of doing so.[939] He is now, however, relying on the power of democratic legislation and calls for faster and more decisive action by the legislature. It is true that, in view of the enormous speed of technological progress and the risks involved, the *laissez faire* argument that innovation should be left unregulated until the technology is mature is going in the wrong direction.[940] What we need today are speedy and forward-looking rules in the form of legislation that are open to development because they are technologically neutral, thus remaining valid in the long term.

Brad Smith seems to be an exception among the grandees of the American tech industry. Can you believe that he really cares about

[938] Smith, Brad and Browne, Carol Ann, Tools and Weapons: The Promise and the Peril of the Digital Age. New York: Penguin Random House US, 2019.

[939] Video: Brad Smith and Jonathan Zittrain on Privacy, Surveillance, and Rebuilding Trust in Tech, 14. November 2014, The Berkman Klein Center for Internet & Society, https://www.youtube.com/watch?v=WY7fuR3eG_0 from 46.00.

[940] Smith, Brad. Tools and Weapons: The Promise and the Peril of the Digital Age. New York: Penguin Random House US, 2019., 295

functioning democracy, the rule of law and the common good? And this not only because the governments of this world are among the largest customers of the now most valuable global corporation, *Microsoft*?

In the European Union, *Microsoft* generates 2 billion euros in licensing revenue per year from public administration alone. The public administrations in Europe have become colonies of *Microsoft*.[941] And in Germany, the leading market for *Microsoft* outside the US, important decisions are pending. Will the federal and state governments continue to rely on *Microsoft*? There are doubts about that. Countries are a long way from the newly discovered necessity of digital sovereignty. 95% of all the world's information is stored on clouds operated by US companies. In view of the widespread use of *Microsoft* Cloud (Azure), the Windows operating system and Microsoft programs for writing (Word), presenting (Power Point), e-mails (Outlook) and video-calls (Skype), it seems rather that Microsoft holds digital sovereignty over Germany's and Europe's offices of public administration.

The German government has already presented a "Strategic market analysis to reduce dependencies on individual software providers"[942]. Once these services are all offered from a *Microsoft* cloud, the European government *Chief Technology Officers* will no longer be able to rely on *Microsoft's* support when operating these services from their own servers from 2026. Migration to *Software as a*

---

[941] Schumann, Harald, Simantka, Elisa, "Europas fatale Abhängigkeit von Microsoft", Tagesspiegel, 13 May 2017, https://www.tagesspiegel.de/gesellschaft/cyber-attacken-auf-staatliche-it-europas-fatale-abhaengigkeit-von-microsoft/19628246.html; The Microsoft Dilemma - Europe as a Software Colony, ARD, Harald Schumann, Arpad Bondy, 19 February 2018, https://www.youtube.com/watch?v=duaYLW7LQvg

[942] "Strategische Marktanalyse zur Reduzierung von Abhängigkeiten von einzelnen Software-Anbietern, Abschlussbericht August 2019", presented to the German Federal Ministry of Interior by Strategy&, which is part of the PwC Network. The Study cannot be found anymore on the website of the Federal Interior Ministry nor the Federal Chief Information officer, but only at https://docplayer.org/156855053-Strategische-marktanalyse-zur-reduzierung-von-abhaengigkeiten-von-einzelnen-software-anbietern-abschlussbericht-august-2019.html

Service[943] from *Microsoft* cloud servers will then become mandatory. This will mean that sovereignty not only over digital, but all thinking and communication in government and administration, will be completely finished. One may perhaps trust an honorable lawyer like *Microsoft's* President Brad Smith. But can states trust American politics, with Trump in charge, and the American stock exchange calling the shots? It is ultimately these that determine whether *Microsoft* has to deliver data stored in Germany or elsewhere to the US on request of US Security Services, or whether it is used for other purposes to make profits.

The US CLOUD Act (Clarifying Lawful Overseas Use of Data Act) shows that the US does not want to accept limits on the direct access of its executive to data produced or stored abroad. The judicial control of such access at home, which has so far taken place within the framework of the mutual legal assistance agreements, is to be eliminated.[944] Brad Smith presents Microsoft's fight against his own government, as heroic and successful, to make people believe that Microsoft is the Switzerland of data. He even presents a Chief Technology Officer of a German federal state (unfortunately anonymously) to bear witness to this fact[945] to convince his own countrymen and customers worldwide. But the reader of his book is haunted by a feeling that things aren't quite right.

On the one hand, the *Microsoft* president argues that technology should be subject to democracy and law, that more legal regulations

---

[943] Software as a Service means that the program itself is no longer sold to customers to be used on their computers, but is only available for use as a service from cloud servers under Microsoft control.

[944] Holzki, Larissa, "IT-Branche: Diese Probleme gibt es, wenn Deutsche demnächst den Microsoft-Clouddienst nutzen." 28 August 2019, https://www.handelsblatt.com/technik/it-internet/it-branche-diese-probleme-gibt-es-wenn-deutsche-demnaechst-den-microsoft-clouddienst-nutzen/24949694.html.

[945] Smith, Brad, and Carol Ann Browne. Tools and Weapons: The Promise and the Peril of the Digital Age. Penguin Publishing Group, 2019, p. 52.

are finally needed in the largely unregulated Internet. But on the other hand, he paints a picture of a company that ultimately always wins in the battle for the interests of its customers, against the American government, in American courts, in American politics, and also in international diplomacy.[946] Does that actually fit together? Probably not. And certainly not with an unpredictable President Trump, who is not afraid to dismantle the rule of law and to rely on power rather than law, both internally and externally. *Microsoft*, even with Brad Smith at the helm, will probably have to accept that it cannot be a state within a state. Not in America, and not in Europe either.

But let us observe further: how, for example, is *Microsoft* using its influence in Washington regarding the negotiations on cross-border access to electronic data as evidence in criminal proceedings between the EU[947] and the US? Does *Microsoft*, together with other GAFAM companies, succeed in enforcing judicial controls and rule of law safeguards against government positions in these negotiations? An increasing commitment by GAFAM to the stronger protection of fundamental rights, the rule of law and democracy alongside technology would in any case be welcome. We need independent US judges to be able to shed light on the darkness of intelligence service data collection activities, also on the basis of complaints from abroad. This is particularly necessary where companies in the US can be served with "gagging orders"[948] under threat of punishment to keep silent about the forced transfer of data to US government agencies, allegedly in the national interest.

---

946 Smith, Brad, and Browne, Carol Ann, Tools and Weapons: The Promise and the Peril of the Digital Age. Penguin Publishing Group, 2019, p. 109ff.

947 The negotiating mandate of the EU Commission in the following documents of the EU Council of Ministers: https://data.consilium.europa.eu/doc/document/ST-9114-2019-INIT/en/pdf and https://data.consilium.europa.eu/doc/document/ST-9666-2019-INIT/en/pdf.

948 Crocker, Andrew, and Mackey, Aaron. "The Failed Fix to NSL Gag Orders: How the Majority of National Security Letter Recipients Remain Gagged After US FREEDOM." Electronic Frontier Foundation, December 13, 2019. https://www.eff.org/wp/the-failed-fix-to-NSL-gag-orders.

The European Court of Justice, in its judgment annulling the EU-US Privacy Shield,[949] which was the legal basis for transfer of personal data of Europeans to US companies for processing in the US, made it clear that a toothless Ombudsperson in the Department of State to deal with complaints on privacy and data protection by EU citizens is not enough. The US must provide non-discriminatory access to the Courts also for non US Citizens, and it must enlarge either ex-post information rights or facilitate standing to sue national security agencies if it wants to stay at the centre of data processing of personal data of people of Europe in the US for profit. The alternative is simple and, contrary to loud claims of lobbyists, not more expensive and certainly not a dramatic break down of economic relations: Processing of data of people of Europe in Europe. There is no right to transfer the data on Europeans to the US and with progress of decentral processing at the edge and infrastructure investments and billions of private investment flowing into new data centers in Europe also no need anymore.

1.4. More use of open source and no state within a state

For a long time, there have been calls for more open source software to be used in public administration. The late conversion to open source at Microsoft through its purchase of *GitHub*,[950] the most important open source platform, is a clever business moves that should enable

[949] Court of Justice of the European Union, PRESS RELEASE No 91/20, "Judgment in Case C-311/18 Data Protection Commissioner v Facebook Ireland and Maximillian Schrems", https://curia.europa.eu/jcms/upload/docs/application/pdf/2020-07/cp200091en.pdf

[950] Microsoft. "Microsoft Acquires GitHub." https://news.microsoft.com/announcement/microsoft-acquires-github/.' "Github-Übernahme: Ein Super Deal - Für Microsoft Und Den Rest - Golem.De." https://www.golem.de/news/github-uebernahme-ein-super-deal-fuer-microsoft-und-den-rest-1806-134769.html

Microsoft to remain powerful even as the move towards open source takes place. IBM followed the trend towards open source with the purchase of *Red Hat*,[951] the most important Linux developer.

But in the end, free societies and free markets should make it difficult for huge corporations like *Microsoft* and the other GAFAM companies to dominate, with or without open source. It is therefore right and proper that the public administration in Germany and Europe should at least partly bid farewell to *Microsoft*, even if *Microsoft* offers more open source products. The principle that state and politics must not become dependent on a corporation must be enforced for the sake of sovereignty. Because without sovereignty there is no freedom and no democracy. As in the case of G5 and Huawei, the same must apply to *Microsoft* and the public sector: Public authorities must not become accomplices of large corporations by encouraging further dependency through their bidding and purchasing behaviour. In their public procurement behaviour, they need to take responsibility for a functioning competition, also in their own interest.

It must also be remembered that *Microsoft* is not as clean-lipped in terms of data protection and democracy as the story its management has painted in dazzling colours suggests. For example, the Dutch and Bremen data protection authorities and the European Data Protection Supervisor have begun investigations into breaches of data protection in *Microsoft* products. It is also unclear how *Microsoft* managed to regain a foothold in the Munich city administration, which had opted for open source.[952] Whether the establishment of

---

[951] RedHat, IBM closes landmark acquisition of RedHat, 9 July 2019, https://www.redhat.com/en/about/press-releases/ibm-closes-landmark-acquisition-red-hat-34-billion-defines-open-hybrid-cloud-future. Grüner, Sebastian, "IBM kauft Red Hat: Die riesige riskante Wette auf die Cloud - Golem.de." 30 October 2018. https://www.golem.de/news/ibm-kauft-red-hat-die-riesige-verzweifelte-wette-auf-die-cloud-1810-137384.html

[952] Mitchell, Courtenay. "EDPS Investigates Contractual Agreements Concerning Software Used by EU Institutions." Text. European Data Protection Supervisor - European Data Protection Supervisor, April 8, 2019. https://edps.europa.eu/press-publications/press-news/press-releases/2019/edps-investigates-contractual-agreements_en

a *Microsoft* research center in Munich has anything to do with this remains an open question.[953] Likewise, many open questions remain about the security of *Microsoft* programs.[954]

Microsoft and *Google* are already working in a core group with Tim Berners Lee,[955] the inventor of the World Wide Web and now professor at MIT in Boston,[956] on a new *private public partnership initiative* with the governments of Germany and France: The *Contract for the Web*.[957] Here again we see GAFAM being built into a position similar to that of the states, raised to the same level as governments and of simulating harmony and common ground where distance and clear separation would be better. The *Contract for the Web* does contain many good things. But the birth defect of GAFAM's privileged position in its creation is inherent in the *Contract*. This is not how democracy can function in the age of the Internet.

1.5. The digital innovation model of the "minimal viable product" cannot be transferred to democracy

The innovation model of the digital has much in common with Schumpeterian creative destruction in that analogue business

953 Investigate Europe. "Why Europe's Dependency on Microsoft Is a Huge Security Risk," April 10, 2017. https://www.investigate-europe.eu/en/2017/why-europes-dependency-on-microsoft-is-a-huge-security-risk/. and Deutschlandfunk Kultur. "Christian Ude über IT-Sicherheit und den Einfluss von Microsoft - 'Das war ein sehr bedeutsamer Fall.'" https://www.deutschlandfunkkultur.de/christian-ude-ueber-it-sicherheit-und-den-einfluss-von.1008.de.html?dram:article_id=412109

954 The Microsoft Dilemma: Europe as a Software Colony, Harald Schumann, Arpad Bondy, ARD-Reportage, 19.02.2018, https://www.youtube.com/watch?v=duaYLW7LQvg

955 Newitz, Annalee. "Opinion | A Better Internet Is Waiting for Us." The New York Times, November 30, 2019, sec. Opinion. https://www.nytimes.com/interactive/2019/11/30/opinion/social-media-future.html

956 MIT CSAIL, "Tim Berners-Lee | MIT CSAIL." https://www.csail.mit.edu/person/tim-berners-lee

957 Contract for the Web. "Groups." https://contractfortheweb.org/groups/. https://contractfortheweb.org/endorse-the-contract/endorse-as-an-individual/

models are completely swept off the market. The digital also replaces hardware-dependent industrial goods, as we have seen in the music industry. So far, there is no objection to innovation in digital technology and the Internet, but the special elements of the digital innovation model we see today go beyond creative destruction.

One reason for the global success of Internet-based digital technologies is the scalability of software over the Internet. Another reason is the possibility of going to market with an initial "minimal viable product."[958] Further optimizing and expanding this product depends on customer reactions. In this way, in the long run, the interaction with customers and users and testing of variants in the market theoretically results in maximum customer benefit with minimum development effort. And the work of optimization is transferred to the customers, reducing costs for the developer.

This innovation model contrasts with the classic development model for industrial products, such as is found in extreme form in the automotive industry. No new Mercedes goes on the market without billions of dollars in research, development and testing, because the product that goes on the market should meet the customer's expectation of perfection from the outset. And it must also comply with extensive legal regulations to obtain approval before it can be put on the road.

The situation is completely different for software and Internet-based services. They are hardly regulated at all.

In addition, the error correction of software already on the market is easier by means of updates via the Internet, and even more so with

[958] Ries, Eric. The Lean Startup: How Today's Entrepreneurs Use Continuous Innovation to Create Radically Successful Businesses. Crown Business, 2011.

"Software as a Service". In theory, this can be further developed in parallel to it being on the market, since the product is never given out of the hands of the manufacturer - unlike the car sold. In the case of a car, it may just be possible to install new software via remote maintenance. But if there are problems with the mechanics or physics, a product recall is necessary, which cause costs and trouble, and therefore, in contrast to updates for software, must remain an exception. The incentive is therefore high to bring only perfected products onto the market.

However, the MVP method is risky. Because it is also possible in software that the product is too far away from the perfection that customers expect.[959]

And the MVP method leads to an underweighting and often complete absence of any consideration of the consequences and impacts of technology or business model, and unfortunately almost any ethical or legal considerations. Legal considerations are kept to a minimum in order to exclude liability risks.

Transferring the tech sector principle of market entry with a *minimal viable product* to legislation and democracy, as Brad Smith suggests, therefore cannot work.[960] On the one hand, it is precisely the MVP principle that has led to innovations in the digital economy being brought to market without any impact assessment in terms of law, ethics, sustainability and democracy. On the other hand, democracy thrives on discourse and majority decisions. For this reason alone, democracy can never move as quickly as digital technological change

---

[959] A cautionary tale for founders looking to launch a Startup with an MVP, Rand Fishkin, Quartz at Work, 16 May 2018, https://qz.com/work/1277369/the-lean-startup-methodology-will-kill-your-reputation-if-youre-not-careful/

[960] Smith, Brad, and Browne, Carol Ann. Tools and Weapons: The Promise and the Peril of the Digital Age. Penguin Publishing Group, 2019, p. 296.

driven by the MVP model. After all, it is normal for companies in a market economy to go bankrupt, for example if they misjudge the "MVP". Market exit is a normal route in the economy. But for democracy and the state, this path must not exist. For these reasons, "MVP" is not an option for legislation, just as other simple transfers of technological and market principles to democracy often do not work.

G2 Active democracy and technology policy

Those who prefer democracy as a control system to capitalism and technology must appear with a claim to design. If they do not do so, the combined forces of technology and capitalism will take the shaping of human life out of the hands of democracy, as they have always tried to do, today more than ever.

Parallel to the political issues we are discussing, work is of course needed on tools for use by individuals, such as security software, virtual private networks, encryption, alternative forms of networking, search and orientation in the sea of data and knowledge or the technical visualization and shutdown of private and state surveillance. What already exists in this space and what tools we still need to support freedom and democracy can be read up on, for example, by civil society network activists.[961] It will also increasingly be a task of allocating public funding to support the development and dissemination of such technologies for democracy, civil and consumer rights. But it is also clear that technical solutions alone are not enough.

Technology, politics and law must go hand in hand; they can support each other to some extent and even substitute for each other in specific cases. But it will not work without self-organization in politics and participation in the political processes of democracy, which in the end - hopefully - will set the 'right' rules. This book is therefore an invitation to all those who work with commitment and expertise on technical solutions to engage more intensively and constructively

---

[961] See, for example, reports in Netzpolitik.org such as: Mühlenmeier, Lennart and Dachwitz, Ingo. "Read unobserved mail: How to protect yourself against tracking pixels in newsletters and the like", 14 July 2019. https://netzpolitik.org/2019/unbeobachtet-mails-lesen-so-schuetzt-ihr-euch-gegen-tracking-pixel-in-newslettern-und-co/ and https://www.eff.org/de/pages/tool

in the political processes. It is necessary to learn how to tolerate frustration that democracy, with its majority principle and its need to organise and convince, needs. Perseverance and tenacity are needed not only among start-up entrepreneurs, developers, programmers and inventors. We also need them in democracy. Political change is not easy to bring about by gamification. It always requires sustained commitment from many, and especially from those with knowledge to bring to the table.

An advertising brochure of an international law firm already identifies more than 450 initiatives worldwide to regulate GAFAM in the public interest.[962] There is no reason to be shy on the questions of how we control these monopolies.[963] The GAFAM companies were and are not shy when it comes to maximising profits by breaking the law and disregarding fundamental values. And some of the most radical initiatives to dismantle and regulate GAFAM so far come from the US itself.

In the following we want to list and justify our proposals.

2.1. Securing independence of journalism and science as a pillar of democracy

The struggle between labour and capital, which has been domesticated in Europe in the form of the social market economy, is taking on

[962] Hogan Lovells, A turning point for Tech, Global Survey on digital regulation, 2019, https://www.hoganlovells.com/~/media/hogan-lovells/pdf/2019/2019_10_30_tmt-final_05319-tmt-study-09-tw-digital.pdf

[963] The authors took important suggestions and encouragement from the work of former Judge at the German Constitutional Court, Professor Wolfgang Hoffmann-Riem, see at https://www.jura.uni-hamburg.de/media/die-fakultaet/personen/hoffman-riem-wolfgang/schriftenverzeichnis-riem-september-2022.pdf, in particular Re:claim Autonomy – Die Macht digitaler Konzerne“, in Rudolf-AugsteinStiftung (Eds.), Reclaim Autonomy – Selbstermächtigung in der digitalen Weltordnung, edition suhrkamp 2714, 2017, S. 121 - 139. https://www.beck-shop.de/hoffmann-riem-materialien-rechtswissenschaftlichen-medien-informationsforschung-big-data-regulative-herausforderungen/product/24069852

new forms in the digital world.[964] Added to this is the new struggle between democracy as a control system and the totalitarian control systems. Political dictatorship, as we see it in China, is now perfected by technology. Digital capitalism that increasingly dominates all areas of politics and individual life has also totalitarian ambitions. It seeks to penetrate the institutions of *checks and balances*, such as the legislators, the free press and the independent sciences, which are conceived of as independent and free in democracy, and leave them dependent, indeed subordinate.

What we need today is exactly the opposite: a strengthening of representative democracy, the independent institutions of journalism as the fourth estate and of science. Journalism that is dependent on the gifts of the digital giants and a science in which the most interesting research and the best individual advancement is only possible if third-party funding is provided by the digital giants or other large corporations, cannot claim the credibility necessary as truth-generating institutions for democracy.

So, let us begin, step by step, to break the comfortable friendship with the digital giants. Let us make it clear that scientists, journalists, publishers and media houses that take money from them cannot be free in the long run. The only exception to this rule is genuine capital foundations, in the management of which the funding companies have no stake and from the income of which journalism or science is permanently financed without any influence from the funding companies. So, let us say yes to capital foundations, but no to donations in the form of operating subsidies, which are temporary and create dependencies. Because at the latest when the operating

[964] On the further decline in payroll and rising capital income in the digital economy, see Seemann, Michael. "Five worrying questions for digital capitalism (Directors Cut)", 10 June 2019, Blog Ctrl+Verlust; http://www.ctrl-verlust.net/fuenf-beunruhigende-fragen-an-den-digitalen-kapitalismus-directors-cut/

aid is used up or expires, you have to go begging again. The GAFAM companies seek this dependency because it promises them influence, at least in the form of the anticipatory holds in the minds of those who know that they will soon have to ask for money again and will therefore curtail criticism.

It is right and proper to invest more public money in research on AI and digital technologies because of their impact on society, the rule of law, democracy and fundamental rights. The establishment with public funds of the Weizenbaum Institute for the Networked Society in Berlin was a step in the right direction in this respect.[965]

However, it is also important to abolish the incentives for non-transparent cooperation between scientists and companies. The pressure in science to acquire third-party funding, regardless of how and from which sources, must be reduced. Strict transparency rules must oblige the scientists and institutes applying for funding to disclose in all their events and publications which companies supports their work, and the conditions for this support. Just as publishers must make visible any newspaper content that is advertising or otherwise purchased, publishers should also become obliged to disclose when they receive money from corporations, for example through the *Google*[966] or *Facebook* journalism initiatives.[967] The dependencies that arise here are less transparent than advertising that a reader sees. What is needed, then, is transparency at least as great as that of advertising in newspapers. At present, there is no transparency as to where EUR 15 million of the *Google* news initiative has gone in Europe.[968]

---

965 https://weizenbaum-institut.de/

966 https://newsinitiative.withgoogle.com/ and https://www.newsgeist.org/exploring-future-news/ It seems the later web site has been taken down. But see https://newsgeist.wordpress.com/

967 https://www.facebook.com/journalismproject

968 Dachwitz, Ingo. "News Initiative: Wohin Googles Millionen für die Medien in Deutschland fließen." netzpolitik.org (blog), September 26, 2018. https://netzpolitik.org/2018/news-initiative-wohin-googles-millionen-fuer-die-medien-in-deutschland-fliessen/

Similarly, a transparency obligation should be introduced for all holders of the title of professor, including honorary and visiting professors. Lawyers in Germany in particular hold such honorary degrees and use them to represent the interests of clients under the flag of academic authority, without disclosing this in their publications and statements. The regulations on academic titles of universities should introduce a corresponding disclosure obligation and the professional ethics of the liberal professions should also be examined to see how the lack of transparency in the protection of interests can be remedied.

More importantly, however, is to find a way to give back to journalism the advertising revenue that *Google* and *Facebook* have taken away from it. It is astonishing how the democracies in the US and Europe have stood by and watched as the GAFAM companies threaten to destroy the fourth power in the state, privately financed journalism, with their enormous profits drained from media. For the cacophony on the Internet is no substitute for the contribution of investigative, reflective and critical journalism to the control of public and private power in free societies.

The Copyright Directive of the EU[969], which regulates, among other things, the ancillary copyright for journalistic products, was the first attempt to repair the imbalance between the profits of the GAFAM companies and the destruction of journalism as a pillar of democracy. As we are now seeing in France, however, *Google is* resisting the application and simply says: we will not pay.[970] It will

---

[969] Directive (EU) 2019/790 of the European Parliament and of the Council of 17 April 2019 on copyright and related rights in the digital single market and amending Directives 96/9/EC and 2001/29/EC, OJ L 130, 17.5.2019, p. 92-125, http://data.europa.eu/eli/dir/2019/790/oj

[970] Kayali, Laura. 2019. 'Google Refuses to Pay Publishers in France'. Politico. 25 September 2019. https://www.politico.eu/article/licensing-agreements-with-press-publishers-france-google/ Fanta, Alexander. "EU-Leistungsschutzrecht droht schon vor dem Start zu scheitern." netzpolitik.org (blog), September 26, 2019.

therefore be necessary to tighten up the copyright directive and make GAFAM liable to pay any links to press content, while at the same time setting up a binding rights collection system. After all, sending publishers to individually negotiate with Google and other GAFAM companies will not work.

Google's business model is based on processing data free of charge from three sources: content available on the Internet, users' personal data and cloud data storage. This model must take a back seat to the interests of democracy in maintaining a free press, the protection of personal data and thus individual freedom and human dignity. If this happens, i.e., if *Google* has to pay for press content, adheres properly to the GDPR and is no longer allowed to record and track users on press websites, *Google* will still remain highly profitable and innovative. But democracy and the rule of law will then have a chance to stay alive through a functioning fourth power.

It would indeed be important to simply prohibit the collection of reader data by third parties on publishing and press pages, insofar as these third parties themselves collect advertising and at the same time collect the personal data of readers. If only the publishers themselves were allowed to insert advertising alongside their content and process reader data with their consent, and if at the same time the GAFAM companies were no longer able to use the content free of charge, the publishers and journalists would once again be at the economic centre of the advertising market. They would receive a larger slice of the advertising cake thus further enabling the plurality of the free, privately financed press.

---

https://netzpolitik.org/2019/eu-leistungsschutzrecht-droht-schon-vor-dem-start-zu-scheitern/. Hanfeld, Michael. "Europa unter Google: Schutzlos." FAZ.NET. https://www.faz.net/1.6406524 GmbH, DWDL de. "'Raubrittertum': Verlage beklagen sich erneut über Google." DWDL.de. https://www.dwdl.de/nachrichten/74653/raubrittertum_verlage_beklagen_sich_erneut_ueber_google/.

At the same time, this could also create new competition between content curators. Because below the GAFAM capitalization threshold of 500 billion euros there is plenty of room for new companies and competition. Let's not forget that obtaining information about what people are reading provides information about political and ideological attitudes. In a free society, this data must not be collected centrally, and certainly not across different media. However, this is exactly what is currently happening with *Google* and *Facebook* and represents a real threat to freedom. The GDPR must be strictly implemented to protect politically and ideologically sensitive data, but more will be needed to save the free press and the freedom of readers.[971]

Google and *Facebook* have already admitted through their "news initiatives" that they are taking too much money away from the privately financed press. They do this by using, free of charge, content generated by the press, processing reader data for their own purposes and advertising on press pages essentially for their own benefit. They return a fraction of the profits they make in this way to the press. They create dependency relationships through new subsidies that flow to the press for digitization projects. These in turn provide them with knowledge about which digitization models work in the press and which do not. In this way, *Facebook* and *Google* acquire knowledge about how they themselves can one day offer digitised press worldwide. No self-respecting democrat could wish for this.

Therefore, it would be better to place the funds of the news initiatives under the management of a foundation controlled by the press itself, which researches models for the future of independent journalism in digital times and at the same time finances their implementation.

---

[971] Cf. Art. 9 GDPR.

On the other hand, it is urgently necessary to ensure that profits from press activity do not accrue to Google and Facebook from the outset. They must remain with the many media outlets whose decline are lamented everywhere, most loudly in America.[972] In addition, thought will also have to be given to increasing support for journalism through foundation models and other measures. In any case,[973] Europe needs a holistic policy for press freedom and press plurality for the sake of democracy.[974]

Moreover, Europe needs initiatives for a new journalism of power control of the digital corporations, as it is planned in America under the heading *Tech Accountability Journalism*, for example by Julia Angwin and *The Markup*.[975] The concept is to mix editorial teams: 50% journalists and 50% data scientists and technicians. This is intended to create a high level of in-house competence of the editorial staff to procure and analyse data themselves. The editorial attention is to be directed mainly at the powerful tech companies but also at digital

---

[972] The death knell for local newspapers? It's perilously close, Margaret Sullivan, Washington Post, 22 November 2019, https://www.washingtonpost.com/lifestyle/style/the-death-knell-for-local-newspapers-its-perilously-close/2019/11/21/e82bafbc-ff12-11e9-9518-1e76abc088b6_story.html; Der Spiegel, Markus Brauck, "DuMont: Vom Zeitungssterben und einem wankenden Geschäftsmodell - Kommentar - Der Spiegel - Kultur." https://www.spiegel.de/kultur/gesellschaft/dumont-vom-zeitungssterben-und-einem-wankenden-geschaeftsmodell-kommentar-a-1255220.html; see also the collection on #Newspaper Death at the TAZ, https://taz.de/Zeitungssterben/!t5028783/

[973] "Journalism Funding in Europe › DAFNE." https://philea.eu/journalism-funding-in-europe/. Klomp, Biba. "Six Ways to Boost Media Philanthropy in Europe." Medium, June 27, 2019. https://medium.com/we-are-the-european-journalism-centre/six-ways-to-boost-media-philanthropy-in-europe-e3ffce78a0c6

[974] What Can Be Done? Digital Media Policy Options for Strengthening European Democracy, Rasmus Kleis Nielsen, Robert Gorwa, and Madeleine de Cock Buning, Reuters Institute, November 2019, https://reutersinstitute.politics.ox.ac.uk/sites/default/files/2019-11/What_Can_Be_Done_FINAL.pdf; see also Nemitz, Paul. Pincer Hold of Artificial Intelligence and Populism? - Democracy and the Fourth Estate in Europe, in: Press and Media Freedom in Europe - a Fundamental Right under Threat ?, Center for Applied European Studies (CAES), January 16, 2019, https://www.frankfurt-university.de/fileadmin/standard/Forschung/CAES/Dokumente/FRA-UAS_CAES_Presse-_und_Medienfreiheit_in_der_EU_ebook_01.pdf

[975] See: "'Big Tech Is Watching You. Who's Watching Big Tech?" The Markup Is Finally Ready for Liftoff'. Sarah Scire Nieman Lab (blog). https://www.niemanlab.org/2020/02/big-tech-is-watching-you-whos-watching-big-tech-the-markup-is-finally-ready-for-liftoff/ and https://themarkup.org/"

technology and AI in governmental activity. Thus, *Tech Accountability Journalism* can become an important factor in controlling these new carriers of technical power in our society.

In Germany, the Netzpolitik.org website in Berlin and its authors partly fulfil this function. A website like Netzpolitik.org in English which has a critical view of EU policies, located directly in Brussels, would be an important step forward in the dialogue within Europe on digital issues but also with the English-speaking world about shaping the digital future.

Also important for securing independent journalism are the new provisions of the draft German State Media Treaty of 5 December 2019,[976] which is to come into force at the end of 2020 or beginning of 2021. In it, media are redefined and thus adapted to the digital age to include platforms. Platforms are no longer to be treated as a technically neutral infrastructure, but as content providers, as intermediaries, and thus made subject to new regulation. It remains to be seen whether this will go far enough, especially with regard to the question of whether high-quality journalistic offerings can be better found on the platforms in contrast to fake news and conspiracy theories. After all, a first step in the right direction, including the transparency of bots, has been taken.

The discussions on the new text, which replaces the regulations from 1991, took six years because of the requirement for unanimity between the 16 federal states, governed by different parties. The

[976] "https://www.die-medienanstalten.de/fileadmin/user_upload/Rechtsgrundlagen/Gesetze_Staatsvertraege/Interstate_Media_Treaty_en.pdf". And in addition: Laufer, Daniel. "Medienstaatsvertrag - Neue Spielregeln für Streamer, Google und Falschmeldungen." netzpolitik.org (blog), December 6, 2019. https://netzpolitik.org/2019/neue-spielregeln-fuer-streamer-google-und-falschmeldungen/. and Frank, Benedikt. "Medienstaatsvertrag: Im Netz wie im Fernsehen?" Süddeutsche.de. https://www.sueddeutsche.de/medien/medienstaatsvertrag-youtube-instagram-fernsehen-1.4708118

thorough debate, including public consultation, and the final consensus are an important sign to Europe: important preparatory work has been done here for new rules that can be an example for Europe as a whole, showing that unanimous agreement is possible in a difficult political environment with difficult partners.

It is good that the German Association for Electrical Engineering (VDE) with its 30,000 engineers as members, has also submitted proposals for strengthening "Trustworthiness in public spaces". In the white paper of the same name, the VDE proposes mechanisms and standards for the tracing of information, the certification of sources based on the existing system of security certification, disclosure obligations for large Internet platforms and forgery-resistant "fingerprints" of information in a blockchain. A system for assigning pseudonyms and a digital dialogue room between Members of Parliament and citizens was also proposed. It is not only the proposals which are encouraging, but also their emergence from association of engineers, who are showing their awareness of shared responsibility for democracy.[977]

Democracy will have to do more to secure the private financing of its fourth estate than we have seen so far in German media policy and at the EU level. Just as resolutely as *Facebook* and *Google* are increasing their profits year after year at the expense of the free press, so must the EU and its member states resolutely ensure that journalism and publishers still have a chance against these giants. At stake is a diverse and locally based press and free, critical journalism. Of course, the publishers and journalists themselves must also be

---

[977] The German Committee for Electrical, Electronic & Information Technologies DKE - VDE is also working on a standardization roadmap for ethics in artificial intelligence https://www.dke.de/de/arbeitsfelder/core-safety/ethik-und-kuenstliche-intelligenz, which should be available in April 2020, and corresponding application rules https://www.dke.de/de/news/2019/referenzmodell-vertrauenswuerdige-ki-vde-anwendungsregel; VDE is developing application scenarios with partners.

innovative. But with the current asymmetry of power relations, appeals alone for faster digitization and more innovation in the press will not be enough. A telling example of this is provided by the Reuters Institute in Oxford, which itself receives millions from *Google* and therefore wisely does not propose anything that could affect *Google's* profitability.

Political action at EU level and in Germany is now required. This starts with the necessary taxation of the digital giants, goes on to copyright and media responsibility for illegal content, and finally to specific rules in the competitive relationship between platforms and the press.

2.2. Uniform legal regulation of transparency of election advertising on the Internet

The rules for electoral advertising on the Internet, their frequency and transparency must be harmonised by law, and in such a way that real comparability between parties facing each other in a concrete election campaign is easily possible, even beyond individual platforms. Furthermore, there must be mechanisms to ensure parties are treated equally and there are no possibilities of circumventing the rules on party funding and campaign costs on the Internet, as we saw in Britain before the Brexit vote.

It is also necessary to record how often advertisements were placed and to which groups they were shown, in general and also specifically by identified candidates and party organizations, to compare the multiplier effects organised within parties through targeted forwarding. Transparency of paid advertisements is only the beginning. The public must also have ways to identify controlled retweet campaigns of party members, and the targeted addressing of certain groups with special content.

A group of prominent scientists and activists has drawn up a catalogue of requirements for election campaign archives, which GAFAM must now introduce.[978] The EU should go the way some US states are going, by introducing legislation.[978a] The central issue of transparency in election advertising cannot be left to voluntary action. The EU needs binding legislation for this purpose, based on the EU code of conduct to combat disinformation as a first step, but going further. The need is justified by the extensive shortcomings of the existing archives of political advertising, which the companies themselves have created, and the higher importance of securing proper democratic election processes.

The embedding of employees of GAFAM companies in the campaign headquarters of political parties is illegal in Europe for data protection reasons,[979] unless a wall is installed within the campaign headquarters between the personal data of the parties and the helpers from GAFAM, thus excluding them from access to party data. The easiest way to do this will be to simply prohibit such direct assistance. In theory, it is conceivable to offset their monetary value against the election campaign costs or, in other words, to book monetary election campaign donations from the company with the high market value of the services of these employees, but this is not sufficient due to the lack of transparency over party financing.[980]

---

[978] Facebook and Google: This is What an Effective Ad Archive API Looks Like, 27 March 2019, https://blog.mozilla.org/blog/2019/03/27/facebook-and-google-this-is-what-an-effective-ad-archive-api-looks-like/ Following Edelson, Laura, Shikhar Sakhuja, Ratan Dey, and Damon McCoy. "An Analysis of United States Online Political Advertising Transparency." ArXiv, 2019. https://arxiv.org/pdf/1902.04385.pdf

See also Fanta, Alexander. "Facebook macht seine Werbung transparenter - aber nur ein bisschen." netzpolitik.org (blog), June 29, 2018. https://netzpolitik.org/2018/facebook-macht-seine-werbung-transparenter-aber-nur-ein-bisschen/

[978a] See now the proposal of the European Commission for this purposes of 25.11.2021, https://eur-lex.europa.eu/legal-content/EN/TXT/?uri=celex:52021PC0731

[979] Seetharaman, Deepa. "How a Facebook Employee Helped Trump Win—But Switched Sides for 2020." Wall Street Journal, November 24, 2019, sec. Tech. https://www.wsj.com/articles/how-facebooks-embed-in-the-trump-campaign-helped-the-president-win-11574521712

[980] Kergueno, Raphael, "Fraud and boats: funding European political parties“, Transparency International EU, https://transparency.eu/boatfraud/

"Parteispenden: Verdeckte Geldflüsse in Millionenhöhe | LobbyControl," May 28, 2018. https://www.lobbycontrol.de/2018/05/parteispenden-verdeckte-geldfluesse-in-millionenhoehe/

2.3. Research and industrial policy: Edge computing and a new mobile operating system as examples of the opportunities offered by 5G and AI

In the long term, Europe's digital sovereignty, a prerequisite for freedom and democracy, can only be preserved if Europe itself remains innovative and thus continues to generate its own prosperity.

The German government and Europe are now relying on European cloud solutions such as Gaia X,[981] among others. One hears and reads more and more about the fact that the pendulum is slowly swinging back, from the centralization of data processing in the cloud back to decentralized processing. Terms such as *Distributed Cloud,* i.e., the distribution of data to many different clouds, close to the point of use, and *Edge Computing,* are moving into the spotlight.[982]

In order to be able to offer more data-intensive services and at the same time minimize data protection risks and data security risks, all GAFAM companies invest in edge computing, i.e., the processing of data close to the customer, ideally with AI applications directly on the smartphone or other portable terminals of the *Internet of Things (IoT). Apple* seems to be the most advanced in this area. But Google, Amazon and Microsoft are also investing in this technology. It could perhaps one day actually solve many data protection problems, by ensuring that personal data remains with the data subject and is processed directly on a device under the control of the data subject.

---

[981] See https://gaia-x.eu/ and Delcker, Janosch and Melissa Heikkila, "Germany, France launch Gaia-X platform in bid for 'tech sovereignty', Politico, 6 April 2020, https://www.politico.eu/article/germany-france-gaia-x-cloud-platform-eu-tech-sovereignty/

[982] Gartner, "Gartner Identifies the Top 10 Strategic Technology Trends for 2020", 21 October 2019, https://www.gartner.com/en/newsroom/press-releases/2019-10-21-gartner-identifies-the-top-10-strategic-technology-trends-for-2020

In the future, *Apple's* mobile phones will use Artificial Intelligence on a chip in the phone to process information directly, such as translating speech, which previously could only be processed after transmission to *Apple's* central data systems.[983] Such technologies could possibly enable the system to react more quickly. They avoid a delay between input and output, the so-called latency. They may also help address some privacy and data security issues.

Technology can work hand in hand with law to solve problems, indeed law and technology can support or even substitute each other in solving problems. All that is needed is a desire to make this work, and an active search for solutions. Correctly used edge technology could be one area in which technology companies and democracy can work together, to ensure personal data no longer leaves the mobile phone and other devices under the control of the user at all and personal data then cannot be viewed from outside, because processing is done directly on the phone, under the control of its owner.

Nevertheless, technical decentralization, currently most advanced in Blockchain technology, should not be confused with a decentralization of power, as is customary in democracy, as long as the entire system is centrally designed and operated, as in *Apple*, and remains closed. A problem of data protection may be solved, but *Apple* will not become less powerful as a result.

Apple's business model was not previously based on aggregating personal information and advertising. However, if the accumulation of personal data on *Facebook* and *Google* (and other centralizers)

[983] Sawhney, Mohanbir, Forbes, "Why Apple and Microsoft Are Moving AI To the Edge". Forbes, 27 January 2020. https://www.forbes.com/sites/mohanbirsawhney/2020/01/27/why-apple-and-microsoft-are-moving-ai-to-the-edge/#5fcf69392570

could be eliminated through the systematic use of Edge Technologies or other means, this would be a real step forward, not only for privacy law, but also for the data sovereignty of citizens and states. Those who lead in Edge Technology may one day be able to play targeted advertising directly with AI from the mobile phone or computer at home, i.e., move the profiling there, and thus avoid the privacy problems of centralization at the operator. With Edge, Apple could then become the main provider of targeted advertising and steal business from the centralizers of personal data, *Facebook* and *Google*.[984]

Let's also not forget that who holds all cards for data processing could change with the rollout of 5G. Because of the high transmission speed and capacity, combined learning from decentral data is becoming possible and efficient for the first time. It is no longer necessary to bring the data together in one cloud system in order to learn using Big Data methods. Edge computing will continue to grow and become more attractive with 5G.[985]

The Spanish digital sociologist Manuel Castells points out that 5G could offer the opportunity to advance concepts that challenge the dominance of the GAFAM companies.[986] The SAP professor for digitization, Key Pousttchi, from Potsdam, says the same. The current dominance of *Apple* and *Google* in mobile phone operating systems will continue into the world of *IoT* unless Europe invests boldly in its own operating system for the mobile communications of the future, which includes Virtual Reality and *IoT* and will operate on 5G and one day the quantum Internet.

---

[984] This was the thought expressed by Michael Veale, a British legal scholar in the field of law and digitization, to Paul Nemitz at a conference in Budapest in 2019.

[985] Forrester Research. "Predictions 2020: Edge Computing Makes the Leap." ZDNet. https://www.zdnet.com/article/predictions-2020-edge-computing-makes-the-leap/ citing report accessed at: https://www.forrester.com/report/Predictions+2020+Edge+Computing/-/E-RES157595#

[986] La Vanguardia. "La revolución 5G, por Manuel Castells," March 29, 2019. https://www.lavanguardia.com/opinion/20190330/461329107516/la-revolucion-5g.html

What Airbus and Galileo were yesterday and are today, joint and successful technological efforts by Europe to maintain independence in strategic areas, the mobile communications operating system could be tomorrow. Europe's strength in the field of the Internet of Things (*IoT*) and virtual reality would be a starting point for this. Already today, and even more so tomorrow, much of the power to control people and systems lies with those who control the mobile operating system. As a new era is starting with AI, 5G, Virtual Reality and *IoT*, European and German technology policy should seize these opportunities together with companies and science.[987]

2.4. Ensure public good orientation in future technical-economic structures: Data trustees and platforms in the public interest

Public interest must in future find greater expression in technical and economic structures. The way to achieve this is through financial incentives, but also through more strategic regulation. For example, the *Data Ethics Commission* of the German Federal Government has suggested that the exercise of data protection rights, but also the benefit from personal data, should be increased at the same time by setting up a group of data trustees[988] who professionally manage data in the interest of the individual. These trustees may well differ in their views of how best to use the data; some could be more oriented towards the monetization of personal data and others more towards the common good, for example using health data to improve hospital

[987] Interview Tobias Schuman, Augsburger Allgemeine, mit Prof. Pousttchi, 30.1.2019, https://www.augsburger-allgemeine.de/wirtschaft/Interview-Bei-der-5G-Infrastruktur-ist-Wettbewerb-sinnlos-id53566751.html

[988] p. 133 ff of the Report of the *Data Ethics Commission*, 2019, https://www.bmi.bund.de/EN/topics/it-internet-policy/data-ethics-commission/data-ethics-commission-node.html

performance. In this way, they can address different preferences of citizens in competition with each other, similar to what happens in the mixed banking system in Germany between private banking, public savings banks and cooperative banks.

Just as banks collect the money of savers for lending or as an investment and thus for the benefit of the economic system as a whole, data trustees and custodians could also aggregate and make data anonymous, thus making it available for the development of AI and other data-driven services, in the interest of all. Europe's mixed banking system could even become data trustees themselves, who would take charge of managing their clients' personal data. After all, if you entrust your money to someone to manage, it is likely that you would also trust them to manage your data. To create trust requires that we get the legal regulatory framework right, which is also necessary in the financial system. Banks certainly have a basic knowledge of how to handle data; the new FinTech Bank start-ups such as *N26* or *Revolut* even more than the traditional banks. In any case, it is interesting to observe that more and more models for data trusts are being worked on worldwide.[989] MIT- Professor Alex Pentland refers to the US *Credit Unions*, which are organized as cooperatives and to which 100 million Americans belong, as possible data trustees of the future.[990]

---

[989] Delacroix, Sylvie, and Neil D. Lawrence. 2019. 'Bottom-up Data Trusts: Disturbing the "One Size Fits All" Approach to Data Governance'. International Data Privacy Law 9 (4): 236–52. https://doi.org/10.1093/idpl/ipz014. https://theodi.org/article/what-is-a-data-trust/ and https://www.slideshare.net/peterkwells/launch-of-odi-2019-data-trust-pilots-work https://mydata.org/ ; https://hello.elementai.com/rs/024-OAQ-547/images/Data_Trusts_EN_201914.pdf https://digitalpublic.io/ https://www.cigionline.org/articles/what-data-trust; https://radicalxchange.org/blog/posts/2019-10-24-uh78r5/

[990] Walsh, Dylan. "How credit unions could help people make the most of personal data", 8. July 2019, MIT Sloane Ideas made to matter, https://mitsloan.mit.edu/ideas-made-to-matter/how-credit-unions-could-help-people-make-most-personal-data

Another public task is to introduce and promote efficient platform models oriented towards public interests, especially in those areas where interest is particularly high, such as education, administration and health. It is precisely in these areas that Europe has traditionally been stronger than the US.[991] So far, these areas have not yet been fully occupied by US platforms. But we must act in haste, as can be seen from the run on the health sector by *Google*. In the end, "software eats it all" may apply to these sectors as well.[992] The question remains: will European companies, including public and non-profit enterprises and the public sector, with the support of politicians, manage to develop and successfully market software themselves? A European approach to digitization will be demonstrated precisely by whether, in these sectors, Europe is capable of developing the traditional strengths of the European social model in the age of AI.

There are already some glimmers of hope for the digitization of public administration: For example, as part of the nationwide programme for the digitization of public administration, a total of 575 administrative services in Germany are to be made fully accessible online by 2022.[993] The work on this is divided between the federal government and 16 federal states. There has already been

---

[991] Nemitz, Paul, and Kwiatkowski, Arndt "Why Europe Needs Public Funding for Platform Development." Delphi - Interdisciplinary Review of Emerging Technologies 2 (January 1, 2019): 95–98. https://doi.org/10.21552/delphi/2019/2/9

[992] Andreessen, Marc. "Why Software Is Eating the World." Wall Street Journal, August 20, 2011, sec. Life and Style. https://www.wsj.com/articles/SB10001424053111903480904576512250915629460

[993] Digitale Verwaltung 2020, Regierungsprogramm 18. Legislaturperiode, https://www.bmi.bund.de/SharedDocs/downloads/DE/publikationen/themen/moderne-verwaltung/regierungsprogramm-digitale-verwaltung-2020.pdf?__blob=publicationFile&v=5

[994] See "Der Senator für Finanzen - ELFE - Einfach Leistungen Für Eltern." https://onlinedienste.bremen.de/Onlinedienste/Service/Entry/ELFE. and Moritz Koch, Dietmar Neuerer, "Digitalstrategie: Das Ende der Papier-Ära: Wie die Bundesregierung den digitalen Aufbruch schaffen will." https://www.handelsblatt.com/politik/deutschland/digitalstrategie-das-ende-der-papier-aera-wie-die-bundesregierung-den-digitalen-aufbruch-schaffen-will/25094234.html

astonishing progress, for example the popular "Einfache Leistungen für Eltern" ("Simple benefits for parents") (ELFE)-App, which was developed in Bremen, and which simplifies access to all state services in the early months after childbirth.[994]

It would be good if the federal and state governments would use their joint gain in competence from such projects and their purchasing power to support the creation of one or two competitive platforms for AI in administrative services. For parallel to the digitization of existing administrative processes, the intelligent use of data in cities and municipalities enables completely new public service projects, greater citizen participation and local democracy. The development of the Smart City should not be left to Google and the GAFAM companies,[995] but must become a central pillar of European democracy policy.[996] It seems that the Hanseatic City of Hamburg, the second richest city in Europe in terms of GDP per capita and ruled by a red-green coalition of Social Democrats and the Green Party, is becoming a front runner in smart city services for participation of citizens and sustainability.[997]

---

[995] Tett, Gillian. "Google's Smart City: Dystopian Nightmare or Model for the Future?," November 13, 2019. https://www.ft.com/content/9fbd70da-05a7-11ea-9afa-d9e2401fa7ca

[996] For example Theo Bass. "Reclaiming the Smart City: Personal Data, Trust and the New Commons." Text. DECODE, July 3, 2018. https://decodeproject.eu/publications/reclaiming-smart-city-personal-data-trust-and-new-commons. https://ec.europa.eu/info/eu-regional-and-urban-development/topics/cities-and-urban-development/city-initiatives/smart-cities_en. Green, Ben, The smart enough city, Boston, 2019, https://mitpress.mit.edu/books/smart-enough-city; https://colab-digital.de/koki/, Bria, Francesca and Amy Lewin, "Barcelona's Robin Hood of data", 16. November 2018, Sifted.eu, https://sifted.eu/articles/barcelonas-robin-hood-of-data-francesca-bria/; Bria, Francesca. "Europa cannot rely on Silicon Valley", Alicia Prager, Euractiv, 8. May 2019, https://www.euractiv.com/section/digital/interview/sam-francesca-bria-europe-cannot-rely-on-silicon-valley/

[997] https://digital.hamburg.de/digitale-stadt/ueber-uns-amt-itd-10494

G3 What can be done to combat the concentration of power in the GAFAM companies?

The World Wide Web, invented by Tim Berners Lee, and the personal computer was intended to empower the individual over the dominant central *mainframe,* which at that time was IBM. The original Internet contained a promise of freedom, and a promise of enabling and promoting democracy.

The EU Commission has already put GAFAM in its place like no other competition authority in the world, through a large number of individual decisions. It has just launched an investigation into the data collection practices of *Google* and *Facebook*. This research is aimed at determining how and for what purpose the companies collect data and whether the vast amount of data from both groups is used to disadvantage competitors or to penetrate further markets, i.e., whether the data is possibly being processed for the purpose of abusing a dominant position.[998]

3.1. Reform competition law and administrative structures

European Competition law allows companies to achieve economic size and even dominance per se. Only abuse of a dominant position can be sanctioned. And the definition of abuse and their finding is constrained to effects on the market, i.e., competitors and consumers. Impacts on Democracy do not allow a finding of abuse.

---

[998] Chee, Foo Yun, "Facebook in EU antitrust crosshairs over data collection", Reuters Technology News, 2. December 2019, https://www.reuters.com/article/uk-eu-facebook-antitrust/facebook-in-eu-antitrust-crosshairs-over-data-collection-idUKKBN1Y625B

There are already doubts as to whether the limited instruments of competition law are sufficient to maintain competition in the market and, above all, the decentralised innovative powers in the economy. This is because so far, no recourse is available against practices such as the systematic buying up of smaller, innovative start-ups[999] or the legal imitation of innovation. A report by scientists to the EU Commission makes initial recommendations for reform and shows how more can possibly be done with existing law, including through new interpretation.[1000]

However, it is all the more true for democracy that the concentration of power will always be a serious problem, even if no abuse within the meaning of competition law can be established.

Competition law as it stands today is therefore not sufficient to address the problem of the concentration of power in GAFAM in such a way to ensure decentralised innovation and democratic compatibility of further developments.

In Germany, the draft bill for a reform of German competition law has been on the table since 7 October 2019.[1001] This draft is intended to enable the *Bundeskartellamt* to determine the market power of platforms more easily, not least because of access to competition-

---

999 https://en.wikipedia.org/wiki/List_of_mergers_and_acquisitions_by_Microsoft; https://en.wikipedia.org/wiki/List_of_mergers_and_acquisitions_by_Alphabet; https://en.wikipedia.org/wiki/List_of_mergers_and_acquisitions_by_Facebook; https://en.wikipedia.org/wiki/List_of_mergers_and_acquisitions_by_Apple; https://en.wikipedia.org/wiki/List_of_mergers_and_acquisitions_by_Amazon

1000 Jacques Crémer, Yves-Alexandre de Montjoye, Heike Schweitzer, Competition Policy for the digital era, European Commission, 2019, https://ec.europa.eu/competition/publications/reports/kd0419345enn.pdf

1001 Draft bill of the Federal Ministry of Economics and Energy of 7 October 2019: Draft of a Tenth Act to amend the Act against Restraints of Competition for a focused, proactive and digital competition law 4.0 (GWB Digitization Act), https://www.d-kart.de/wp-content/uploads/2019/10/GWB-Digitalisierungsgesetz-Fassung-Ressortabstimmung.pdf

relevant data. Under certain circumstances, the refusal of access to data is to be regarded as an abuse of market power. The disclosure of data may become mandatory, although this must be done in accordance with the GDPR. It is also important that the draft creates a legal base for cooperation and exchange of information between the competition authorities, the data protection authorities and other regulators provides. It expressly authorises and encourages the exchange of information between them. All in all, the draft is a first, important step towards using competition law to deal with the data power of GAFAM. Even if further steps towards structural regulation are necessary,[1002] independent of individual case decisions such as in competition law, this draft, if implemented, will be a pioneer in the regulation of digital market power worldwide.[1003]

The power of GAFAM and thus the dependence on a few large corporations must not be repeated in the next technological leap that is imminent with the Internet of Things, AI and 5G as well as quantum computers.

With regard to the platform economy, we will have to rethink regulation in order to reduce these positions of power, e.g., through more competition. Otherwise, they will represent an even bigger danger to democracy, and the functioning of the market in our SME-dominated European economy will be called into question. We need basic principles of platform neutrality, interconnectivity and the separation of activities in order to secure democracy, fundamental rights and the rule of law, as well as a functioning market, innovative capacity, digital sovereignty and social justice.

[1002] Piétron, Dominik and Marita Wiggerthale. "New competition rules for the platform economy", Netzpolitik.org, 6 December 2019, https://netzpolitik.org/2019/neue-wettbewerbsregeln-fuer-die-plattformoekonomie/#spendenleiste

[1003] This was given as the opinion of the international law firm Hogan Lovells in its autumn 2019 antitrust law radar, https://www.hoganlovells.com/~/media/germany_folder-for-german-team/newsletter/kartellrechts_radar_herbst_2019.pdf

Fragmented regulators must learn to deal with platforms operating in many markets simultaneously. It will be necessary to consider whether the current set up which sees multiple, structurally-fragmented economic regulators is appropriate, or whether it should be abandoned. This is because in the digital platform economy in the age of AI, problems clearly defined by sector or topic – like those that characterised the economy of the last century - are becoming increasingly rare. If platforms become multilateral and use general-purpose technology such as AI, this should also be reflected in the structure of the regulatory authorities. This is especially true since the subject of regulation, data and AI, will also become a common topic for all supervisory authorities. It will become increasingly difficult to explain why public prosecutors' offices in many EU Member States are organised in a unitary manner, but the supervisory authorities for markets and public interest, such as data protection, competition, consumer protection, banking and telecom regulation, are fragmented into many small authorities, which are less flexible at adaptation, even though they are increasingly all confronted with the same limited number of powerful platforms and groups.

The insights today of the tech giants into our thinking and communication goes too far. The concern about the security of data in the cloud provided by American companies, especially after what we have learned from Edward Snowden, but also the desire for industrial and political independence, is driving European[1004] and German[1005] cloud initiatives. However, these are primarily designed for business and science. For the time being, the individual remains alone with the GAFAM terms and conditions, which give *Google* and

[1004] European Commission, "The European Cloud Initiative." Text. Shaping Europe's digital future -, April 19, 2016. https://ec.europa.eu/digital-single-market/en/%20european-cloud-initiative

[1005] Grüll, Phillip, and Samuel Stolton. "Altmaier Charts Gaia-X as the Beginning of a 'European Data Ecosystem.'" Www.Euractiv.Com (blog), June 5, 2020. https://www.euractiv.com/section/data-protection/news/altmaier-charts-gaia-x-as-the-beginning-of-a-european-data-ecosystem/

Co. insight and usage rights to all communication. This is where the data protection authorities are called upon to finally enforce the GDPR more rigorously.

3.2. The principles of the constitutional state as a model for the control of technical power

Let us continue with Popitz, who addresses the question on what political control of technical action should look like:

> *"What success - even to a limited extent – can such an enterprise achieve? We have barely begun to try for it. I do not intend to conclude with a flowery "if we don't . . " statement. But even a sober reflection reveals that a control over technical action, understood as a control over enormous power potentials that are increasing tremendously further, is unthinkable without difficult changes, difficult to imagine, comparable with the conceptual and institutional innovations that have put into being the modern constitutional state."*[1006]

Popitz recommends that we examine the innovations of the modern constitutional state in terms of its capability to control power to see whether they could also control and limit technical power in such a way that safeguards democracy, the rule of law and fundamental rights.

Here we can borrow from the principle of separation of powers, which is implemented in many constitutional democracies in two ways, namely in the form of a horizontal and a vertical separation. The horizontal separation of powers describes the functional division of state power between the legislative (legislation), executive

[1006] Popitz, Heinrich. Phenomena of Power: Authority, Domination, and Violence. Columbia University Press, 2017, p8

(government) and judicial (courts). The fourth power is the free independent press, which controls power through critical reporting, inquiries and investigations. Free and independent science also plays a critical function in holding the state accountable.

The vertical separation of powers describes the division of responsibilities between the EU, the federal government, the Länder (federal states) and the municipalities. No level may "rule through", no one has all the power.

It is very important that the whole system, both the separation of powers and the entire substantive law, is supervised by the institutions themselves and by the citizens, who have the possibility to bring violations of the rules of the Constitution before the courts, which are independent and have decision-making powers. However, before and in addition to any legal action taken, a parliament can investigate the activities of a government, set legal limits on it and even vote it out, to give but a few examples. The government is therefore accountable to both the parliament and the judges. The public also controls and uncovers. The press, as the fourth estate, traditionally plays a major role in this, as do civil society organizations. Their findings flow into the work of parliament, the government and sometimes also the courts and can influence elections. It is therefore essential for safeguarding democracy that attacks through technical manipulation or dominant economic influence on the public and civil society are prevented and that the platforms are held accountable.

3.3. Separation of powers in the modern constitutional state as a model for the containment of digital power in the age of the AI

The probability increases that with the further development of AI and quantum computers, technical power will become even further entrenched, possibly even setting itself goals and becoming "autonomous" in some areas. In view of this, is it conceivable to introduce a general principle of sharing technical power by law, just as the constitutional state shares power strictly vertically and horizontally?

This idea may sound revolutionary at first sight. However, it is already reality in the field of media law. In order to ensure the plurality of public opinion, which is essential for democracy, media law in many countries limits the technical reach of magazines, radio and television stations and/or the market share that a media group can hold, and this without proof of abuse, as is necessary in general competition law.

As the BVerfG (The German Constitutional Court) has stated in several decisions, the emergence of predominant power able to influence public opinion must be prevented by measures to ensure diversity. In principle, this is a horizontal separation of powers according to functions, i.e., within the press, which is, however, in Germany still overlaid by the vertical separation of powers between the Federal Government and the Länder.[1007]

A very topical question is whether social networks and search engines, which have become a very important source of political information, should consequently be treated like media within the framework of media concentration law and thus be subject to the separation principle of media concentration law.

44% of Americans already get their political information from *Facebook*. And in Germany, *Facebook*, *Instagram* and *Google are* already competing with the traditional main sources of News in TV and

---

[1007] 4th Radio Decision of the First Senate of the German Constitutional Court, 4 November 1986 case 1 BvF 1/84, published in BVerfGE 73, 118-205.

printed press, namely the *Tagesschau, heute* and the *Bild* newspaper for the top positions as sources of political information - among people up to 30 years old, the platforms rank highest. So why not subject them to media concentration law?

Nothing really prevents us from adopting this approach. Especially because these platforms are not passive, but rather create content actively in some cases. And they certainly arrange and edit it for display, i.e., by automating editorial functions or, more recently, having them performed by journalists on *Facebook*.[1008]

They exercise power of opinion by deciding what information appears on the first hit pages. Therefore, their algorithms of automatic editing and curating are not, and cannot be, value-neutral, since they influence essential areas of public communication.

Because the GAFAM networks and the search engines themselves aggregate political news and disseminate it according to users' personal profiles and their own priorities, they themselves have become part of the electronic public sphere. Therefore, the media law that protects the function of the public sphere in democracy must also be applied to them, at the very minimum in the form of the media concentration law.

A concrete example helps to illustrate the point: If one day *Facebook* or *Google* would like to buy the German publishing house Springer, in which the US investor KKR already has a stake, in order to integrate its advertising portals but also its journalistic products such as *Bild*, *Welt*, and in Brussels *Politico Europe*, then the law on media concentration currently does not apply. This cannot be right. After all, *Facebook*

[1008] Palmer, Annie. "Facebook Is Hiring Journalists to Curate Its News Tab." CNBC, August 20, 2019. https://www.cnbc.com/2019/08/20/facebook-is-hiring-journalists-to-curate-its-news-tab.html

and *Alphabet*, alongside *Google* News and *YouTube*, are already very important news sources in Germany. In fact, they compete with newspapers and television for advertising, but also for attention.[1009] In the discussion on the question of whether Facebook should distribute political ads and whether it has a role in verifying advertising claims, Facebook described itself as a medium of the press.[1010]

The new German draft State Media Treaty, which replaces the State Treaty on Broadcasting, is a step in the right direction in that it requires these intermediary platforms to disclose how news is aggregated and how it is presented to users. But media concentration law still needs updating.[1011]

With a view to regulations aimed at ensuring diversity of opinion, a regulation for the so-called media intermediaries, i.e., the large platforms, would be conceivable, which would not only have to ensure that serious news sources can be found. Rather, in order to safeguard diversity, consideration should also be given to safeguarding the general public. One option would be to require windows on search and social networking sites to display a uniform, non-personalised selection of news of the day in news feeds to

---

[1009] See in detail Lobigs, Frank and Neuberger, Christoph, Meinungsmacht im Internet und die Digitalstrategien von Medienunternehmen, Gutachten für die Kommission zur Ermittlung der Konzentration im Medienbereich (KEK), 2018, https://www.kek-online.de/fileadmin/user_upload/KEK/Publikationen/Gutachten/Meinungsmacht_im_Internet_ALM51_web_neu.pdf p. 69ff. Chiavegato, Alexian. n.d. 'Council Post: The Attention Economy's Impact, And Why Publishers Must Shift Their Strategy'. Forbes. https://www.forbes.com/sites/forbescommunicationscouncil/2020/02/12/the-attention-economys-impact-and-why-publishers-must-shift-their-strategy/

[1010] Goodman, Ellen, and Karen Kornbluh. "How Facebook Shot Themselves in the Foot in Their Elizabeth Warren Spat | Ellen Goodman and Karen Kornbluh." The Guardian, October 15, 2019, sec. Opinion. https://www.theguardian.com/commentisfree/2019/oct/15/facebook-elizabeth-warren-regulation

[1011] The law on media concentration is no longer up to date, interview with the chairman of the Commission on Concentration in the Media (CEC), Prof. Georgios Gounalakis, medienpolitik.net, 26 August 2018, 'https://www.medienpolitik.net/2018/08/medienpolitikdas-medienkonzentrationsrecht-ist-nicht-mehr-zeitgemaess/'; on the activities of the CEC see https://www.kek-online.de/

all users. It should also be possible to diversify these locally. The selection and compilation of sources could be based on competition between reputable, professional media sources, under the control of a self-regulatory body of the press or the independent decentralised state media authorities (*Landesmedienanstalten*) as they exist in every German land. This would be a modern variant of the existing "window regulation" in the German Interstate Broadcasting Treaty, which ensures diversity in private television.[1012]

In America, Facebook itself has chosen the news sources it distributes for payment, including the right-wing extremist propaganda website Breitbart.com. In addition, Facebook retains the personalization with its own algorithms. Both are detrimental to the democratic public. The example confirms that this matter must be regulated by democracy and cannot be left to the GAFAM companies.

In addition to the goal of ensuring diversity, which has the objective of avoiding predominant power of opinion and which is as important today as it was in the age of mass media, the goal of securing a public sphere that can be viewed and reviewed by all must be added. The radical personalization policy of the GAFAM business model must limited when it comes to the public exchange of information. Otherwise, this common public sphere and thus the most important prerequisite of a democratic society may be lost.

3.4. From net neutrality to platform neutrality

But back to the principles of the state, the separation of powers and the principle of neutrality towards those subject to the law. Net neutrality is a new expression of the principle of neutrality as regards transmissions on the Internet. But net neutrality alone, plus

---

[1012] See § 30 f of the former RSTV (Interstate Broadcasting Treaty).

transparency obligations for intermediaries, as now laid down in the German Interstate Media Treaty, will not suffice to ensure the functioning of the democratic public in an age of concentration of power in the GAFAM companies.

It will probably be necessary to introduce a political neutrality requirement for platforms with market power. The current discussion in the US about the political one-sidedness of the platforms foreshadows the problems of the concentration of power among the platforms in a democratic society.

For example, a Democratic Party presidential candidate, Elizabeth Warren, announced the break-up of *Facebook* as a political goal, and Facebook then took her message down. It is impossible to imagine what it would mean if *Facebook* boss Mark Zuckerberg were to decide to support only other candidates by giving them,[1013] but not Elizabeth Warren, access to raw data about the citizens of America collected free of charge on *Facebook* and only they are allowed to distribute their content on *Facebook*.[1014] Or indeed if Zuckerberg was to stand for the Presidency, as has sometimes been the speculation.[1015] But as if the raw data was not enough, *Facebook* could use its power over behavioral predictions, after all 200 trillion predictions per day, and thus provide a *mega-microtargeting machine* for Warren's political opponents that would be far greater than *Cambridge Analytica*. This must not be allowed to happen with such dominant companies as *Facebook* and *Google*. But no law yet exists to prevent this.

1013 Lima, Cristiano. "Facebook Backtracks after Removing Warren Ads Calling for Facebook Breakup." Politico. 11 March 2019. https://politi.co/2TKyxX1

1014 Lima, Cristiano, "Facebook backtracks after removing Warren ads calling for Facebook breakup", Politico. 11. March 2019, https://www.politico.com/story/2019/03/11/facebook-removes-elizabeth-warren-ads-1216757

1015 Gilbert, Ben. n.d. 'President Trump Says Mark Zuckerberg Running for President "Wouldn't Be Too Frightening"'. Business Insider. https://www.businessinsider.com/president-donald-trump-facebook-mark-zuckerberg-running-for-president-2020-1

We therefore need a legally established political neutrality obligation for large platforms and search engines to complement net neutrality. Of course, control of the political neutrality of platforms must not be placed in the hands of the government, for example a "Ministry of Truth". But the platforms can and should be obliged to participate in the forums of self-organization of the press, as with Germany's Press Council. It should be the responsibility of these forums, which are far removed from the state, to monitor the neutrality of the platforms.

3.5. From the separation of powers in the state to the separation principle for GAFAM platforms

Furthermore, we will have to think about a general principle of separation from certain services in order to prevent a concentration of political power in the corporations by such an analogy to the separation of powers. This is not a new response to significant problems posed by monopolies, and dates back to at least the Glass-Steagall Act of 1933, separating commercial and investment banking in response to the stock market crash of 1929.

The basic principle must be the separation of data control and system control, and in addition, there must always be newly developed interruptions and discontinuities built into the overall system in order to prevent total control or excessive concentration of power. In Germany after the Second World War, due to the terrible experience with the abuse of highly centralized power under fascism, a political system was developed in which, in addition to the separation of the power of the state into executive, judicial and legislative branches, a more extensive vertical distribution of responsibilities was established. Thus, the distribution of power on different levels, as federal, state and local government, as well as the control of power in the economy, became a central element of the reason of state and democratic capability of the second German Republic.

In the future, the control of technical - systemic power and the separation of "powers" of technical and economic power will have to be a central task of state-driven, democratic action. For the combined technical systems of the GAFAM companies, which are, theoretically, equipped with absolute control and manipulation power, a dynamic doctrine of system separation and data separation must be developed. The capacity for innovation and democracy and the freedom of the individual must be secured anew again and again, since enormous technical progress and the concentration of wealth always pose a new threat to these fundamental values.

This will not only be achieved through competition law and ex-post decisions on individual cases based on criteria of illegality. Rather, this requires forward-looking, preventive regulation that establishes the separation principle in a technology-neutral manner and without hindering connection and complementary innovation. There must be an incentive for economic actors to think about the possible consequences of innovation on the distribution of power in state and society from the outset when innovating technology and business models. The freedom and fundamental rights of individuals, the legal system and the rule of law, the functioning of democracy and ecological sustainability must be factors for consideration. If the basic structure of democracy today is determined by personal data, the Internet, Artificial Intelligence, then the basic principles of *checks* and *balances* and the separation of powers must be transferred to these technical systems.

The idea of digitising democracy has largely failed. Too many problems have arisen in the attempts that have been made so far, for example by spreading liquid democracy, using blockchain to fix politics and in electronic voting. The decline of the Pirate party in Europe and *Die*

Piraten in Germany followed the disappointment of digital democracy dreams. The bottom line is that in developed democracies the Internet is currently not a boon to democracy, but a danger.

We therefore need a new regulatory theory of the Internet and the key technologies and business activities based on the Internet.

The new theory of regulation must be based on a meta-theory of the relationship between technology and democracy that clearly identifies and further analyses the dangers of technological absolutism.

It must then learn from previous experience with legal rules for controlling power in complex networked systems:

- Theory and practice of horizontal and vertical separation of powers in the state and independent institutions in democracy.
- Systemic regulation of areas particularly relevant to democracy, such as election advertising during election campaigns, party and campaign financing, regulation of internal plurality and against concentration of press and media outlets.
- Theory and practical experience of unbundling and interconnectivity obligations in relation to network infrastructure and services provided in strategic areas such as general competition law, railways, telecommunications and electricity.

However, raising the prospect of separation means we are entering a difficult debate. Because the GAFAM companies are potentially in competition in so many areas, it should be noted that a separation principle that is too simple can lead to a reduction of this potential competition. Example: If we were to introduce a rule that those who collect data about people on the Internet using social networks or search engines may not also collect personal data through free email services and messengers, we would reduce competition in the area of messengers and email by excluding *Gmail*, *WhatsApp* and *Facebook Messenger*.

Better, in relation to this example, would be an intelligent disconnection rule that only regulates the collection of personal data, but not the provision of services themselves. This would mean that those who already collect personal data intensively in the area of social networks or search engines should not at the same time collect personal data in other areas such as individual communication via messenger and e-mail. The services may be offered, but without any collection and analysis of the content of the communication and its metadata.

The reflection on a separation principle is manifest in different jurisdictions, and for different services. In the US, there are already bills in Congress that stipulate that social networks and search engines may not provide banking services.[1016] We could also imagine that cloud services of third-party data storage could be separated from other data collection activities in order to eliminate the risk of combining data from the three major sources: personal data, corporate data and Internet content, such as *Google* does it.

Consider another example that falls within the competence of the federal states in Germany: should we not ensure that those companies that provide digital services in schools or universities are not the same as those that collect data about our children in other areas of life in their leisure time or later in their professional life, and thus can compile a comprehensive life profile?

[1016] see: Keep Big Tech Out of Finance Act draft: "H.R.4813 - 116th Congress (2019-2020): Keep Big Tech Out of Finance Act." Legislation, November 13, 2019. 2019/2020. https://www.congress.gov/bill/116th-congress/house-bill/4813. ; Fortune. "One Issue Bringing Democrats and Republicans Together? Banning Big Tech from Banking." https://fortune.com/2019/11/19/big-tech-financial-regulations-banking/.; Markets Media. "QUICK TAKE: Will Keep Big Tech Out of Finance Act Happen?," August 1, 2019. https://www.marketsmedia.com/quick-take-will-keep-big-tech-out-of-finance-act-happen/

The purpose of such a separation would be to structurally prevent GAFAM from being able to create personality profiles from childhood on, which would serve as a basis for lifelong *profiling*. The temptation to do this, for example at *Microsoft*, is enormous. But founder Bill Gates' foundation project to collect data with the educational technology initiative *InBloom* at schools in the US was rightly defeated because of parental mistrust.[1017] A childhood profile would be an ideal basis for later offerings and *profiling* in the area of games and professional networks as well as practically all other areas of the digital world. *Microsoft* with its many acquisitions, including the popular game Minecraft and the professional database *LinkedIn*, is already in a great position to build this type of cradle to grave surveillance technology.

In any case, it would make sense to contrast the strongly centralizing tendencies of the digital technology and business models of the GAFAM companies with a general principle of separation as a basic idea for avoiding the concentration of power in society. In this way, the concrete manifestations of the fantasies of omnipotence in thinking and acting about digital systems in these companies could be countered with concrete measures. We need this new way of thinking about the separation principle of democracy for the digital-informational complex, because the categories of competition law alone, which only have the market in mind, and media concentration law, which is intended to protect against media concentration, are not enough to protect us as a society and as individuals against the concentration of power in the digital world which is robbing us of democracy and freedom.

[1017] Bulger, Monica Patrick McCormick, Mikaela Pitcan, "Data&Society, "The Legacy of Inbloom", 02 February 2017 https://datasociety.net/pubs/ecl/InBloom_feb_2017.pdf

3.6. Interconnectivity obligation for messengers and social networks to break the network effect

While the separation principle is intended to prevent the omnipotence of GAFAM, an interconnectivity obligation can serve to reduce the *winner-take-all effects* in the platform economy and thus create new opportunities for competitors to enter the market.

In her doctoral thesis of 2016, Inge Graef discussed interconnection rules for telecommunications as an example, but she doesn't identify a need for interconnection rules for social networks in view of the "dynamic market", also due to concerns about the alleged negative effect on innovation.[1018] Three years later, there is no sign of the dynamic market for social networks, the power of *Facebook* has only increased. Negative effects for innovation through the obligation of interconnectivity in telecommunications are not discernible.

The obligation to interconnect will lead to the creation of a new uninterrupted connection of services with each other, both between services of the same function and across functions, e.g. between social networks, search, cloud services and messenger services and many new combinations of interconnection.[1019] This increases

---

[1018] Graef, Inge. Data as Essential Facility: Competition and Innovation on Online Platforms: Thesis, 2016, p.154 , https://core.ac.uk/download/pdf/34662689.pdf.

[1019] Further approaches to the commitment to interconnectivity and to enabling dynamic data portability: "Stigler Center (University of Chicago) Report on Digital Platforms." Antitrust Digest by Pedro Caro de Sous (blog), July 12, 2019. https://www.chicagobooth.edu/research/stigler/news-and-media/committee-on-digital-platforms-final-report. Yale School of Management. "Prof. Fiona Scott Morton Leads Call for Greater Competition among Digital Platforms," May 21, 2019. https://som.yale.edu/news/2019/05/prof-fiona-scott-morton-leads-call-for-greater-competition-among-digital-platforms.

GOV.UK. "Unlocking Digital Competition, Report of the Digital Competition Expert Panel." https://www.gov.uk/government/publications/unlocking-digital-competition-report-of-the-digital-competition-expert-panel. https://ec.europa.eu/competition/publications/reports/kd0419345enn.pdf; https://www.bundesregierung.de/breg-de/service/publikationen/ein-neuer-wettbewerbsrahmen-fuer-die-digitalwirtschaft-1669480

competition between providers, as network effects no longer occur only at one company in a closed system, which in turn becomes the starting point for achieving dominance in subsequent areas. Rather, the benefits of the network also serve all competitors, at least as far as networking with the members of the network is concerned. Connecting always should follow the rules of the network, which is the starting point of the of the party seeking a new connection.

The interconnectivity can be demonstrated by the following hypothetical model: *Facebook* (FB) maintains its current business model. In addition, two competitors form. One is a commercial network of a publishing house (V). For this network, participants can choose whether they want to participate "for free", using their user data, or for a monthly or annual fee, in which case the data is not used. To make its network attractive, V plays out its own exclusive content on the network in our example. A third network is self-organized by users, without any advertising and free (OW). It does not collect personal data. Content is posted by members or Wikipedia alone. It is financed like Wikipedia, by voluntary membership and donations.

Now when a user of V or OW connects to a friend who is on FB, that friend will see all content posted, rated or forwarded by the user and all friends of the user, and vice versa. If the user has chosen an ad-supported model, he/she will continue to receive advertising from V. If he/she has chosen the paid model without advertising, he will not receive any advertising, even if he/she is connected to users of FB. FB may process personal data and usage data of users of networks V and OW only to the extent permitted by them.

It may be correct to initially make the interconnectivity obligation asymmetric,[1020] i.e., to oblige only the GAFAM platforms to open up, but not the smaller new networks. However, this needs to be carefully considered. After all, the interest in the smaller platforms increases with the fact that users of different smaller networks can also interconnect. But this is only possible if there is an obligation to interconnect even between smaller networks.

The decisive factor will probably be that networking will always only take place according to the rules of the network on which the user is registered, and that smaller networks will in any case be allowed to reserve their exclusive content or functionalities for their users, which go beyond a minimum level to be defined by law. They should not be obliged to open these to users originating from other networks. The centrality of the initial profile of each platform will ensure that the interconnectivity obligation actually results in a wider range of services, and better design and content. We consider it conceivable that precisely the combination of social network and exclusive content could become an attractive model for the press, if these conditions are met.

[1020] in detail, and on already existing open source systems of networking between networks: Piétron, Dominik, Digital Sovereignty through Interoperability, FES WISO Direkt 24/2019, https://library.fes.de/pdf-files/wiso/15852.pdf

G4 Ethics and law to ensure that the AI serves the public good

It is astonishing how long the advocates of the right to regulate AI have been on the defensive. After all, there is a long history of technology regulation by law. Every architect must learn the building code during their studies and work according to its legal rules, which give form to the public interest.[1021] Every car on the road must pass through a type of approval process for safety reasons. The legal requirement to wear seat belts, which industry and automobile clubs fought fiercely against, cut the number of road deaths in two. Critical technologies such as nuclear power and biotechnologies were regulated as a matter of course, not least because responsible scientists recognized not only the benefits but also the potential dangers. Time and again, society has experienced that it is the provision of rights, that serves the interests of the general public with regard to critical technologies.

Closer to the present, the work on the ethics of AI has already identified many very important challenges that AI poses to the rule of law, democracy and individual rights. It has also led to numerous catalogues of ethical rules for AI and autonomous systems. By 2017, Alan Winfield already counted ten codes of ethics for AI worldwide[1022], and by 2019 a global inventory counted 80 codes.[1023]

1021 "We all have a role to play: Games and AI Ethics," keynote, IEEE Conference on Computational Intelligence in Games (CIG), NYU, Brooklyn, NY, 22–25 August 2017. http://www.cs.bath.ac.uk/~jjb/ftp/Bryson%20CogX%20Ethics%20Keynote%20London%20June%202018%20Ethics.pdf for a general introduction also see: Bryson, Joanna J. "The Artificial Intelligence of the Ethics of Artificial Intelligence: An Introductory Overview for Law and Regulation." The Oxford Handbook of Ethics of Artificial Intelligence, 2019. https://researchportal.bath.ac.uk/en/publications/the-artificial-intelligence-of-the-ethics-of-artificial-intellige

1022 A summary of robotics and AI ethics: Part 1 Principles, 23.12.2017, https://robohub.org/a-round-up-of-robotics-and-ai-ethics-part-1-principles/

1023 Mulgan, Geoff, 'AI Ethics and the Limits of Code(s)'. Nesta, 16 September 2019, https://www.nesta.org.uk/blog/ai-ethics-and-limits-codes/

Recent additions to this list are the Declaration of the European Group on Ethics in Science and New Technologies of March 9, 2018,[1024] on Artificial Intelligence, Robotics and Autonomous Systems, the Ethics Recommendations *of the EU High Level Group on trusted AI,*[1025] the recommendations of the German *Data Ethics Commission,*[1026] and the recommendations of the *Institute of Electrical and Electronics Engineers (IEEE)* on *Ethically Aligned Design.*[1028]

There is no shortage of proposals for ethical principles for AI. And it is possible to give machines ethical rules that are operationalized in programs, alongside transforming these ethical principles, often criticized as doing little to actually resolve some of the issues created by AI, into hard law. It is not possible for AI systems to give themselves values or develop a morality that is supposedly superior to the human one. Values and value decisions are and remain a human matter.

4.1. The principle of essentiality helps to determine the necessary legal regulations

Now that all this material is on the table, it is time to move on to the crucial question in democracy, namely which of the challenges

---

1024 European Commission, "European Group on Ethics in Science and New Technologies (EGE)." https://op.europa.eu/en/publication-detail/-/publication/dfebe62e-4ce9-11e8-be1d-01aa75ed71a1/language-en/format-PDF/source-78120382

1025 European Commission, Ethics Guidelines for a trustworthy AI, High Level Expert Group on AI, 8 April 2019, https://ec.europa.eu/digital-single-market/en/news/ethics-guidelines-trustworthy-ai.

1026 Opinion of the German *Data Ethics Commission*, 23 October 2019, https://www.bmi.bund.de/EN/topics/it-internet-policy/data-ethics-commission/data-ethics-commission-node.html. https://www.bmi.bund.de/EN/topics/it-internet-policy/data-ethics-commission/data-ethics-commission-node.html

1028 IEEE, Ethically Aligned Design, First Edition, 2019, https://standards.ieee.org/content/dam/ieee-standards/standards/web/documents/other/ead1e.pdf?utm_medium=undefined&utm_source=undefined&utm_campaign=undefined&utm_content=undefined&utm_term=undefined for more background see: https://ethicsinaction.ieee.org/ and for more information https://ethicsstandards.org/p7000/.

of AI can be safely and in good conscience left to ethics and which challenges of AI must be addressed by rules that are enforceable and based on democratic processes, i.e., laws.[1029]

In answering this question, responsible politics will have to consider the principle of essentiality, which has long governed legislation in constitutional democracies. This principle stipulates that any matter that is essential because it either concerns the fundamental rights of the individual or is important to the State must be regulated by a parliamentary law with democratic legitimacy.[1030]

Another important element to consider in answering this question will be whether, after the experience of the lawless Internet, our democracies can once again take the risk of a new, all-pervasive and decisive technology like AI that remains unregulated and is therefore likely to have significant negative effects. Furthermore, unlike the Internet, AI today is not, from the outset, an *infant industry* driven mainly by academics and idealists. Rather, it is largely developed and deployed under the control of the powerful Internet technology companies.

[1029] For introduction see Drexl, Josef , Wirtschaft, Gesellschaft und Recht im digitalen Wandel, 7 December 2019, 'https://www.badw.de/fileadmin/user_upload/Files/BADW/0Jahrfeier/Festvortrag_2019_Josef_Drexl.pdf' and Hildebrandt, Mireille. Law for Computer Scientists and Other Folk. Oxford, New York: Oxford University Press, 2020, https://www.cohubicol.com/assets/uploads/law_for_computer_scientists.pdf ; with detailed proposals Mario Martini, Blackbox Algorithmus - Grundfragen einer Regulierung Künstlicher Intelligenz, 2019;

[1030] Jo Eric Khushal Murkens considered the UK High Court ruling on Brexit decisions which are not solely in the hands of the Government but which give Westminster a say an application of the principle of essentiality, see Brexit Decision of the High Court: A Lesson in Constitutional Law for the British Government, in: Constitutional Blog - On Constitutional Issues, 3rd ed. November 3, 2016, available at https://verfassungsblog.de/the-high-courts-brexit-decision-a-lesson-in-constitutional-law-for-the-uk-government/ ; on the principle of essentiality in German, US and EU law, see Saurer, Johannes. EU Agencies 2.0: the new constitution of supranational administration outside the EU Commission, in: Comparative Administrative Law: Second Edition 2017, edited by Susan Rose-Ackerman, Peter L. Lindseth, Blake Emerson, p. 619.628; other recent examples of the case law of the European Court of Justice are the judgements in cases C-355/10, ECLI:EU:C:2012:516, European Parliament v. Council, par. 64ff, with further precedent, available at https://eur-lex.europa.eu/legal-content/en/TXT/?uri=CELEX:62010CJ0355, and C-293/12 and 594/12/, ECLI:EU:C:2014:238, Digital Rights Ireland, par. 54ff, available at http://curia.europa.eu/juris/documents.jsf?num=C-293/12.

It is these powerful corporations that have already proven that they cannot be trusted to pursue a public interest on a large scale without being guided by the hard hand of the law and its rigorous enforcement. We need regulation that sets limits and even provides direction for innovation that is in the public interest. Indeed, some representatives of these companies have recently come to this conclusion themselves and have called for legislation on AI.[1031]

4.2. Rule of law, fundamental rights and democracy by design in AI

The design of AI for autonomous development and its very broad application may well lead to far more catastrophic effects than the unregulated Internet has already done. Looking only at the experience with the Internet and the capabilities of AI and its potential widespread use, there are good reasons to establish a precautionary legal framework that lays down the basic rules necessary to protect the public interest in the development and deployment of AI.

The triad of human rights, democracy and the rule of law are the core elements of Western, liberal constitutions, the "trinitarian formula" of our constitutional law.[1032] These principles are the supreme law - all actions of government, legislators and also social reality are measured against them. In view of the foreseeable spread of Artificial Intelligence and other new technologies based on the Internet, it is

[1031] Bass, Dina, "Microsoft Says AI Advances Will Require New Laws, Regulations", Bloomberg, 18 January 2018, https://www.bloomberg.com/news/articles/2018-01-18/microsoft-says-ai-advances-will-require-new-laws-regulations.

[1032] Term coined by Kumm, Matthias. "The Cosmopolitan Turn in Constitutionalism": An Integrated Conception of Public Law' [2013] 20 Indiana J Global Legal Studies 605; on the role of this formula in and for international law see Mattias Kumm, "Constituent Power, Cosmopolitan Constitutionalism, and Post-Positivist Law" [2016] 14 Int J Constitutional L 697.

legitimate and necessary to ask how these new technologies should be designed to support the maintenance and strengthening of the constitutional "Trinitarian formula".

In many areas of life, AI will decide according to certain rules or prepare decisions that were previously made by people. So, as AI increasingly contains and sets the rules by which we live, we must get used to the fact that AI must always be treated like the law itself. And it is normal for a law to be checked before it is passed to see if it violates higher law.

The test to which any law must be subject is whether it is in conformity with fundamental rights and the Constitution, whether it has been adopted by a legitimate procedure, whether it conflicts with other existing laws, is sufficiently clear and comprehensible and lays down obligations that are proportional to the purpose pursued. It is this test that the AI, which contains or even sets rules for society and applies them through decisions or in any case prepares decisions, must also pass. The AI will only pass this test if the principles of democracy, the rule of law, respect for fundamental rights and compatibility with the law as a whole are integrated into the AI from the outset, i.e., from the beginning of the programme development. This must be done in such a way as to ensure that this compatibility is also guaranteed for subsequent individual decisions taken by the AI, even if the programme learns and mutates. This can be achieved, for example, by setting absolute or relative corridor values for individual decisions, which must never be exceeded even if the programme mutates. Even if AI cannot apply principles per se, rules can still be written into the algorithms, which concretize the principles for many, if not all, use cases in such a way that an automated application of the rules is possible and leads to the application of the principle. It will be the task of AI science and research to continue on this path of operationalizing principles into rules applicable in AI. In his book "Human Compatible, Artificial Intelligence and the Problem of Control", Stuart Russell has described a path and at the same time its limits: AI must be programmed to return questions back to humans for deliberation over new value decisions that have to be made.[1033]

So, the answer proposed here is: we need a new culture of technology and business model development for the age of AI, which we call the *Rule of Law, Fundamental Rights and Democracy by the Design in AI*. The reference to "by design" refers both to policy making for the legal rules of the Internet and new technologies such as Artificial Intelligence and the Internet of Things, and to the technology and business models themselves.

The principle of the *Rule of Law, Fundamental Rights and Democracy by the Design in AI* is necessary because the capabilities of AI combined with other technologies surpass anything that has been seen before in terms of potential societal impact. AI, combined with large amounts of data and a range of devices and sensors connected to the Internet of Things, will ultimately determine the core functions of society. From education, health care, science and business to law, security and defence, political discourse and democratic decision-making. All of this will be subject to the ever-increasing control of AI and other future technologies.

4.3. Binding to law and the principle of legality as a principle of control of technical power

We live in a world that is at least as much influenced by technology as by law and democracy. Just as people form the law and the law

---

[1033] Russell, Stuart. Human Compatible: AI and the Problem of Control. 1st edition. Allen Lane, 2019, p. 172ff.

forms people's behaviour, we have to get used to the fact that this relationship is reciprocal. Every technology has lived by being shaped by law, and it is high time that Silicon Valley and the digital Internet industry also accepted the need for democracy. At a time when the Internet and AI are becoming ubiquitous, the non-regulation of these ubiquitous and often decisive technologies by law would mean the end of democracy.

However, there are two particularities in relation to the combined importance of the Internet, AI and quantum computers and the power associated with them, which justify asking a question that has already been a hot topic in the work on data protection reform and GDPR: Can there actually be complete entrepreneurial freedom for the companies that master and develop these technologies, i.e. above all the GAFAM groups, according to the principle: What is not forbidden is allowed in a free society?

Or should legislatures around the world consider whether a principle of legality must be introduced, as in the GDPR, i.e., the operation of AI or quantum computers should only be permitted if there is a legal basis for it in the law and the conditions for its lawful operation, set out in the legal base, are complied with?

To be clear: At present, every architect and citizen who wants to build a house needs a permit to be issued, after a public authority or a body delegated by it has checked that the building plan is in line with the law. And every new car which comes on the street needs a type approval. Only software and AI need nothing like that. As with the GDPR, one could for low-risk AI refrain from such prohibitions with the possibility to grant a permit, and simply, as in GDPR, say: Here are the legal conditions for putting AI on the market, document that you fulfil them, no permit or *ex ante* check by government needed. But if masses of people are impacted, still no permit and no *ex ante* checking? After all, a building permit is also required for the construction of a house because very often the interests of third parties, namely neighbours, but also public interests, such as town planning, can be affected. The same applies to the construction of any factory or workshop, however small it may be. And a business must also be registered and pass a reliability check.

For the processing of personal data, such authorization was also required earlier, also to protect the citizens as the party concerned. But that is no longer the case today, so there is no longer a data protection law that requires permission.

However, the principle of legality still applies: personal data may only be processed if one of the seven reasons for lawful processing under the GDPR is present and the rules of the GDPR are complied with in all other respects. The person responsible for processing the data checks whether this is the case, and also documents that this check has taken place.

Third party interests are always involved in the processing of personal data and the processing of personal data has become a mass phenomenon. AI will become a mass phenomenon, as widespread as electricity (which also requires a licence to produce it). AI is a *general purpose technology*, which has an impact on society as a whole and which can be used for all kinds of purposes. Moreover, as we have shown, this technology also has the potential to exercise power in many ways. It would therefore probably be right, at the very least, to lay down a principle of legality for both AI and quantum computers, in other words to define conditions of lawful operation, as has been done in the GDPR for the processing of personal data. AI could then be used without the prior approval of an authority. But anyone using AI would have to ensure that the conditions for legal use were met.

And for critical applications of AI and quantum computers, such as those with significant fundamental rights implications or relating to important community concerns, the principle of official approval should be introduced, as is common practice in building law and for new cars. This is also what the *Data Ethics Commission* of the German Federal Government has recommended, in the form of a criticality pyramid.[1034]

Another prerequisite for the acceptance of AI is clear liability rules in case these systems cause damage. There is no doubt that AI, especially if it develops independently in certain areas of application, raises questions of civil liability. Whether current law is sufficient to cover these questions of responsibility or whether new law is needed is currently being examined by the EU Commission on the basis of an expert report.[1035] Proposals by the EU Commission are expected in 2020 and are already included in the report of the German *Data Ethics Commission*.[1036]

Certainly, as a citizen and consumer, you should also be able to expect that whenever you interact with Artificial Intelligence, whether in writing or orally, you will be made aware that it is AI, and that you are not interfacing with a human, unless it is obvious from the outset. In the administrative law of many Member States, including Germany, it is already obligatory to mark an automatically issued administrative act as such. That is not too much to ask for, nor is it a great burden, even if the transparency obligation for AIs is extended to their use by private individuals.

[1034] *Data Ethics Commission*, pp. 163-182.

[1035] See details at http://ec.europa.eu/newsroom/just/item-detail.cfm?item_id=615947.

[1036] Data Ethics Commission, p. 219 ff.

4.4. The GDPR regulates AI when processing personal data

The GDPR applies to AI when it processes personal data. The GDPR contains important rights for users in connection with the processing of their personal data, as well as obligations on the part of those responsible for processing, which determine the way in which AI is developed and applied in the processing of personal data. The principles of privacy and data protection *by design and by default* as laid down in the GDPR will certainly become very important for AI, as will the limitations of automated processing and the associated rights of individuals concerned to obtain meaningful information about the logic, meaning and intended consequences of processing personal data with AI. A new law is not necessary in this respect.

On the other hand, in democratic discourse it is important to know whether the counterpart in the discussion is a human or a machine. If machines could participate in political discourse without being identified as such or even be allowed to embody people without sanction, this would lead to a distortion of the discourse that is unacceptable in democracy. No law currently assures us that we will be fully and objectively informed in this regard, or forces us to clarify when machines enter into dialogue with us in a political context. Since a transparent political discourse among people is the key to democracy, the principle of essentiality requires that we must have transparency underpinned by law as to whether a machine or a person is speaking. In transparent machine language and, above all, the embodiment of another human being by machines should be sanctioned, and those who maintain large infrastructures of political discourse should be held responsible for ensuring that there is full transparency with regard to machine created language and messaging in their infrastructures.

This necessitates a new law that encourages GAFAM companies to enforce the obligation to inform each individual when using machine language, including within their systems. This does not mean that anonymous speech will become impossible, because it is only a matter of indicating that a machine is talking, not who is behind it and who has controlled and programmed it. Nor does it mean that freedom of speech is restricted. Because the content of what the machine then says remains unchanged. Only now we know that, for example, the enthusiastic, millionfold repeated praise for a populist candidate or the millionfold false assertions about a candidate from the democratic spectrum, come from a machine.

The American voters did not know that possibly billions of tweets with political content were produced by machines during the 2020 election campaign. This probably swung the election in Trump's favor.[1037] The draft of the new German media state treaty is a step in the right direction towards transparency, but it is not enough. For it only obliges transparency on the intermediaries covered by the State Treaty, but not on general transparency in the use of AI. And the transparency obligation of the GDPR in the case of automated processing of personal data applies only to decisions and decision-like acts and only when personal data are processed, which is also not enough.

For the collection, processing and use of data on human beings, a strict regime of limitation must prevail. In contrast to the US, Europe, through the GDPR, already restricts the collection of personal data, and this is essential, because theoretically, due to increasingly cheap technologies, the collection of personal data will become much easier and more effective. In the US, there are almost no restrictions

---

[1037] Snyder, Timothy. The Road to Unfreedom: Russia, Europe, America. Random House, 2018. Chapter 6

on collection at all; only restrictions on use are being considered. The bank robber may steal the money, but how he spends it, we regulate. This cannot work.

4.5. We need a holistic total accounting for surveillance

In the technical world, freedom can only be preserved if a systems-wide view always takes the place of the evaluation of individual measures, because the enormous potential for knowledge, prognosis and manipulation arises from the networking and connection of data and systems.

From the jurisprudence of the German Federal Constitutional Court derives the principle that a holistic total accounting of surveillance must be carried out at any time new measures of surveillance are introduced, to determine the corresponding need of higher protection of privacy and personal data.[1038] This approach aims to cover the ever-denser network of surveillance of humans according to different legal bases and by means of different technical systems in its entirety. The aim is to have a detailed, but also holistic view of data collection which will enable better assessment of the proportionality of new legal bases and systems for surveillance, but also of existing systems whose intensity of surveillance increases with further developments in technology.

We need a rigorous application of this principle, both with regard to the increasingly efficient technologies for collecting personal data, in particular those related to AI, *IoT*, cameras and sensors, and with regard to models of organizational data accumulation by

---

[1038] See for examples: https://digitalcourage.de/ueberwachungsgesamtrechnung/sammlung; https://epicenter.works/thema/ueberwachungsgesamtrechnung

private and public bodies. The concentration of power over personal data in a few companies and the state, which are then also able to create comprehensive profiles and, with the help of AI, behavioral predictions about each individual, must be included in such an overall calculation, since this results in new restrictions of freedom. The US *National Security Commission on AI*, headed by Eric Schmidt, former head of *Google*, has pointed out the scope for AI to be used as a tool to undermine freedom and privacy on a broad scale.[1039]

Since this concentration of personal data is also an element of the concentration of economic power, data protection authorities could cooperate constructively with the cartel authorities and in Germany the Monopoly Commission, which annually reports on tendencies of economic concentration, in this area. The draft of the German 10th Amendment to the Act against Restraints of Competition (GWB) of October 2019 provides the basis for this and further operational cooperation.[1040] The Monopolies Commission's regular report on concentration trends in the economy should include a chapter on the concentration of personal data in certain companies and, as a kind of data concentration aggregate, could also be a model for monitoring the aggregate of all data held by tech companies and their surveillance capacities.[1041] Just as the Monopolies Commission, independent of the Federal Cartel Office, uses rigorous scientific methodology to assess tendencies towards market concentration in the economy, an independent data and monitoring accounting commission, independent of the data protection authorities, could

---

[1039] National Security Commission on AI, Interim Report, November 2019, https://epic.org/foia/epic-v-ai-commission/AI-Commission-Interim-Report-Nov-2019.pdf and https://www.nscai.gov/

[1040] Draft of the Federal Ministry of Economics and Energy of 7 October 2019: Draft of a Tenth Act to amend the Act against Restraints of Competition for a focused, proactive and digital competition law 4.0 (GWB Digitization Act), https://www.d-kart.de/wp-content/uploads/2019/10/GWB-Digitalisierungsgesetz-Fassung-Ressortabstimmung.pdf

[1041] https://www.monopolkommission.de/en/

share the findings from the regular monitoring of surveillance accounting. Until we have such a new commission, data protection authorities in Germany and across Europe should take this task into their own hands.

4.6. Non-discrimination and consumer protection

Considering the much-discussed cases of discrimination by AI, one can probably say in Europe that the body of legal rules, in particular those of EU law in the field of non-discrimination and consumer protection, is fully applicable to AI. In this respect, the situation differs considerably from that in the US.

However, there is a lack of enforcement of these protections against companies. Resolving this will involve better organization of enforcement and better equipping of the responsible authorities, including with the technology of AI, which will be necessary to detect discrimination by AI.

In principle, sanctions will also have to be tightened. Because in the market, nefarious actors simply divide the probability of being caught by cost of the sanction to determine if compliance is economically rational. If the probability of being caught is small, which it is today, the sanction must be high in order to develop the motivational effect necessary for compliance, called deterrence. This should apply across the board to prevent people making an economic calculus about whether to break the law, as they do already in many areas, including tax evasion. The abused of dominant position, in case of GAFAM, should be found in competition law if as part of their business model on data and technology, they systematically not comply with the rules applicable to their business and technology. As demonstrated above, because of their size, they intimidate the fragmented sectoral and special regulators, as well as competitors who would normally bring complaints, and this needs a hard response in the most effective area of enforcement existing today, namely competition law. In EU law, competition law provides a maximum sanction of 10% of turnover while GDPR only provides 4%.

G5 The future of AI law

There are a number of approaches that can be helpful in developing a general AI law. The *Data Ethics Commission* of the German Federal Government has meanwhile presented a detailed catalogue of recommendations to the European and German legislators.[1042]

These are just a few of the fundamental issues that should be considered when drafting a general regulation on Artificial Intelligence. In this area of very rapidly advancing technology, it is not now a question of enacting technology-specific, small-scale rules. Because these could be obsolete again tomorrow. Rather, it is a question of laying down some basic legal rules in binding form, which are formulated in a technology-neutral and future-oriented manner, that is to say, rules which retain their regulatory content when technology and business models are developing rapidly, enabling dynamic interpretation and which can be updated by interpretation in the light of new circumstances rather than having to be replaced with new legislation.

The law has the great advantage in that it is written for people who can think for themselves. This is an important distinction between the law and technical code. You can interpret and apply the law again and again in line with new challenges, provided that the legislator makes good use of the open texture of the language and does not get lost in the details or even buzzwords of the times, as is often demanded by lobbyists. This is possible without calling into question legal certainty today.

[1042] Data Ethics Committee, 163-197.

That law in the technical age must not be rewritten in the way computer programs receive updates. Computer programmes need new versions because the code was written for the dumb - namely for machines - who, unlike humans, are not in a position to constantly reinterpret the language of the law. The interaction of the open texture of human language with the brain as the source of reason within the framework of linguistic communication enables a form of dynamic adaptive intelligence. This is far from being achieved by the combination of code - data - computer and is probably also fundamentally inaccessible to machines. This is, by the way, particularly noticeable in the fact that intelligent, understanding of speech and interpretation still cannot be automated and the applications don't do much more than basic *Speech to Text and Text to Speech* as well as translation. The open texture of language, written for interpretation by humans who think themselves, is a strength to be used in the formulation of technology neutral law, not a weakness.

5.1. The EU GDPR is an example of technology-neutral and innovation-open law

All the arguments now being put forward against AI legislation were put forward against data protection legislation, in the years before 1995, when the first Directive on the protection of personal data was introduced in the EU, and again during the period running from 2012 to 2016, when the four years of negotiations on the GDPR were taking place.

None of these arguments convinced the legislator - and rightly so:

The assertion that law cannot develop as fast as technology and business models is contradicted by technology-neutral law in Europe. The GDPR is a modern example of technology-neutral law whose meaning and relevance changes with the progress of technology, including AI.

For example, the answer to the question of what personal data is that enables a person to be identified will change when new self-learning algorithms for tracking and identifying people online and offline are implemented with the use of AI. An AI analysis tool was recently released by *Facebook's* Director of Machine Learning, based in Paris. This tool uses 5000 locations on the human body in motion to identify body shapes to match fashion, offering a great new innovation for the fashion industry.[1043] But, it comes with a darker side, it will most likely enable the identification of an individual through their unique body shapes and movement, providing an additional surveillance tool alongside facial identification. The fashion tool produces personal data. AI-supported pattern recognition methods in Big Data can allow the identification of individuals from some anonymized data. This only works, however, if sufficient data is available and the progress of technology turns the former anonymization into a pseudonymization. The frivolous assertion that anonymization is impossible today is just as wrong as the assertion that any cryptographic encryption is broken.

The assertion that a law is fundamentally not precise enough to regulate complex technologies, and that a law that lies beneath the detail, precision and usability of good code is not a good law and should therefore not be passed by the legislature, is another fallacy of the world view propounded by GAFAM. By definition, a law passed in the democratic process requires compromise. The GDPR was negotiated between the co-legislators of the EU, with almost 4000 amendments laid by the European Parliament. Compromise texts of laws that have been drafted through a democratic process

---

[1043] See details at: http://densepose.org/

are the noblest expression of democracy. And a democracy that has lost the ability to negotiate compromises is in crisis. Compromise texts of laws in democracy usually fulfil their function, because the compromise reached, ideally after long public deliberation, represents an advance towards a consensus – which never can be fully reached - on the rules we want to live by; yet it is an accepted compromise because it was reached democratically.

These compromise laws are not written to be applied by machines and automation - like code. Laws are created to be applied by people who can justify their own decisions. And in case of dispute, they are there to be interpreted by reasonable judges.

It is this process of openness of the law and the legal process for later interpretation by wise and independent judges with the help of independent science[1044] that gives the law the flexibility to cope with the demands of constant technical innovation and constantly evolving business models. This can be done without having to rewrite the laws like a code that has to be constantly revised after version 1.0.

To put it bluntly, the requirement that the law must either be as precise as the technical code or be updated as quickly as the code is simply undemocratic, as this ignores the need for democratic reflection and compromise and the time needed for democratic debate and due process of law. Moreover, this demand is based on a fallacy that the law can act as a technical signal language, and this approach seeks to transfer the alleged unambiguity of code to natural language. It would require lawmakers to spend their entire

---

[1044] In a society increasingly dominated by private sector markets and technologies, it is essential to increase the share of public funding for science in order to safeguard its orientation towards truth and public interests and its independence, and to avoid it being dominated by private interests.

time reviewing technical code, rather than speaking to constituents about the important principles that should govern the code. Due to the basic function of language as a world-opening device for human beings, this is fundamentally not unambiguous in the sense of a technical code, which, however, is not its deficiency but its superiority. It is the openness of language that makes technology-neutral and constantly updated legislation possible at all through human interpretation.

Incidentally, lobbyists had no problem criticising successive drafts of the GDPR with contradictory arguments. The criticism began with the assertion that the excessively open wording did not provide sufficient legal certainty today. The next day this changed to complaints that the wording was too prescriptive and not open enough to create flexibility for the future. These contradictory assertions showed that they had only one aim, namely, to avoid the law, to discredit it in whatever way possible.

Companies have no problem with the fact that a code of ethics lacks democratic legitimacy and cannot be enforced. But that is actually the main advantage of the law compared to ethics and self-regulation, given the concentration of power in the hands of large companies: The law has democratic legitimacy and can be enforced, even against powerful mega-corporations. Law creates a credible threat that sanctions can be imposed if companies breach the law. This creates a level playing field for all. The level playing field created by the law is also beneficial because it provides guidance to innovators by creating incentives to focus innovation on the public interest.

In the world of technological domination and populist attempts to undermine democracy, we must strengthen democracy by restoring the law to its noble function To express the will of the majority in a form that is binding on everyone. This will, through law, can be translated into reality by the exercise of public and private powers of enforcement, even in the face of opposition and non-compliance.

It is a matter of breaking the central claims of the impossibility of democratic regulation and design of the Internet, digitalization and AI, as anchored in the minds of many, such as those of John Perry Barlow and the *post-privacy ideology*, because only in this way will freedom and democracy survive in the age of AI.

The GDPR has shown that Europe can set global standards with democratic laws that are bound by fundamental rights. The claim by neo-liberalism and technological ideology, innovation, growth and freedom are best served without any democratic laws affecting technology is simply false.

The opposite is correct: software-based services on the Internet are economically most advantageous if they follow a uniform, standardized business model worldwide. This is why global corporations such as *Microsoft*, for example, now promise to comply with the GDPR worldwide.[1045] If they did not do so, they would have to abandon the worldwide uniformity of their offer, which would result in higher costs, or discontinue their offers in the EU internal market. Both would reduce revenues.

Global financial markets are gradually realising that AI cannot be scaled up worldwide without built-in basic principles of law, democracy and without respect for fundamental rights such as data protection and non-discrimination. Apps and technologies that are being developed and operated in China for face recognition, for example, are simply illegal here in Europe and to some extent in

---

[1045] Microsoft on the Issues. "Microsoft's Commitment to GDPR, Privacy and Putting Customers in Control of Their Own Data," May 21, 2018. https://blogs.microsoft.com/on-the-issues/2018/05/21/microsofts-commitment-to-gdpr-privacy-and-putting-customers-in-control-of-their-own-data/

the US. It therefore comes as no surprise that already between 2017 and 2018 private investment in AI in China fell by 30%.[1046] Since 2018 investments fell even further. Only in dictatorships will AI Made in China be scalable, but not in the free world.[1047]

Just as the greening of General Electric and of industry in general has come about because environmental legislation has created incentives for innovation towards environmental sustainability, the EU's GDPR will now drive innovation for a way of collecting and processing personal data that respects individual rights and the importance of privacy in democracy. The same will happen with regard to well-regulated AI. And just as the European and German economy has achieved market leadership in many areas of sustainable technology and environmental technology, thanks to regulatory pressure, so will it be successful worldwide with high data protection standards and AI, in which the basic values of the Enlightenment and constitutional democracies are incorporated from the outset.

5.2. What is illegal for humans cannot be legal for AI

A first way of guiding the general law of AI may be to simply ask whether there are acts that may be illegal when performed by a natural person but legal when performed by the AI. If, on an objective level, the answer to this question is no, and we agree that this should continue to be the case, it becomes important to codify the principle that an act carried out by the AI is illegal if the same act carried out by a human being would be.

[1046] Lucas, Louise. "China's Artificial Intelligence Ambitions Hit Hurdles," November 15, 2018. https://www.ft.com/content/8620933a-e0c5-11e8-a6e5-792428919cee

[1047] Nikkei Asian Review. "Why China's AI Players Are Struggling to Evolve beyond Surveillance." https://asia.nikkei.com/Spotlight/The-Big-Story/Why-China-s-AI-players-are-struggling-to-evolve-beyond-surveillance

Such a simple codification would maintain the rule of law in the age of AI while providing clear guidance for AI developers and users.

5.3. Principles of technology regulation, pharmaceutical law and chemicals regulation

A second line of thought is to consider whether the regulatory principles that exist in certain areas of law should be generalised for AI. For example, in most areas of sensitive human-technology interaction, i.e., where risks to humans are associated with technology, is a far-reaching obligation to test products and procedures for type approval as a prerequisite for market access. In addition, there is an obligation to monitor the effects of the application of the technology or the medicinal products on humans over the lifetime of the product. These rules apply, for example, in the Medicines Acts, medical device regulation or in chemicals regulations. The aim of these regulations is to prevent damage and to safeguard public interests. The diesel scandal in the car industry, showed the importance of considering the public interest of environmental protection, which was ignored by the fraudulent technology manipulations of VW and other car companies. AI is a prime candidate for similar procedures and obligations, both at a general level and with specific adaptations when developed or applied to specific areas. When we order the testing of cars, medicines and chemicals for harmful effects and sometimes also prescribe precise standards for this purpose, compliance must be demonstrated before these products can be marketed. There is nothing shocking about taking the same approach for AI. These laws steer innovation in the right direction, towards the public good (if they are not circumvented by criminal innovation, as was the case with the car companies in the emissions scandal), they are not hostile to innovation. Nothing else could be said about such rules on AI. And just as the developers of new chemicals or drugs have to follow their products even after the development and market launch has been completed in order to understand their changing interactions with humans and the environment, so AI has to be pursued dynamically.

5.4. A renaissance and intensification of technology assessment on three levels

A third way is to return to the proven principles of technology assessment and to systematically apply the latest state of the art in technology assessment to AI. The renaissance of technology assessment, which has been a good tradition of parliaments in Europe and the US since the 1970s, would correspond to the urgently needed intensified dialogue between democracy and technology. It would also help to promote a general culture of responsibility in the tech industry which is both binding and flexible enough to take account of and promote new technological developments.

While in Europe parliamentary technology impact assessment has become routine, based on the responsibility principle of Hans Jonas, who regarded investment in such impact assessment as a key element of the precautionary approach, in the US the *Congressional Office of Technology Impact Assessment* was closed during the Reagan administration and never re-opened - an early victory for the anti-science movement. Hillary Clinton had promised in her election campaign to re-establish this office or a similar process. The *American Academy for Arts and Sciences* preserves the memory of this important institution. There are ongoing demands in the US to reopen this office.[1048]

[1048] Bipartisan Policy, "Congress Needs the Office of Technology Assessment to Keep up with Science and Technology | Bipartisan Policy Center." 25 July 2019. https://bipartisanpolicy.org/report/congress-needs-the-office-of-technology-assessment-to-keep-up-with-science-and-technology/

In Europe, however, the principles and methods for assessing the impact of technologies in the short and long term have proved their worth in terms of informing policy-makers and legislators. The *European Parliamentary Technology Assessment* (EPTA) network brings together the methods and individual impact assessments for parliaments in Europe in a common database, which already contains several preliminary studies on the capabilities and impact of AI.[1049]

What is necessary to increase confidence in technology in the age of Artificial Intelligence, where high technology is increasingly colonising every aspect of life, are binding impact assessments for new technologies at three levels:

First, as mentioned above, parliamentary technology assessment at the level of policy-making and legislation to determine whether essential interests are affected by the technology in question and what legislation needs to be adopted to ensure the public interest in this context. This impact assessment should ideally take place before high-risk technologies are used. The decisions on the consequences resulting from the risk assessments carried out by experts are in the hands of governments and legislators and, at EU level, in the hands of the Commission and the Council and Parliament as co-legislators.

Second, at the level of developers and users of this technology. There is already an obligation to carry out a data protection impact assessment when the AI processes personal data in the context of automated decision making.[1050] For AI, it would certainly also be

---

[1049] This is available at: http://www.eptanetwork.org to distinguish it from the Technology Assessment, the Legislative Impact Assessment, which assesses the impact of draft legislation, originally in terms of costs for companies, but now also in terms of social impact; see details at https://ec.europa.eu/info/better-regulation-toolbox_en

[1050] Article 35 (3) GDPR;

justified to extend the obligation to carry out an impact assessment to all aspects of democracy, the rule of law and fundamental rights. This should at least be done when the AI has the potential to be used in the exercise of public authority, in the democratic and political sphere or in the provision of services of general interest, whether or not personal data are processed. The importance of these impact assessments at the developer and user level would be that they would underpin public knowledge and understanding of AI, which currently suffers from a lack of transparency about the capabilities and hence its impact. They would also help companies, their managers and the engineers who develop the new technologies and their applications to face up to the power they wield. They would help to create a new culture of technological responsibility for democracy, the rule of law and fundamental rights.[1051]

The standards for such an AI impact assessment, which must be carried out at the latest before a new AI programme is published or marketed to customers, would have to be laid down by law, as we see in the GDPR for the specific case of privacy impact assessments.[1052]

As with the GDPR, compliance with the standards for impact assessment would have to be monitored by the authorities. Non-compliance should be punished by sufficiently dissuasive sanctions. In cases where AI is to be used in the exercise of public authority or for broad public use, the impact assessment would have to be made available to the public, and in high-risk cases the authority using

[1051] Practical examples of elements of a possible in-house technology assessment for AI are the questionnaire developed by the EU High Level Group on AI, see Ethics Guidelines for a Trustworthy AI of 8 April 2017, p. 32ff, https://ec.europa.eu/futurium/en/ai-alliance-consultation.1.html; and the guide developed by the Omidyar Network and the Institute of the Future, available at https://ethicalos.org/wp-content/uploads/2018/08/Ethical-OS-Toolkit-2.pdf

[1052] See Articles 35-36 of the DPA and recitals 74-77,89-92,94-95; see also the guidelines for the Article 29 Working Party's privacy and data protection impact assessment of 4 October 2017, EN/17 WP 248 rev. 01, available at https://ec.europa.eu/newsroom/document.cfm?doc_id=47711

AI would have to carry out its own complementary assessment and present a risk mitigation and contingency plan.

Without distinguishing between the private and public sectors, the most comprehensive plan to date for the establishment of an EU agency for ex-ante certification and registration and a legal framework for essential rules on research, development and use of AI and robotics is contained in the European Parliament resolution of 16 December 2017.[1053]

Third, individuals affected by the use of AI should have a statutory right to be informed about the functioning of AI, the logic behind it and the impact of its use on the interests of the data subject. The impact of the use of AI on individuals must be made, based on new law, even if the AI does not involve the processing personal data. A new law is necessary only to cover this constellation, as where personal data is processed by AI, there is already a right to such information under the GDPR.[1054]

In this context, the claim of the tech giants should be rejected that explanations as to how AI works and how it arrived at decisions cannot be given. There have already been many studies and applications giving examples of the interpretability of AI.[1055] Since

---

[1053] See in particular the sections on "Ethical principles" (par. 10ff) and "A European Agency" (par. 15ff) and the Annex to the resolution with recommendations on the content of the legislative proposal to the European Commission, European Parliament resolution of 16 February 2017 with recommendations to the Commission on Civil Law Rules on robotics (2015/2103(INL), https://eur-lex.europa.eu/legal-content/EN/TXT/?uri=CELEX:52017IP0051

[1054] Articles 13-15 in conjunction with Art. 22 GDPR;

[1055] Miron, Marius. 2018. 'Interpretability in AI and Its Relation to Fairness, Transparency, Reliability and Trust', https://mariusmiron.com/blog/interpretability/; Hamon, Ronan Junklewitz, Henrik Sanchez, Ignacio, "Robustness and Explainability of Artificial Intelligence", 2020 - European Commission JRC, 2020. https://publications.jrc.ec.europa.eu/repository/bitstream/JRC119336/dpad_report.pdf. And JP Hall (2018) 2020. Jphall663/Awesome-Machine-Learning-Interpretability. https://github.com/jphall663/awesome-machine-learning-interpretability.

the duty to state reasons is part of the rule of law, at least when the authorities act in the exercise of official authority, the simple reality is: programs of AI that do not provide reasons and whose decisions cannot be explained by humans cannot be used in the exercise of official authority, since the authority could not fulfil its duty to state reasons by using such programs.[1056]

A new intensity of three-stage technology impact assessment is a necessary component of a new intensity of dialogue between technology and democracy.[1057]

If the debate on AI results in this new responsibility of engineers and technology companies for democracy, fundamental rights and the rule of law, then AI has already created important conditions of trust which it needs in order to find broad acceptance in society.

5.5. Implement the recommendation of the *Data Ethics Commission* of Germany

For the presentation of the report of the *Data Ethics Commission* to the German Federal Government on 23 October 2019,[1058] the Director General for the Digital European Single Market in the EU Commission, the Italian Roberto Viola, travelled from Brussels to Berlin especially

---

[1056] Pasquale, Frank A. 2017. 'Toward a Fourth Law of Robotics: Preserving Attribution, Responsibility, and Explainability in an Algorithmic Society'. SSRN Scholarly Paper ID 3002546. Rochester, NY: Social Science Research Network. https://papers.ssrn.com/abstract=3002546

[1057] Nemitz, Paul. Constitutional democracy and technology in the age of artificial intelligence, 376 Philosophical Transactions of the Royal Society A: Mathematical, Physical and Engineering Sciences, 2018, http://doi.org/10.1098/rsta.2018.0089

[1058] See the English translation of the report at https://www.bmi.bund.de/EN/topics/it-internet-policy/data-ethics-commission/data-ethics-commission-node.html. https://www.bmi.bund.de/EN/topics/it-internet-policy/data-ethics-commission/data-ethics-commission-node.htm; the author Paul Nemitz is a Member of the Commission of 16 which wrote this report and adopted it in consensus

for this occasion. Michael O'Flaherty, the Irish Director of the European Fundamental Rights Agency, came from Vienna.

The presentation of the *Data Ethics Commission*'s report was a European event – the co-chair of the *Data Ethics Commission,* the physician Prof. Dr. Christiane Woopen, is also chair of the independent European Group on Ethics (EGE), which is based at the EU Commission. The other, Prof. Dr. Christiane Wendehorst, is President of the European Law Institute, a Europe-wide association of lawyers who draft model laws for the EU.

In fact, the document with its 75 proposals on 240 pages contains much of interest to the EU Commission. From the risk of inadequate regulation of algorithmic systems and AI to questions of civil liability in the event of damage; the interoperability obligation of *messengers* and social networks to additions to the General Data Protection Regulation. It also extended from questions of rights to digital heritage to a legal framework for trustees of personal data.

These are topics that are also discussed in Brussels, and on which work is being done to some extent. Ursula Von der Leyen, President of the EU Commission, announced in her political guidelines for the new European Commission taking office in 2019 that she will present a legislative proposal within 100 days of taking office for a coordinated European approach on the human and ethical implications of Artificial Intelligence.[1058a]

It would be a mistake to dismiss the report of the *Data Ethics Commission* as typically German. The report breathes the air of

---

[1058a] p. 13, A Union that strives for more, My agenda for Europe, by candidate for President of the European Commission Ursula von der Leyen, Political Guidelines for the next European Commission 2019 - 2024, https://commission.europa.eu/system/files/2020-04/political-guidelines-next-commission_en_0.pdf

European enlightenment, of the European categorical imperative, of the ability to reach a consensus typical of the EU: always act in such a way that the principle underlying your actions could be a general law in Europe, with which Europe would function well.

The work of the *Data Ethics Commission*, with the presidents of employers and consumer protection organizations on board, has shown that the will to talk and understand each others' concerns among democrats is very strong.

The *Data Ethics Commission*'s signal to politicians is: we have reached an agreement. It has presented a unanimous report on ethics and law in the age of Artificial Intelligence. Others can do it too, whether in Berlin or Brussels.

Reaching unanimity was not easy. But it's important. Because on the other side of the Atlantic, in the US, agreement in politics on digital matters is obviously not possible at present. And in China, the Communist Party is the only ruling party. Europe's strength is that there is no absolutism here, not that of the party as in Beijing, not that of technology as in Silicon Valley. Consensus is one of the productivity factors of our democracy in Europe. And it is good for technical and economic innovation, which always needs broad social acceptance and support.

It is already the most powerful, richest Internet companies in the world today, which are now also pushing the next power technologies with huge investments: AI and quantum computers. *Microsoft* has just put a billion dollars into the Open AI project. *Facebook* produces 200 trillion behavioral predictions about people per day with its AI. *Google* announces quantum superiority.

The new technologies are in the hands of the already most powerful tech companies in the world, the GAFAM companies, who will increase their already enormous power as exponentially as they represent an exponential development of technology.

This is the background against which the report of the *Data Ethics Commission* is to be read, whether in Brussels or Berlin, whether in government, parliament or universities.

The sociologist Heinrich Popitz wrote about man's unlimited capacity for innovation, that we do not know where our innovation might take us. Because innovation has not yet ended, Popitz predicted that the control of technical power generated by innovation would become the most important task of democracy. One of California's leading Artificial Intelligence professors, Stuart Russel, has just published his new book, Human Compatible: AI and the Problem of Control. The control of technical power has long since ceased to be an issue solely for the market and competition law. This was pointed out by professor and former constitutional judge Hoffman-Riem when he presented the report of the *Data Ethics Commission*. The regulation of digital and AI is now also about individual freedom, a functioning public sphere, democracy and fundamental rights.

On the road to digitization, we, on this continent of wars and populism, must cultivate the factors that contribute to our productivity: freedom and openness to innovation, democracy and the rule of law. In the face of the pincer grip between infantile dreams of omnipotence in Silicon Valley and Beijing and rising authoritarian populism, our democracy in Europe is facing a test, not only in the regulation of digital and AI.[1059]

[1059] The text of this chapter was already published in a slightly abridged form in the Tagesspiegel Background of 8 November 2019, https://background.tagesspiegel.de/digitalisierung/regulierung-des-digitalen-europa-vor-der-bewaehrungsprobe

G6 Net neutrality and data protection: US fragmented, Europe uniform: The strength of the EU internal market

In the US, there is no longer a nationally enforced regulation on net neutrality after President Trump, who put his man in the telecom regulator's (FCC) executive chair, took power. The abolition of net neutrality was described as a step towards the end of democracy.[1060] However, the American states and cities have taken various initiatives to restore net neutrality, partly in the form of regulation and partly by setting up local networks. The picture of the variety of different regulations in individual states and cities is now extremely complex,[1061] similar are the rules governing the protection of personal data and privacy. There are only a few sectors in the United States, such as students, credit data and health data, that have some degree of data protection under federal law. In these areas individual states that have adopted very different regulations.[1062] We think it is fair to say that at federal level it was mainly economic interests of a very different kind that prevailed.

In the EU, net neutrality has been introduced and initially secured with the REG 2120/2015.[1063] This was a success of the enormous civil

---

[1060] Collins, Keith. 'Net Neutrality Has Officially Been Repealed. Here's How That Could Affect You.' The New York Times, 11 June 2018, sec. Technology. https://www.nytimes.com/2018/06/11/technology/net-neutrality-repeal.html

[1061] Landgraf, Greg . "The State of Net Neutrality," American Libraries Magazine. 2 January 2019. https://americanlibrariesmagazine.org/2019/01/02/state-net-neutrality-roundup/

[1062] Brumfield, Cynthia. "11 new state privacy and security laws explained: Is your business ready?", CSO, 8. August 2019, https://www.csoonline.com/article/3429608/11-new-state-privacy-and-security-laws-explained-is-your-business-ready.html

[1063] Regulation (EU) 2015/2120 of the European Parliament and of the Council of 25 November 2015 on measures regarding access to the open internet and retail charges for regulated intra-EU communications and amending Directive 2002/22/EC and Regulation (EU) No 531/2012, consolidated version: https://eur-lex.europa.eu/eli/reg/2015/2120/2018-12-20

society commitment[1064].

A 2019 report by the EU Commission on the implementation of the Regulation confirms that some of the GAFAM companies and the major telecoms would have liked it otherwise.[1065] However, as in all areas of the digital world, the battle between different interests continues. The introduction of the 5G standard makes it technically even easier to prioritise and forward certain content from certain senders. This could undermine net neutrality. We must be vigilant here.[1066]

Both in terms of net neutrality and data protection, we will have to take care in the medium term in Europe that the GAFAM companies do not succeed in softening up the rules and their enforcement in individual Member States by way of a *divide et impera*, divide and rule, and then, by way of the process of habituation described by Shoshana Zuboff. Continuous illegal practices, if tolerated, can overturn the attitude of the public and governments to these fundamental issues one after the other.

Uniform application of common EU law will thus become a central challenge to ensure that GAFAM actually comply with the law everywhere in their technology development and business models. For just as a development from mass production to individualization can be observed in the history of industry, the progress of digital technology and control by systems of AI also allows for an increasingly differentiated business model. Once it is no longer true that a globally

---

1064 EDRi. "Net Neutrality vs. 5G: What to Expect from the Upcoming EU Review?," 5 December 2018. https://edri.org/net-neutrality-vs-5g-what-to-expect-from-the-upcoming-eu-review/

1065 European Commission. "Commission Report on Open Internet." Text. Shaping Europe's digital future, April 30, 2019. https://ec.europa.eu/digital-single-market/en/news/commission-report-open-internet

1066 EDRi. "Net Neutrality vs. 5G: What to Expect from the Upcoming EU Review?," 5 December 2018. https://edri.org/net-neutrality-vs-5g-what-to-expect-from-the-upcoming-eu-review/

uniform business model is the most efficient and profitable for digital business models because it is cheaper to control the differentiation of offers by AI, there is a danger that the GAFAM companies will systematically exploit weaknesses in the law and its enforcement in the various EU states to offer business models throughout the EU at the lowest denominator identified within the EU.

The previous concentration of the European branches of the GAFAM companies in Ireland was motivated, in addition to receiving the benefits of English as national language and the low wages, by the tax regime, which made special arrangements possible to reduce group taxes. The same applies to the location of the European tax headquarters of *Microsoft* and *Amazon* in Luxembourg.[1067] This loophole has now been largely closed by state aid decisions of the EU Commission.[1068] *Microsoft* and *Amazon* chose Luxembourg, which was always "ready to talk" when it came to tax and data protection. According to press reports, *Microsoft* has also been very active in lowering the level of protection afforded by the new data protection law of its state of domicile in the US, Washington State,[1069] despite an otherwise general assertion that the company is committed to a high level of data protection.

The search for the most favourable regulatory environment for profitability will become easier for groups in times of AI. This must be matched by increased solidarity among Member States and

[1067] The Silicon six and their 100 Billion $ global Tax Gap, Fair Tax Mark, December 2019, https://fairtaxmark.net/wp-content/uploads/2019/12/Silicon-Six-Report-5-12-19.pdf; and: Taylor, Chloe. "Silicon Valley Giants Accused of Avoiding over $100 Billion in Taxes over the Last Decade." CNBC, 2 December 2019. https://www.cnbc.com/2019/12/02/silicon-valley-giants-accused-of-avoiding-100-billion-in-taxes.html

[1068] Süddeutsche Zeitung, Konzern will Steuern zahlen, 2 January 2020, https://www.sueddeutsche.de/wirtschaft/google-konzern-will-steuern-zahlen-1.4741449

[1069] Scott, Mark. "How Lobbyists Rewrote Washington State's Privacy Law." Politico, April 26, 2019. https://www.politico.eu/article/how-lobbyists-rewrote-washington-state-privacy-law-microsoft-amazon-regulation/

common EU discipline. This common discipline is also necessary in the equipment and organization of the authorities responsible for the enforcement of regulation. The organization and equipment of the authorities must not become the Achilles' heel of European law. The GDPR must finally be rigorously implemented throughout Europe. Already there are complaints about the lack of enforcement of the GDPR. People get the impression that what the legislator has produced in six years of hard work is simply not properly implemented. In particular, the collection of personal data is not sufficiently restricted.[1070]

The *Data Ethics Commission* of the Federal Government has called for a structural reform of data protection in Germany, which is fragmented among the federal states. The authorities must be equipped in such a way that they can deal with the conflict that inevitably arises when the law is enforced against large companies.[1071]

The authorities responsible for enforcing the GDPR need at least three times as many staff as they currently have. This is because they fulfil a central task in the age of Artificial Intelligence, quantum computers and the mass collection and processing of data: they protect the freedom of the individual and the functioning of democracy. From a ridiculed secondary function in the huge administrative apparatuses of governments, poorly equipped and in Europe often banned from the capital, traditionally without any real experience of law enforcement, have now become authorities with a central task in this new age. They must be staffed and equipped accordingly.

1070 Vinocur, Nicholas. "'We Have a Huge Problem': European Regulator Despairs over Lack of Enforcement." Politico, December 27, 2019. https://www.politico.eu/article/we-have-a-huge-problem-european-regulator-despairs-over-lack-of-enforcement/ Hare, Stephanie. "These New Rules Were Meant to Protect Our Privacy. They Don't Work | Stephanie Hare." The Guardian, November 10, 2019, sec. Opinion. https://www.theguardian.com/commentisfree/2019/nov/10/these-new-rules-were-meant-to-protect-our-privacy-they-dont-work

1071 *Data Ethics Commission*, point 3.2.4, p. 103.

We cannot rant about a quantum leap in technology and business models, praise the GDPR in soapbox speeches as an important achievement, but leave the data protection authorities starved of resources. What we need now are data protection authorities as genuine investigative authorities, with expert staff at the top, who have experience in competition law, public prosecutions or in tax investigations, meaning: who have tough enforcement experience.[1072]

The US Federal Trade Commission (FTC) which is also concerned with privacy and data protection, has issued a decision against Facebook, which has been criticised as still too favourable in many cases, according to which Facebook must pay a fine of $5 billion for data protection offences.[1073] This is about 10% of Facebook's turnover in 2018. The GDPR only allows up to 4% of a company's global turnover to be imposed as a fine in Europe. In view of the many violations of the GDPR in Europe, it is now urgent to make use of this possibility in order to give companies clear incentives to finally comply with the GDPR. And it is above all the task of the data protection authorities in Ireland and Luxembourg to pursue complaints against GAFAM companies with vigour, in the interests of all EU citizens. They can no longer see themselves as authorities that safeguard national interests; they are taking on European responsibilities. To this end, they should make use of active administrative assistance from their colleagues in other Member States and also involve seconded personnel from other Member States in their investigations and exchange relevant information. If they fail to do so, and if they do not fulfil their European responsibility to

---

[1072] Recommendations 1,2, 4 of the *Data Ethics Commission* of the Federal Government of 23 October 2019, p. 17 of the report; in detail point 3.2.4. of the report, p. 103. https://www.bmi.bund.de/EN/topics/it-internet-policy/data-ethics-commission/data-ethics-commission-node.html

[1073] https://www.ftc.gov/news-events/news/press-releases/2019/07/ftc-imposes-5-billion-penalty-sweeping-new-privacy-restrictions-facebook and https://fortune.com/2019/07/20/facebook-5-billion-ftc/

apply the GDPR consistently, the authorities of other Member States will be forced to take more and more GDPR urgency decisions against companies based in Ireland and Luxembourg in order to protect EU citizens as a whole.[1074]

The advantages for companies from the GDPR *One Stop Shop*, whereby they always deal directly only with the authority of their head office, would thus be lost, as would confidence in the European orientation of the supervisory authorities in these Member States. Either way, the GDPR will have to fully develop its effect in protecting the informational self-determination of individuals in the world of data collected by *IoT* and analysed and used by AI.

[1074] Articles 61 (8) and 66 GDPR.

G7 Protection of whistleblowers and support for the self-organization of GAFAM staff

In its previous mandate, the European Commission proposed as an important step a directive to strengthen the protection of whistleblowers. A few days before the new EU Commission took office, the European Parliament adopted this directive on 23 October 2019.[1075]

It will be possible to further develop the increased protection of whistleblowers, which the German newsmagazine Der Spiegel described as a fifth power in democracy,[1076] given its importance for the toolbox of democracy and the rule of law. In the United States, hundreds of millions of dollars are paid on a legal basis to whistleblowers,[1077] whether they come from businesses or administrations, who uncover illegal activities at their workplace. In an age of increasing complexity due to digitization and AI, we will need more of these instruments, as it is almost impossible to understand these complex systems from the outside and to check that they are functioning lawfully.

[1075] Directive (EU) 2019/1937 of the European Parliament and of the Council of 23 October 2019 on the protection of persons reporting infringements of Union law, OJ L 305, 26.11.2019, p. 17-56, http://data.europa.eu/eli/dir/2019/1937/oj

[1076] The Fifth Power, Lothar Gorris, Rene Pfister, Marcel Rosenbach, Christoph Scherman, Der Spiegel of 16 November 2019, No. 47, https://magazin.spiegel.de/SP/2019/47/166982304/index.html?utm_source=spontm_campaign=centerpage

[1077] https://www.whistleblowers.org/know-your-rights/whistleblower-rewards-an-international-framework-for-detecting-corruption-and-fraud/; Stephen M. Kohn, Founding Director, National Whistleblower Center, Effective Whistleblower Protections, Testimony before the European Parliament, November 21, 2018, https://www.europarl.europa.eu/cmsdata/157293/7%20-%2002%20Kohn-Testimony-Europe-FINAL.pdf

It will also be important to strengthen the self-organization and assumption of responsibility of employees of the GAFAM companies, even if they work as "freelancers" in the Gig Economy, by means of labour law. In the US, a dangerous trend is emerging where GAFAM company employees are removed who call for colleagues to be unionized[1078] or who stand up for human rights[1079] or ethics[1080] or call out a misogynistic culture.[1081] This must not happen in Europe. In Germany and other EU member states, *Amazon* has been fighting for years against the normality of a works council and the organization of employees in trade unions. In Austria, there is still no works councils at *Amazon*; nearly all employees are only temporary workers.[1082]

The countervailing power of trade unions in these companies, including the so-called "freelancers", needs to be strengthened by law, as does the protection of whistleblowers. The new possibilities of co-determination of workers and owners with the help of digitization and AI in the companies also need support for further development.

1078 Meredith Whittaker: 'The tech industry at large has a culture of retaliation, Colm Gorey, Siliconrepublic, 29. September 2019, https://www.siliconrepublic.com/machines/meredith-whittaker-google-retaliation-ai-now-institute. Netzik, Ann-Kathrin. "Eine ganz gewöhnliche Firma", Die Zeit, 6. November 2019, https://www.zeit.de/2019/46/google-mitarbeiter-proteste-vertrauen-mitbestimmung-drohnen

1079 Tiku, Nitasha. 'A Top Google Exec Pushed the Company to Commit to Human Rights. Then Google Pushed Him out, He Says.' Washington Post. 2 January 2020. https://www.washingtonpost.com/technology/2020/01/02/top-google-exec-pushed-company-commit-human-rights-then-google-pushed-him-out-he-says/

1080 Thompson, Nicholas, "Tristan Harris: Tech Is 'Downgrading Humans.' It's Time to Fight Back." Wired. 23 April 2019. https://www.wired.com/story/tristan-harris-tech-is-downgrading-humans-time-to-fight-back/. and Newton, Casey. "The Leader of the Time Well Spent Movement Has a New Crusade." The Verge, April 24, 2019. https://www.theverge.com/interface/2019/4/24/18513450/tristan-harris-downgrading-center-humane-tech

1081 GeekWire. "Analysis: Why Unionization in Tech Could Actually Gain Traction in 2020," January 4, 2020. https://www.geekwire.com/2020/analysis-unionization-tech-actually-gain-traction-2020/

1082 Futurezone, 14.6.2019, https://futurezone.at/netzpolitik/ministerium-wird-nach-kritik-an-amazon-arbeitsbedingungen-aktiv/400524313

G8 Development of control technologies for data processing and AI

Currently, inspectors involved in the approval of medical devices that work with AI, for example, are not at all able to test data and algorithms themselves.[1083] It is clear that there is an urgent need to develop control methods and control technologies.

At the same time, more public resources must be invested in the development of control technologies for all AI and data-intensive technologies. Without such technologies, which must then be widely used by law enforcement agencies, it is not possible at all to check, for example, whether personal data have been deleted from a controller's or processor's IT system or which personal data are transferred from a mobile phone and to what location for storage, let alone the functionality and legality of the operation of a complex AI program.

Just as the large data controllers and processors are already equipping themselves with control and data management systems, the corresponding, more advanced forensic technologies for the enforcement agencies must also be developed and acquired. In addition, for the control of AI it will be necessary to install parallel logging and control systems from the beginning and to provide interfaces for them.

There is a need for a legal obligation not only to provide for interfaces in technology, but also to grant state and law enforcement authorities

---

[1083] Vollmer, Sebastian, Bilal A. Mateen, Gergo Bohner, Franz J. Király, Rayid Ghani, Pall Jonsson, Sarah Cumbers, et al. 2020. 'Machine Learning and Artificial Intelligence Research for Patient Benefit: 20 Critical Questions on Transparency, Replicability, Ethics, and Effectiveness'. BMJ 368 (March). https://www.bmj.com/content/368/bmj.l6927

access to data and algorithms when the public good requires. To this end, the appropriate, forward-looking legal basis for the regulation of AI must be created, especially for large systems, such as navigation systems that are not currently subject to any authorization at all, but whose operation or malfunction can nevertheless have enormous public interest implications.

A new methodology, indeed a new field of research, is needed for the predictive behavioral control of Artificial Intelligence, as the newly recruited director of the Berlin Max Planck Center for Man and Machine, Iyad Rahwan, formerly at the Boston MIT, has explained.[1084] The view of the data and the programming of the algorithm must be combined with the view from the outside on the behaviour of the AI in test situations. This observational psychology for machines must complement the view inside on the code and the data with the view from the outside in dynamic test situations. Only in this way can the function and effect of the AI in the context of its application really be understood.

[1084] Rahwan, I., Cebrian, M., Obradovich, N. et al. Machine behaviour. Nature 568, 477–486 (2019) https://www.nature.com/articles/s41586-019-1138-y

G9 Duty to tell the truth: Put an end to GAFAM's lies to governments and parliaments

Society is so dependent on the systems and knowledge of the GAFAM companies that reasonable regulation and government is no longer possible at all without the knowledge of these companies.

Unfortunately, however, experience shows that GAFAM groups often tell parliaments and governments half-truths or at least avoid providing complete information. Either they do not appear at hearings at all, invoke trade secrets or instruct the representative on site to simply answer "I don't know" questions and promise to provide information later, but then do not provide very meaningful information. Mark Zuckerberg did this before the European Parliament[1085] and the US Congress.[1086]

While in European competition law the European Commission can enforce information and also sanction false information, the European Parliament and the European Commission have not yet had such instruments at their disposal in the legislative procedure.

However, it is precisely in the legislative process that the right information is decisive for good legislation. Anyone who wants

---

1085 Stupp, Catherine. "MEPs Outraged over Zuckerberg's EU Parliament Show." Www.Euractiv.Com (blog), May 23, 2018. https://www.euractiv.com/section/data-protection/news/meps-outraged-over-zuckerbergs-eu-parliament-show/. and Kraus, Rachel. "EU Lawmakers Raked Mark Zuckerberg over the Coals — but, as Usual, It Didn't Matter." Mashable. https://mashable.com/2018/05/22/zuckerberg-facebook-eu-hearing-highlights/

1086 Frenkel, Sheera. "Facebook Gives Lawmakers Follow-Up Answers, but Not Much Is New." *The New York Times*, June 11, 2018, sec. Technology. https://www.nytimes.com/2018/06/11/technology/facebook-lawmakers-follow-up-answers.html

good legislation must therefore demand a duty to be placed on corporations to provide accurate, and complete, information, which should be enforced if necessary. Non-observance should be sanctioned even during the simple legislative procedure.

Data protection authorities are also dependent on accurate information. In future, they should issue more requests for information in the form of a formal instruction to the GAFAM companies so that they can then impose appropriate sanctions in the event of non-compliance with the instruction, which includes false information. The GDPR permits such a procedure.[1087] This could also accelerate the hitherto lengthy procedures for enforcing data protection law.

As a result, the duty of truthfulness should be enforced in data protection and in legislative procedures in the same way as in competition law, as there is no good reason to differentiate here in matters of the duty of truthfulness.

[1087] Art. 58 (1) in conjunction with Art. 83 (2) and (5) GDPR.

G10 The financial markets: From ESG to democracy and rule-of-law-abiding AI as an investment object

In the financial markets, environmentally and socially oriented ESG[1088] investment funds have been contributing to the reorientation of the economy and society towards sustainability. Some of these funds, which promise to invest in ecologically and socially sustainable economic activity, are launched with the claim that this type of investment even generates higher yields over the long term. However, many investors in these funds are also prepared to forego a certain proportion of the return if they know that their investments will finance less or no external costs in the form of pollution and exploitation of workers.

The assessment of the "G" component of ESG, namely "governance", is already concerned with legal compliance, and the protection of privacy and personal data. Some analysis houses are already evaluating data protection practices in the GAFAM companies in detail today and highlight the risks of inadequate data protection for investors' earnings.[1089] In most analysis houses and funds, however, this analysis has so far played only a minor role.[1090]

Investors now have the choice of a large number of environmentally oriented investment funds. It would be good if there were soon

---

[1088] Environmental, social, governance.

[1089] Sustainalytics, Managing data privacy risk: comparing the FAANG+ stocks, Moin Syed, Matthew Barg. Martin Vezer, Doug Morrow, June 7, 2018, https://www.researchgate.net/publication/326065806_Managing_data_privacy_risk_comparing_the_FAANG_stocks

[1090] see for example https://www.refinitiv.com/content/dam/marketing/en_us/documents/methodology/esg-scores-methodology.pdf, p. 16; https://www.esade.edu/itemsweb/biblioteca/bbdd/inbbdd/archivos/Thomson_Reuters_ESG_Scores.pdf, p. 10 and 11

an equally wide choice of funds that rely on a high level of data protection in the data economy and an AI equipped with the principles of the *rule of law, democracy and fundamental rights by design*. Already in America, the large pension funds and other major investors are being asked not to invest in companies that provide the technology that facilitates the oppression of the Uighurs in China.[1091]

After years of work, markets and politicians have only just reached a first rough agreement on the definition of common testing standards in the field of environmental sustainability.[1092] Finding a common standard for funds that concentrate investment in companies where data protection is taken seriously and AI is developed with due regard for law and values should in any case be easier for data protection, as the GDPR is a clear audit standard.

For AI, the regulation announced by the EU Commission will also be important for this purpose. It will give the financial markets a common, binding orientation, just as the GDPR already does for data protection.

In the US, "Ranking digital Rights" was founded in 2013 by Rebecca MacKinnon. It has taken the first steps towards a quality index[1093] that assess the extent to which companies help to promote free expression and privacy on the Internet, without, however, strictly checking the full application of the GDPR. Other forums are also already discussing how financial markets can contribute to a more sustainable use of data.[1094]

1091 Pletka, Danielle, and Derek Scissors. "Opinion | Stop Investing in China's Brutality." The New York Times, December 5, 2019, sec. Opinion. https://www.nytimes.com/2019/12/05/opinion/stop-investing-in-china.html

1092 Finke, Björn. "Kriterien für grüne Fonds." Süddeutsche.de. https://www.sueddeutsche.de/wirtschaft/nachhaltigkeit-am-finanzmarkt-kriterien-fuer-gruene-fonds-1.4713070.

1093 https://rankingdigitalrights.org/

1094 World Economic Forum. "Here Are 4 Ways Investors Can Influence More Secure and Responsible Innovation." https://www.weforum.org/agenda/2018/12/four-ways-investors-influence-more-secure-responsible-innovation/

The $5 billion fine imposed by the FTC on *Facebook,*[1095] the threat of a fine of up to four percent of global sales in the GDPR,[1096] but also the downward stock market swings such as Facebook's following various scandals of non-compliance with data protection[1097] and also the decline in private investment in AI in China[1098] are clear indications that aligning investments in the data economy and AI with strict data protection and responsible handling of AI can help to avoid losses and increase profits. Conversely, the rise of *Microsoft* to become the most valuable company in the world, with its promise to provide data protection in accordance with the GDPR to all customers worldwide, shows that investing in data protection is worthwhile.[1099] This applies to the marketing of its own services and products as well as to gaining the trust of investors in the financial markets.

[1095] Federal Trade Commission. "FTC Imposes $5 Billion Penalty and Sweeping New Privacy Restrictions on Facebook," July 24, 2019. https://www.ftc.gov/news-events/press-releases/2019/07/ftc-imposes-5-billion-penalty-sweeping-new-privacy-restrictions

[1096] Article 83 GDPR.

[1097] Timberg, Craig and Elizabeth Dwoskin, "How years of privacy controversies finally caught up with Facebook", Washington Post, 27. July 2018, https://www.washingtonpost.com/technology/2018/07/26/how-years-privacy-controversies-finally-caught-up-with-facebook/; Gecgil, Tezcan, InvestorPlace Contributor Mar 18, 2019, "One Year After Cambridge Analytica: Where Is Facebook Stock Now?". https://investorplace.com/2019/03/one-year-after-cambridge-analytica-where-is-facebook-stock-now/

[1098] Lucas, Louise. "Cold Water Hits China's AI Industry," July 9, 2019. https://www.ft.com/content/973bfc08-a15f-11e9-a282-2df48f366f7d., China Money Network. "Next 10 Years: Predictions From 10 Experts on China's Artificial Intelligence Future." https://www.chinamoneynetwork.com/next-10-years-predictions-from-10-experts-on-chinas-artificial-intelligence-future

[1099] Microsoft on the Issues. "Microsoft's Commitment to GDPR, Privacy and Putting Customers in Control of Their Own Data," May 21, 2018. https://blogs.microsoft.com/on-the-issues/2018/05/21/microsofts-commitment-to-gdpr-privacy-and-putting-customers-in-control-of-their-own-data/

G11 The future of control of technical power in the age of AI

Thinking in categories of control, power and domination, as presented in this book, is an achievement of civilization. For without thinking in these categories a critique of power and domination is not possible. But this is a prerequisite for human autonomy, as far as it is possible, and also for self-determination in democracy.

In the technical discourse on ethics and the future of AI, these categories are often ignored today. There is then talk of "humility"[1100] towards man's own control abilities in a complex system. There is also the goal of harmony in a system that optimizes man and machine in interaction and mutual complementarity, without, however, reducing the parameters of optimization to those which are countable and measurable.[1101]

It is true that man and women must empower himself and herself and equip themselves with instruments in order to be able to exercise control in the highly efficient system of technology, such as AI and quantum computers. Humility before the limits of our own natural abilities is the driving force behind the advancement of the science of behavioral control of "autonomous" systems[1102], as is now to be done at the Max Planck Institute in Berlin under the direction of Prof. Iyad

[1100] Karachalios, Konstantinos, and Ito, Joichi. "Human Intelligence and Autonomy in the Era of 'Extended Intelligence,'" n.d., 5. https://globalcxi.org/wp-content/uploads/CXI_Essay.pdf

[1101] Ito, Joichi. "Resisting Reduction: A Manifesto." Journal of Design and Science, November 1, 2017. https://doi.org/10.21428/8f7503e4

[1102] Rahwan, Iyad, Manuel Cebrian, Nick Obradovich, Josh Bongard, Jean-François Bonnefon, Cynthia Breazeal, Jacob W. Crandall, et al. "Machine Behaviour." Nature 568, no. 7753 (April 2019): https://doi.org/10.1038/s41586-019-1138-y

[1103] Max Planck Institute for Human Development, "New Director: Iyad Rahwan's Research Focuses on the Societal Challenges of Digitization." https://www.mpib-berlin.mpg.de/press-releases/new-director-iyad-rahwan

Rahwan.[1103]
This new type of behavioral control of self-learning and evolving machines, which complements the internal view of the program with an external view based on the view of the behavioral therapist or psychologist or psychotherapist in humans, will require a new scientific system. This must become an essential part of the risk and impact assessment of novel AI systems.

It is certainly also true that the world cannot and must not be reduced to something countable, and certainly not to something measurable in money or economic categories, as Joichi Ito writes in his Manifesto *Against Reductionism.*[1104]

But the progress of civilization, according to Norbert Elias, consists in the progress of self-regulation and self-control of human beings and man-made systems, such as states, companies and technologies, through their own insights. Even at an advanced age, Elias gave practical proof of this thesis at the International Congress of Sociologists in Hamburg in 1988 in his lecture on the development of accident fatalities per kilometre driven in Berlin since the turn of the century.[1105] The authors attend this lecture together as students.

Even in the face of a technology that can do much, and perhaps one day almost everything, that can be planned and calculated better than humans, we must not accept the transition from human self-regulation to third-party control by such systems. Yes, technology changes us, yes, we become more and more dependent on it, yes, it

---

[1104] Ito, Joichi. "Resisting Reduction: A Manifesto." Journal of Design and Science, November 1, 2017. https://doi.org/10.21428/8f7503e4

[1105] He cited the example again in an interview in Spiegel 1988, Festerberg, Von N and M. Screiber, "Spiegel Gespräch: Wir sind die späten Barbaren", Der Spiegel, 23 March 1988. https://www.spiegel.de/spiegel/print/d-13529892.html. see also Dunning, Eric, and Jason Hughes. 2013. Norbert Elias and Modern Sociology: Knowledge, Interdependence, Power, Process. A&C Black, p.50

itself exercises power and power is exercised over it.
But the progress of civilization, as Norbert Elias defined and recognized it, is finished when we give up the claim to control technical power, as Popitz formulated it. Instead, it is important to continue to take the "initiative to use our reasonable freedom", as Jürgen Habermas writes in his Magnum Opus. With Habermas, we believe it is important "- like Kant in his time - to continue to encourage the contemporaries with arguments (...) to make autonomous use of their reason and to create their practical social existence."[1106]

This is true on a small scale and in concrete terms, for every program and every device. But it also applies above all on a large scale, namely the basic principle that criticism and questioning of power, including and especially technical power, is a basic prerequisite for freedom and democracy. We must be able and willing to understand all power, including and especially the technical power of large systems such as the Internet, IoT, search engines, social networks and Artificial Intelligence. And we must be able to test them, according to our standards of legitimacy, we must be able to regulate them and also to put them in their place or shake them off if we decide.

These are the standards we must set for the future technologies of power. We must maintain our ability to act according to these standards. That is the task of responsible science and politics.

[1106] Habermas, Jürgen. Auch eine Geschichte der Philosophie: Band 1: Die okzidentale Konstellation von Glauben und Wissen. Band 2: Vernünftige Freiheit. Spuren des Diskurses über Glauben und Wissen (Berlin: Suhrkamp Verlag AG, 2019), p. 13.

G12 From a Hippocratic oath for engineers to chartered "Engineers for Democracy"

It is sometimes claimed that a Hippocratic oath for engineers is what we need now. One option presented is to draw from the examples of doctors and lawyers who have strict codes of professional ethics, which are enforced by regulatory bodies. It is true that the voluntary commitment of the technical intelligence, the technical elite to the constitution, fundamental rights, democracy and ethics could certainly help. If, in addition, at least one compulsory lesson per week on these topics were added to the study programme - even better. Otherwise, however, the parallel with doctors and lawyers is misplaced. An oath by engineers to the public interest alone will certainly not solve the problems that the unleashed power of the GAFAM has created.

First, medical and legal professions are also regulated by law. Here too, pure professional self-regulation has not proved successful. It may well be that professional representatives still sing the song of self-regulation. But these professions are too important. In a democracy they cannot be left to regulate to the professions alone, as in the Middle Ages they were left to guilds. Incidentally, in the ideal world, they are nevertheless liberal professions with a special legal fiduciary relationship with their clients and patients, which is why they can look back on a long history of *regulated self-regulation*.

Second, the job description of the IT engineer is different. It lacks both the special fiduciary duty to users and the independence of the liberal profession, as well as, third, the history of professional self-regulation. In other words, none of the three elements that form the basis of the liberal professions' sometimes legitimate claim to self-regulation are present in the case of engineers.

For other reasons, the press also enjoys a special status as the fourth power in democracy. Because we want it to be independent and far removed from the state in democracy, because its task is to control and criticise power, it should largely, though not completely, regulate itself. But again, it is only the democratic and binding law on the press that creates this status and ensures the independence of the press. But journalists are a largely self-regulating profession, based on privileges which they are granted in law.

However, these models, for the reasons outlined, cannot be transferred to engineers and *Big Tech* without further ado.

G13 The new responsibility of the technical intelligence and engineers

The nature of the new digital networks and technologies of AI as domination technologies of a new kind gives a new responsibility to those who develop, operate and maintain them.

To recall the buried sources of the human being and thinking, especially for those who belong to the technical intelligence and design the AI of tomorrow as engineer, programmer, data scientist or entrepreneur, is an important concern of this book. It is therefore necessary to enter into the history of people's reflections about themselves as individuals, but also about the forms of coexistence, i.e., the history of democracy and the various forms of dictatorship, in sum the different ways in which people have exercised power over other people. And in this context, it is necessary to rediscover the sources of the self, of human self-consciousness and the foundation of human dignity and freedom of action.

We want to help ourselves and the technical intelligence, against the background of our criticism of the present and the historical achievements of people's thinking about society, to recognize the responsibility for people and the self-determination of people in democracy and to draw the consequences of responsibility and commitment in the age of the AI.

This book is an invitation to technical experts, engineers, programmers, software designers not to only stay in their profession when it comes to developing knowledge about the risks and opportunities from the age of AI, but to get involved in democracy and politics. Without the contribution of technical intelligence, democracy will not be able to manage the complex issues of regulating new technologies in such a way that the good aspects of technology are preserved and strengthened, but the bad aspects, such as the theft of freedom from individuals and the destruction of democracy, are avoided. Without the participation of technical intelligence in democratic processes and the associated learning about freedom and self-organization of people and the value of individual freedom, fundamental rights and democracy, technical intelligence will move itself further and further away from freedom and democracy. With such a technological design that is detached from democratic processes, one would encourage populists and autocrats who want to bet on the weakness of democracy and bring it about.

Probably in no other country in the world is technical intelligence trained in front of such a powerful historical backdrop to understand its new responsibility in the age of Artificial Intelligence as in Germany. After the fall of the Nazi regime, there was a widespread conviction in Germany that the training of engineers and technical intelligence in general needed to be value-oriented. After all, the fall of the Weimar Republic and the rise of Hitler, up to and including the concentration camps and the extermination of the European Jews, was pursued by unscrupulous policymakers and implemented with the help of an obedient technical elite. German Engineers laid the railroad tracks to the concentration camps without asking questions about what would happen in the camps, Adorno observed critically after the war in his pamphlet on the education after Auschwitz.[1106a]

For this reason, many universities in post-war Germany introduced compulsory courses in politics, history and law for engineers. In

---

[1106a] Education after Auschwitz, https://archive.org/details/education-after-auschwitz-theodore-adorno ; german original Theodor W. Adorno: Erziehung nach Auschwitz. (1966) In: Erziehung zur Mündigkeit, Vorträge und Gespräche mit Hellmuth Becker 1959 – 1969. Published by Gerd Kadelbach. Frankfurt am Main 1970, p. 92–109

many cases these were abolished with the student revolution of 1968, as students protested against the content of these courses: they considered them too conservative.

Political scientist and sociologist Eugen Kogon was one of those who tirelessly worked to teach democracy and fundamental rights at the faculties of engineering. His 1976 book, the title of which can be translated as "The Engineers' Hour: Technological Intelligence and Politics" was a plea for humanitarian values in an increasingly technically centralized world and a wake-up call for commitment to democracy. Kogon's study was optimistic. An overwhelming majority of the engineers interviewed were prepared at the time to accept responsibility for the societal and political consequences of technological progress.[1107]

Concern about the political attitude of the technical elite had diminished as time has accrued distance from the Nazi dictatorship and the atomic bomb dropped on Hiroshima. Dürrenmatt's "The Physicists" had become compulsory reading at school.[1108] The student revolt and the new environmental movement, including the fight against nuclear power, attracted many engineers, not least those who developed new sustainable environmental technologies and made them an economic success.

The Green Movement and later the Green Party in Germany, in parallel with the development of sustainable technologies, contributed significantly to the success of both. For the technological developments for sustainability and resource efficiency showed that sustainable policy was feasible. The green technology entrepreneurs also brought important organizational and management talent to the movement. At the same time, they were supported by green

---

[1107] Kogon, Eugen. Die Stunde der Ingenieure: Technologische Intelligenz u. Politik. Düsseldorf: VDI-Verlag, 1976, p.362.

[1108] Dürrenmatt, Friedrich. The Physicists: A Play in Two Acts. Translated by James Kirkup. London: Samuel French Ltd, 2010.

policies. The fact that it was possible to build sustainable technology made the Greens as a political movement interesting for the technical intelligence, in contrast to the social-democrats, which had nothing to offer for ecologically committed engineers and also nothing to build for other engineers.

The success of the Greens is based to a large extent on this symbiosis of technical innovation and the economic success of the entrepreneurs of sustainability, a symbiosis of technological development and politics that the Social Democratic Party in Germany still lacks today, and which may partly explain its decline. Engineers can build technology for an ecological future. But a future of social justice, can it be built by engineers?

Over the last 70 years, the political and economic power of engineers has increased enormously, but their involvement in the institutions of democracy has decreased considerably. Today, there are eight engineers, one physicist, one microbiologist and one chemist in the US Congress, but 218 lawyers and 208 businessmen.[1109] The proportion of engineers, mathematicians and physicists in the Bundestag has decreased from 13.2% in the 12th term in 1990 to 7.2% in the 19th term in 2017.[1110]

Nevertheless, the tradition of scientific technology assessment[1111] and the ethics of engineers[1112] has developed in Europe. The European

[1109] Manning, Jennifer E. "Membership of the 115th Congress: A Profile," 20 December 2018, https://fas.org/sgp/crs/misc/R44762.pdf

[1110] German Bundestag, DHB, chapter 3.11 Occupational structure, 20.11.2018, page 10, https://www.bundestag.de/resource/blob/273350/e521f1d217d7cd471e8ec50217d1502a/Kapitel_03_11_Berufsstruktur-pdf-data.pdf

[1111] Grunwald, Armin, and Carsten Orwat. "Technology Assessment of Information and Communication Technologies." Chapter. Encyclopedia of Information Science and Technology, Fourth Edition. IGI Global, 2018. https://doi.org/10.4018/978-1-5225-2255-3.ch370.

[1112] compare the Ethical Guidelines of the German Informatics Society, 2018, 'https://gi.de/fileadmin/GI/Allgemein/PDF/GI_Ethische_Leitlinien_2018.pdf; VDI, Ethical Principles of the Engineering Profession, 2002, https://www.engineerseurope.com/sites/default/files/EN_VDI_Ethische_Grundsaetze_des_Ingenieurberufs.docx

Parliament has proposed a specific ethical code of conduct for robotic engineers.[1113]

For engineers and computer scientists who develop, operate and maintain AI and other forms of new digital power technologies, it will be necessary to do more than just appeal to goodwill. As a first step, every computer scientist and engineer in this field could be required to attend a seminar with lawyers or philosophers during their studies and to solve some tasks in interdisciplinary groups. Some universities in the US already offer such mixed seminars. For example, Joichi Ito, the former director of the MIT Media Lab, and the jurist Jonathan Zittrain from Harvard University teach together.[1114] In this way, both professors and students learn to communicate between the disciplines. The work in the German *Data Ethics Commission* has shown that this mutual understanding is one of the most difficult but also most important prerequisites for meeting the new challenges posed by AI and other digital power technologies. In this way, lawyers, philosophers, and other social and cultural scientists should also be brought closer to the understanding of technology that is necessary for coping with the future.

Then, as is already the case today for teachers and lawyers, a state examination should be introduced for engineers and computer scientists. The aim should be to check whether the graduates have acquired sufficient knowledge in their training to be able to carry

---

1113 European Parliament, Civil regulations in the field of robotics, European Parliament resolution of 16 February 2017 with recommendations to the Commission on Civil Law Rules in the field of robotics, P8_TA(2017)0051 (2015/2103(INL), https://eur-lex.europa.eu/legal-content/EN/TXT/?uri=CELEX:52017IP0051

1114 The Ethics and Governance of Artificial Intelligence Professor Joichi Ito, Director of the MIT Media Lab, co-taught with Professor Jonathan Zittrain, Harvard Law, MIT Spring 2018, https://dam-prod.media.mit.edu/x/2018/07/30/Syllabus%20Ethics%20and%20Governance%20of%20AI%20.pdf.

1115 See: Turner, Jacob. Robot Rules: Regulating Artificial Intelligence. 1st ed. 2019 edition. New York, NY: Palgrave Macmillan, 2018.

out their profession in the age of AI and digital power in a way that reconciles power and responsibility, democracy, fundamental rights and the rule of law.[1115] Just as the exercise of certain functions in public administration requires the ability to hold the office of judge, so in technical functions it will also be necessary to determine those for which a qualification as a "democracy engineer or computer scientist" is required, i.e. a special state examination of knowledge about the interaction of technology, democracy, fundamental rights and the rule of law.

The struggle for dominance between the totalitarian technocapitalism of American influence, the autocratic digital state capitalism of Chinese provincialism and democracy will ultimately be decided by the commitment of the Technical Intelligence. If it is alone on the side of capitalism and technology, it is more likely that democracy will end and be replaced by autocratic and dictatorial systems. Donald Trump and Boris Johnson are on the way to a large-scale attempt to decouple liberal democracy and economic growth, towards a "*Meritocratic Technolopis*", as described by the head of Johnson's Brexit campaign, Dominic Cummings.[1116]

The price for this path will be the loss of freedom, humanity and self-determination, for the members of the technical intelligence itself. The human being as the centre of all our endeavours would then be replaced by a totalitarian technological-economic system, into whose systemic imperatives the human being would have to completely integrate himself and against which he would be able to assert neither the dignity of man nor claims to freedom.

[1116] The Economist, "An optimistic Eurosceptic", 21 January 2016, https://www.economist.com/britain/2016/01/21/an-optimistic-eurosceptic.

In order for this not to happen, it is important to invite the technical experts to also become 'engineers' of democracy and freedom,[1117] and create a democratic technical intelligence which, in everything it does, especially with regard to personal data and algorithmic systems such as Artificial Intelligence, always keeps in mind individual freedom, fundamental rights and democracy. Continuing the line of learning by taking responsibility for technology and its consequences in a design of digital technologies, networks and AI that is oriented towards democracy and fundamental rights is the great new future task of Technical Intelligence. It will only be able to fulfil this task if it becomes more involved in the institutions of democracy, the parties, the parliaments and civil society and helps to maintain the creative power of living democracy - also and especially in the age of global networks and dominant corporations with fantasies of technical omnipotence.

[1117] In a conversation with Paul Nemitz, the American political scientist Henry Farell coined the term "Democracy Technician", by which he does not mean technicians who organize democracy according to technical rules, but conversely who align technology with democratic values.

G14 Afterword: The book, a work of friendship

"The Human Imperative" contains the thoughts of two friends on this topic. It reflects our professional experience as EU official and journalist, but also our 35 years of discussions as lawyer and philosopher. What we have written down here is our personal opinion, not necessarily that of our employers or clients.

The register in which we conduct these conversations already indicates that we are not *digital natives*. We are expressly committed to our background as migrants from an analog age. When we started the debates in the kitchen of the shared flat in Hamburg St. Pauli, we pulled the telephone behind us on long cables and typed our homework on mechanical typewriters, which at the time of our final exam had become electrical and thus contained 4 lines of memory for corrections. We looked for the sources for our papers in the greasy card index boxes of the university library. Our network consisted of real friends.

During our research for this book, we found a historical bookmark in one of the books from that time: a flyer from the Federal Statistical Office in Wiesbaden on the 1987 census, under the heading "So zählt unser Land" ("This is how our country counts"), which listed the benefits of the then upcoming census ("Misinvestments in the billions are avoided"). Nevertheless, we did not open the shared flat door to the official census taker at that time.

Today, we want to live in a digital world where we can search for information without being searched, where we can read, without being read, where we can network without being caught in the net, neither by our own government nor by foreign powers or private companies.

It is our common conviction that enlightenment and humanism are not at an end, but rather only at the beginning. Just as the rash and fashionable proclamation of an end to history after the European freedom revolution of 1989 was wrong, so too the swan songs to humanism and Homo Sapiens are just as misplaced today as they are to democracy. We take Mark Twain, who, when he found his own obituary in the newspaper, wrote laconically: "The news of my death is greatly exaggerated."[1118]

We owe the creation of this book to many people. We would like to explicitly mention our families, who have been patient with us over the past year and had to endure long absences. We thank our teachers for everything we could learn from them, especially Herbert Schnädelbach. We thank our editors Alexander Behrens and Uwe Optenhögel, Bonn, who made this book possible. We thank our diligent and inspiring translator and discussion partner Nyasha Weinberg, London. We thank as well as our friends and conversation partners for numerous suggestions. We have tried to mention them as completely as possible in footnotes and bibliography. Where this has not been done, we hope we will be forgiven. This is also part of the Human Imperative.

[1118] New York Journal, 1895-1901, June 2, 1897.

Bibliography

Agrawal, Ajay; Gans, Joshua; Goldfarb, Avi, (Ed.). 2019. The Economics of Artificial Intelligence, An Agenda, National Bureau of Economic Research, Chicago

Andrews, Lori. 2011. I know who you are, and I saw what you did, Social Networks and the Death of Privacy, New York

Arendt, Hannah. 1970. On Violence. Houghton Mifflin Harcourt, New York

Arendt, Hannah. 1994. Critical Essays. Ed. By Lewis P. Hinchman and Sandra K. Hinchman, New York

Arendt, Hannah. 2018. The Freedom to Be Free: From Thinking Without a Banister. Knopf Doubleday Publishing Group, New York.

Arendt, Hannah, and Margaret Canovan.1998. The Human Condition: 2nd Revised edition edition. Chicago: University of Chicago Press.

Aristotle. The Politics. 2000. Rev edition. Penguin, Harmondsworth, England, New York,

Arute, Frank, Kunal Arya, Ryan Babbush, Dave Bacon, Joseph C. Bardin, Rami Barends, Rupak Biswas, et al. "Quantum Supremacy Using a Programmable Superconducting Processor." Nature 574, no. 7779 (October 2019): 505–10. https://doi.org/10.1038/s41586-019-1666-5.

Barbrook, Richard, Cameron, Andy, "The Californian Ideology." Text. Mute. Mute Publishing Limited, September 1995. https://www.metamute.org/editorial/articles/californian-ideology.

Bell, Daniel. 2000. The End of Ideology: On the Exhaustion of Political Ideas in the Fifties: With 'The Resumption of History in the New Century'. Harvard University Press Cambridge, London.

Bloch, Ernst, 1995. The Principle of Hope, MIT Press, Cambridge, Mass.

Bloch, Ernst. 1986. Natural Law and Human Dignity. MIT Press, Cambridge, Mass.

Bostrom, Nick. 2014. Superintelligence, Paths, Dangers, Strategies, Oxford

Brockman, John. 2019. Possible Minds, 25 Ways of Looking at AI; New York

Bridle, James. 2018. New Dark Age: Technology, Knowledge and the End of the Future. Verso Books.

Brynjolfsson, Erik, Daniel Rock, and Chad Syverson. 'Artificial Intelligence and the Modern Productivity Paradox: A Clash of Expectations and Statistics. ' Working Paper. Working Paper Series. National Bureau of Economic Research, November 2017. https://doi.org/10.3386/w24001.

Buchanan, Mark. 2002. Small Worlds and the Groundbreaking Science of Networks. New York.

Castells, Manuel. 2000. The Rise of the Network Society: Economy, Society and Culture v.1: The Information Age: Economy, Society and Culture Vol 1. 2nd Revised edition. Oxford

Cunningham, Anne C. 2016. Drones, Surveillance, and Targeted Killings. Greenhaven Publishing LLC, New York.

Crémer, Jacques, Yves-Alexandre de Montjoye, Heike Schweitzer. 2019, Competition Policy for the digital era, A report. European Commission, available: https://ec.europa.eu/competition/publications/reports/kd0419345enn.pdf

Davies, Simon: Privacy, A personal Chronicle, Washington 2018

Dennett, Daniel C. 1996. Darwin's Dangerous Idea: Evolution and the Meaning of Life. Simon and Schuster. New York.

Dawkins, Richard. 1989. The Selfish Gene. Oxford University Press.

Diamandis, Peter H., and Kotler, Steven. 2016. Bold: How to Go Big, Create Wealth and Impact the World. Simon and Schuster. New York.

Dijk, Jose van; Poell, Thomas; de Waal, Martijn. 2018. The Platform Society, Public Values in a Connective World. Oxford.

Dumaine, Brian. 2020. Bezonomics: How Amazon Is Changing Our Lives, and What the World's Companies Are Learning from It. Simon and Schuster. New York.

Fry, Hannah. Hello World. How to Be Human in the Age of the Machine. 2018, Norton& Company. New York.

Frankfurt, Harry G. 2005. On Bullshit. Princeton University Press, Princeton, NJ.

Foer, Franklin. 2017. World Without Mind. The Existential Threat of Big Tech. Penguin Press. London.

Franck, Georg, 1999, The Economy of Attention, on the attention economy in the digital version,

Freedberg, Sidney J. Jr.: How Ai could change the art of war, 25.4.2019 Networks and Cyber https://breakingdefense.com/2019/04/how-ai-could-change-the-art-of-war/

Fromm, Eric. 2000. The Art of Loving: The Centennial Edition (A&C Black).

Freud, Sigmund, 1929. Civilization and Its Discontents. World Heritage Encyclopedia. Available under: http://www.stephenhicks.org/wp-content/uploads/2015/10/FreudS-CIVILIZATION-AND-ITS-DISCONTENTS-text-final.pdf

Grunwald, Armin, and Carsten Orwat. "Technology Assessment of Information and Communication Technologies." Chapter. Encyclopedia of Information Science and Technology, Fourth Edition. IGI Global, 2018. https://doi.org/10.4018/978-1-5225-2255-3.ch370.

Habermas, Jürgen. 1984. Theory of Communicative Action, Volume 1: Reason and the Rationalization of Society: Reason and the Rationalization of Society Vol 1. New Ed edition. Beacon Press, Boston. https://teddykw2.files.wordpress.com/2012/07/jurgen-habermas-theory-of-communicative-action-volume-1.pdf

Habermas, Jürgen. 2015. The Structural Transformation of the Public Sphere: An Inquiry Into a Category of Bourgeois Society. John Wiley & Sons. Hoboken.

Habermas, Jürgen. 2014. Europe: The Faltering Project. John Wiley & Sons, Hoboken

Habermas, Jürgen. 1986. Theory and Practice. Translated by John Viertel. 1 edition. Polity. Cambridge.

Habermas, Jürgen. Further reflections on the public sphere, in Calhoun, Craig J. 1992. Habermas and the Public Sphere. MIT Press. Cambridge Mass.

Habermas, Jürgen. "Political Communication in Media Society: Does Democracy Still Enjoy an Epistemic Dimension? The Impact of Normative Theory on Empirical Research1," Communication Theory 16, no. 4 (2006): 411–26, https://doi.org/10.1111/j.1468-2885.2006.00280.x. p.422

Haagerup, Ulrik. 2014. Constructive News, Why Negativity destroys the media and democracy, Copenhagen

Hawamura, Katsutoshi. 2005. Kant's criticism of the Golden Rule// Kant between West and East. To commemorate the 200th anniversary of Kant's death and 280th birthday. Ed. by Prof. Dr. Vladimir Bryuschinkin. Vol.2. Kaliningrad.

Harari, Yuval Noah. 2016. Homo Deus: A Brief History of Tomorrow. Random House, New York.

Harari, Yuval Noah. 2018. 21 Lessons for the 21st Century. London.

Hegel, Georg Wilhelm Fredrich. 1991. Elements of the Philosophy of Right. New Edition. Cambridge: Cambridge University Press.

Hildebrandt, Mireille: Law for computer Scientists, 2019, https://lawforcomputerscientists.pubpub.org

Horkheimer, Max, and Theodor W. Adorno. 2002. Dialectic of Enlightenment. Stanford University Press, https://monoskop.org/images/2/27/Horkheimer_Max_Adorno_Theodor_W_Dialectic_of_Enlightenment_Philosophical_Fragments.pdf

Humboldt, Wilhelm von. 1999. "On Language": On the Diversity of Human Language Construction and Its Influence on the Mental Development of the Human Species. Cambridge University Press

Ito, Joichi. "Resisting Reduction: A Manifesto." Journal of Design and Science, November 1, 2017. https://joi.ito.com/weblog/2017/12/20/resisting-reduction.html

Irving, John Good, "Speculations Concerning the First Ultraintelligent Machine." The Edward A. Feigenbaum Papers - Spotlight at Stanford. 1966, Accessed July 16, 2020. https://exhibits.stanford.edu/feigenbaum/catalog/gz727rg3869.

Jeffries, Stuart. 2017. Grand Hotel Abyss: The Lives of the Frankfurt School. Reprint edition. Verso, London, New York.

Jonas, Hans. 1985. The Imperative of Responsibility: In Search of an Ethics for the Technological Age. University of Chicago Press.

Jonas, Hans. 2001. The Phenomenon of Life: Toward a Philosophical Biology. Northwestern University Press

Jordan, Tim. 1999. Cyberpower: An Introduction to the Politics of Cyberspace. Routledge

Kant, Immanuel. 1997: Critique of Practical Reason. Cambridge University Press, Cambridge

Kant, Immanuel. 1879. Critique of Practical Reason and Other Works on the Theory of Ethics. Longmans, London. https://oll.libertyfund.org/title/abbott-kant-s-critique-of-practical-reason-and-other-works-on-the-theory-of-ethics

Kant, Immanuel. 1981. Grounding for the Metaphysics of Morals, trans. J.W. Ellington, Hackett Publishing Company, Inc.; Indianapolis, Indiana.

Kaplan, Jerry. 2016. Artificial Intelligence: What Everyone Needs to Know. Oxford University Press

Karachalios, Konstantinos, and Ito, Joichi. "Human Intelligence and Autonomy in the Era of 'Extended Intelligence,'" n.d., 5. https://globalcxi.org/wp-content/uploads/CXI_Essay.pdf

Kelsen, Hans. 2000. What Is Justice? Justice, Law, and Politics in the Mirror of Science: Collected Essays (The Lawbook Exchange, Ltd.).

Kirkpatrick, David. 2011. The Facebook Effect: The Inside Story of the Company That Is Connecting the World. Simon and Schuster. New York.

Kurzweil, Raymond. 2006. The Singularity Is Near. Duckworth. London.

Kurzweil, Ray. 2013. How to Create a Mind: The Secret of Human Thought Revealed. Penguin Books, London

Lanier, Jaron. 2014. Who Owns the Future? Export ed. edition. New York: Simon & Schuster, New York.

Lessig, Lawrence. 2012.Republic, Lost: How Money Corrupts Congress - and a Plan to Stop It. Reprint edition. Twelve. London:

Lévi-Strauss, Claude. 1966, The Savage Mind, Weidenfeld and Nicolson. London.

Lévi-Strauss, Claude. 1983, Structural Anthropology, Chicago

Levy, Steven. 2011. In the Plex: How Google Thinks, Works, and Shapes Our Lives. Simon and Schuster. New York.

Levitsky, Steven, and Daniel Ziblatt. 2018. How Democracies Die: What History Reveals About Our Future. UK US Canada Ireland Australia India New Zeland South Africa: Viking, New York.

Lovelock, James. 1995. Gaia: A New Look at Life on Earth. Oxford University Press.

Lovelock, James, 2019. Novacene, The Coming Age of Hyperintelligence, London.

Luhmann, Niklas. 2017. Trust and Power. Cambridge.

Lyotard, Jean-François, 1990, Pacific Wall, Culver City.

Madary, Michael, and Metzinger, Thomas K. "Real Virtuality: A Code of Ethical Conduct. Recommendations for Good Scientific Practice and the Consumers of VR-Technology." Frontiers in Robotics and AI 3 (2016), https://doi.org/10.3389/frobt.2016.00003

Margetts, Helen, et al., 2016. Political Turbulence, How Social Media Shape Collective Action, Princeton

McLuhan, Marshall, 2001, Understanding Media: The extensions of man, London and New York,

McLuhan, Marshall, Fiore, Quentin, 2008, The Media is the Massage, Penguin UK

Marcuse, Herbert. 2013. One-Dimensional Man: Studies in the Ideology of Advanced Industrial Society. Routledge

Masters, Blake, and Peter Thiel. 2015. Zero to One: Notes on Start Ups, or How to Build the Future. 1st edition. London: Virgin Books

Moglen, Eben: Snowden and the future, 2013, available under: http://snowdenandthefuture.info/snowdenandthefuture-westward.pdf,

Moore, Martin / Tambini, Damian (Ed.). 2018: Digital Dominance, The Power of Google, Amazon, Facebook, and Apple, Oxford

Morozov, Evgeny. 2013. To Save Everything, Click Here: The Folly of Technological Solutionism. New York, PublicAffairs.

Nagel, Thomas. 1974. What is it like to be a Bat? https://www.jstor.org/stable/2183914

Nemitz, Paul. Democracy through law: The Transatlantic Reflection Group and its manifesto in defence of democracy and the rule of law in the age of "artificial intelligence", 01 December 2021, European Law Journal, https://onlinelibrary.wiley.com/doi/full/10.1111/eulj.12407 and https://doi.org/10.1111/eulj.12407

Nemitz, Paul. Constitutional democracy and technology in the age of artificial intelligence, 376 Philosophical Transactions of the Royal Society A: Mathematical, Physical and Engineering Sciences, 2018, http://doi.org/10.1098/rsta.2018.0089

Nilsson, Nils J. 2009.The Quest for Artificial Intelligence. A History of Ideas and Achievements. New York, Cambridge University Press

Pasquale, Frank A. November 8, 2017 "The Automated Public Sphere." SSRN Scholar Paper. Rochester, NY: Social Science Research Network, https://papers.ssrn.com/abstract=3067552

Parker, Walter C. 2005. „Teaching Against Idiocy ", Bloomington: Phi Delta Kappan.

Pico de la Mirandola, 2016. Oration on the Dignity of Man, London

Pascal, Blaise. 2013. Thoughts: Selected and Translated. Reissue Edition. Cambridge; New York: Cambridge University Press

Popitz, Heinrich. 2017. Phenomena of Power: Authority, Domination, and Violence. Columbia University Press.

Popper, Karl. 2002. The Open Society and its Enemies, Volume 1 The Spell of Plato, Routledge, London.

Rahwan, Iyad, Manuel Cebrian, Nick Obradovich, Josh Bongard, Jean-François Bonnefon, Cynthia Breazeal, Jacob W. Crandall, et al. "Machine Behaviour." Nature 568, no. 7753 (April 2019), p. 477, https://doi.org/10.1038/s41586-019-1138-y

Rapp, Friedrich, Analytical Philosophy of Technology, Boston Studies in the Philosophy and History of Science (Springer Netherlands, 1981), https://doi.org/10.1007/978-94-009-8412-7

Ries, Eric. 2011. The Lean Startup: How Today's Entrepreneurs Use Continuous Innovation to Create Radically Successful Businesses. Crown Business.

Rid, Thomas. 2016. Rise of the Machines. The lost history of cybernetics, Scribe, London.

Rismal, Nina. 2017: Ends of Utopian Thinking, Dissertation, Cambridge.

Rotenberg, Marc. 2019. The AI Policy Sourcebook 2019; Electronic Privacy Information Center, Washington.

Sandel, Michael J. 2007. The Case Against Perfection: Ethics in the Age of Genetic Engineering. Cambridge: Harvard University Press.

Scheler, Max, Eugene Kelly, and Anthony J. Steinbock. The Human Place in the Cosmos. Translated by Manfred S. Frings. Evanston, Ill: Northwestern University Press, 2009.

Schnädelbach, Herbert, Philosophy, A Basic Course, Volume 1, p. 92. Translation quoted in Schweppenhäuser, Gerhard. 2009. Theodor W. Adorno: An Introduction. Duke University Press.

Schnädelbach, Herbert. 2009. Philosophy in Germany 1831-1933, Cambridge.

Schmidt, Eric and Cohen, Jared. The New Digital Age: Transforming Nations, Business and Our Lives, New York, Vintage, 2014,

Schirrmacher, Frank. 2015. EGO: The Game of Life. 1 edition. Cambridge, UK ; Malden, MA: Polity,

Smith, Brad, and Browne, Carol Ann. 2019. Tools and Weapons: The Promise and the Peril of the Digital Age. Penguin Publishing Group,

Snyder, Timothy. 2018. The Road to Unfreedom: Russia, Europe, America. Random House,

Snow, C. P. 1993. The Two Cultures. Cambridge University Press.

Staab, Philipp. 2019. Digital Capitalism: Market and Rule in the Economy of Scarcity, Berlin

Tegmark, Max. 2017. Life 3.0: Being Human in the Age of Artificial Intelligence. 01 edition. Allen Lane,

Toffler, Alvin. 1980. The third wave, An Entrepreneur's vision of the future, William Morrow, New, New York

Turner, Jacob. Robot Rules. 2018. Regulating Artificial Intelligence. 1st ed. 2019 edition. New York, NY: Palgrave Macmillan.

Turing, Alan 1936. On Computable Numbers, with an Application to the "Entscheidungsproblem", London, available under: https://www.cs.virginia.edu/~robins/Turing_Paper_1936.pdf

Turner, Fred. 2006.From Counterculture to Cyberculture: Stewart Brand, the Whole Earth Network and the rise of Digital Utopianism, Chicago

Turner, Frederick J., 1921. The frontier in American History, New York, available under: https://www.gutenberg.org/files/22994/22994-h/22994-h.htm

Vamvacas, Constantin J. 2009. The Founders of Western Thought – The Presocratics, Springer, New York

Vernant, Jean-Pierre. 1984. The Origins of Greek Thought. Cornell University Press

Vollmer, Sebastian, Bilal A. Mateen, Gergo Bohner, Franz J. Király, Rayid Ghani, Pall Jonsson, Sarah Cumbers, et al. 2020. 'Machine Learning and Artificial Intelligence Research for Patient Benefit: 20 Critical Questions on Transparency, Replicability, Ethics, and Effectiveness'. BMJ 368 (March). https://doi.org/10.1136/bmj.l6927

Walsh, Toby. 2062: The World That AI Made. La Trobe University Press, 2018

Walter, Chip. 2020. Immortality, Inc. New York, Simon and Schuster

Weber, Max. 2005. The Protestant Ethic and the Spirit of Capitalism. Routledge, Oxfordshire

Weizenbaum, Joseph. 1993. Computer Power and Human Reason: From Judgment to Calculation. Penguin Books.London

Wiener, Norbert. 1961.Cybernetics or Control and Communication in the Animal and the Machine. Second edition edition. Cambridge, Mass: MIT Press

Wulf, Andrea. 2015. The Invention of Nature: The Adventures of Alexander von Humboldt, the Lost Hero of Science. First Edition. London: John Murray

Yao, Xinzhong. Yao, Hsin-chung. 2000. An Introduction to Confucianism, Cambridge University Press

Zeilinger, Anton. 2010. Dance of the Photons: From Einstein to Quantum Teleportation. New York: Farrar Straus Giroux,

Žižek, Slavoj. 2018. Like A Thief In Broad Daylight: Power in the Era of Post-Human Capitalism. 01 edition. Allen Lane

Zuboff, Shoshana. 2019. The Age of Surveillance Capitalism: The Fight for a Human Future at the New Frontier of Power, Public Affairs, New York